Nanodiamonds

Dean Ho
Editor

Nanodiamonds

Applications in Biology and Nanoscale Medicine

Springer

Editor
Dean Ho
Northwestern University
Evanston, IL
USA
d-ho@northwestern.edu

ISBN 978-1-4419-0530-7 e-ISBN 978-1-4419-0531-4
DOI 10.1007/978-1-4419-0531-4
Springer New York Dordrecht Heidelberg London

Library of Congress Control Number: 2009933602

Printed on acid-free paper

Springer is part of Springer Science+Business Media (www.springer.com)

To my parents, Sarah, friends, family
and colleagues for their inspiration
and encouragement

I would like to gratefully acknowledge
my colleagues who have contributed to
this book, as well as the National Science
Foundation, National Institutes of Health,
Wallace H. Coulter Foundation,
V Foundation for Cancer Research,
and Northwestern University,
for supporting our work.

Preface

Nanodiamonds represent an emerging class of materials with important implications at the intersection of biology and medicine. Their consistent dimensions, unique surface properties, facile processing parameters and scalability, innate biocompatibility, and applicability as imaging/diagnostic and therapeutic platforms make them an ideal foundation for theranostic approaches. Their introduction to the field of nanomedicine couldn't come at a more important time. The optimized monitoring and treatment of physiological disorders such as cancer and inflammation, as well as the addressing of key domains such as regenerative medicine necessitates the development of technologies such as nanodiamonds which can be engineered to meet the aforementioned as well as a spectrum of additional medical needs.

A growing international community of researchers has convened towards the production of this book by sharing their multi-faceted strategies that have forged the role of nanodiamonds in impacting biology and medicine. As such, this book unites the expertise of pioneering efforts in the fundamental fabrication and characterization of nanodiamond particles and ultrananocrystalline diamond thin films, fluorescent nanodiamonds for biological labeling and cytotoxicity analysis, protein capture, and drug delivery particle and device design. Furthermore, the coalescence of nanodiamond platforms with technologies for cellular interrogation and nanomanufacturing are introduced to illustrate the impact of this versatile platform towards a broad spectrum of applications.

The reader will be introduced to a very diverse set of methodologies that span the disciplines of chemistry, physics, materials science, bioengineering, and beyond. These contributions are this intended to formulate a foundation for a roadmap that spans synthesis and characterization through translational application.

Prof. Dean Ho Biography

Dr. Dean Ho is an Assistant Professor in the Departments of Biomedical Engineering and Mechanical Engineering in the Robert R. McCormick School of Engineering and Applied Science, and Member of the Robert H. Lurie Comprehensive Cancer Center of Northwestern University where he directs the Laboratory for Nanoscale Biotic-Abiotic Systems Engineering (N-BASE).

Dr. Ho investigates the scalable fabrication of nanomaterial-based devices for applications in chemotherapy, anti-inflammation, and regenerative medicine. His research has garnered news coverage on the CNN homepage, Nature, Reuters, Yahoo, ABC News, CNBC, MSNBC, The Chicago Tribune, USA Today, and United Press International. Additionally, Dr. Ho was featured in the National Geographic Channel program 'Known Universe'. He is the Editor-in-Chief of the Journal of the Association for Laboratory Automation and an associate editor of both the Journal of Biomedical Nanotechnology and Journal of Nanotechnology Law and Business.

Dr. Ho has authored more than 100 peer-reviewed journal and proceedings publications in the areas of nanomedicine and drug delivery, and has delivered several national and international invited talks. Dr. Ho is a recipient of the National Science Foundation CAREER Award, Wallace H. Coulter Foundation Early Career Award in Translational Research, V Foundation for Cancer Research V Scholar Award, John G. Bollinger Young Manufacturing Engineer Award of the Society of Manufacturing Engineers, and Distinguished Young Alumnus Award of the UCLA School of Engineering and Applied Science.

Contents

Contributors

Shashishekar P. Adiga
Argonne National Laboratory, Argonne, IL, USA

Ashfaq Adnan
Northwestern University, Evanston, IL, USA

Alexander Aleksenskiy
Russian Academy of Sciences, St. Petersburg, Russia

Marina Baidakova
Russian Academy of Sciences, St. Petersburg, Russia

Seunghyun Baik
Sungkyunkwan University (SKKU), South Korea

Amanda Barnard
Commonwealth Scientific and Industrial Research Organisation (CSIRO), Australia

Ted Belytschko
Northwestern University, Evanston, IL, USA

S.-J. Cai
National Dong Hwa University, Hualien, Taiwan

C.- C. Chang
National Chiao Tung University, Hsin-Chu, Taiwan

Huan-Cheng Chang
Institute of Atomic and Molecular Sciences, Academia Sinica, Taiwan

J.- I. Chao
National Chiao Tung University, Hsin-Chu, Taiwan

Mark Chen
Northwestern University, Evanston, IL, USA

C.- L. Cheng
National Dong Hwa University, Hualien, Taiwan

C.- Y. Cheng
National Dong Hwa University, Hualien, Taiwan

C.-D. Chu
National Dong Hwa University, Hualien, Taiwan

P.- H. Chung
National Dong Hwa University, Hualien, Taiwan

Suzanne A. Ciftan Hens
International Technology Center, Raleigh, NC, USA

Larry A. Curtiss
Argonne National Laboratory, Argonne, IL, USA

Horacio Espinosa
Northwestern University, Evanston, IL, USA

Dieter M. Gruen
Argonne National Laboratory, Argonne, IL, USA

Michelle Hallikaine
Northwestern University, Evanston, IL, USA

Houjin Huang
Northwestern University, Evanston, IL, USA

Eung-Soo Hwang
Seoul National University (SNU) South Korea

Seyoung Im
Korea Advanced Institute of Science and Technology (KAIST), South Korea

Sheraz Kalim
University of California-Los Angeles, Los Angeles, CA, USA

Moon Ki Kim
Sungkyunkwan University (SKKU), South Korea

Taesung Kim
Sungkyunkwan University (SKKU), South Korea

Young-Jin Kim
Sungkyunkwan University (SKKU), South Korea

Adrian M. Kopacz
Northwestern University, Evanston, IL, USA

Robert Lam
Northwestern University, Evanston, IL, USA

Jessica Lee
Northwestern University, Evanston, IL, USA

Junghoon Lee
Seoul National University (SNU) South Korea

K.- K. Liu
National Chiao Tung University, Hsin-Chu, Taiwan

Wing Kam Liu
Northwestern University, Evanston, IL, USA

Kian Ping Loh
National University of Singapore, Singapore

Owen Loh
Northwestern University, Evanston, IL, USA

Eugenio Oñate
International Center for Numerical Methods in Engineering (CIMNE), Barcelona, Spain

Eiji Ōsawa
NanoCarbon Research Institute, Shinshu University, Japan

Vladimir Osipov
Russian Academy of Sciences, St. Petersburg, Russia

E. Perevedentseva
National Dong Hwa University, Hualien, Taiwan

Jason Reed
University of California-Los Angeles, Los Angeles, CA, USA

George Schatz
Northwestern University, Evanston, IL, USA

Joanna Schmitt
Veeco Digital Instruments, Inc., Tucson, AZ, USA

Olga A. Shenderova
International Technology Center, Raleigh, NC, USA

Michael A. Teitell
University of California-Los Angeles, Los Angeles, CA, USA

Albert To
Northwestern University, Evanston, IL, USA

J.- S. Tu
National Dong Hwa University, Hualien, Taiwan

Yonhua Tzeng
National Cheng Kung University (NCKU), Taiwan

Alexander Vul'
Russian Academy of Sciences, St. Petersburg, Russia

Weng Siang Yeap
National University of Singapore, Singapore

Chapter 1
Single-Nano Buckydiamond Particles

Synthesis Strategies, Characterization Methodologies and Emerging Applications

Eiji Ōsawa

Abstract This Chapter presents a brief overview of a long history and recent rapid developments of the so-called *detonation nanodiamond*. Although this new version of artificial diamond was discovered as early as 1963, it became a victim of secret research under a military regime, during which there was no exposure to scientific community, and virtually no meaningful progress was made until about 1990. Confusion continued until 2005, when the primary particle was isolated for the first time in dispersed form. The dark age of detonation nanodiamond was briefly summarized in Sect. 1.1. Bitter experience on the hazard of secretive custom in technology prompted us to disclose details of isolation procedure, to which the Sect. 1.2 is devoted. Nevertheless, unexpected difficulties in nanoparticles prevented us to make fast progress in its development. Only in 2007–2008, an illuminative guide was presented by theoreticians regarding geometrical and electronic structures in the primary particles of detonation nanodiamond, which solved most of its mysterious behaviors that we encountered in the past. This lucky incidence is mentioned in Sect. 1.3 together with our provisional extention of the theory. Three potential *Applications* are reviewed in the light of the new theoretical model in the penultimate section. Section 1.5 refers to future directions of production and applications.

1.1 Introduction

In recent years, we are often shown in TV on how accurate are the missile attacks, which usually end up with black smokes coming up from the exact target position. I always wonder "how many people watching the TV realize that about half of the smoke is diamond?" Smoke means that explosives used in

E. Ōsawa (✉)
NanoCarbon Research Institute, AREC, Shinshu University, Ueda, 386-8567, Japan

D. Ho (ed.), *Nanodiamonds: Applications in Biology and Nanoscale Medicine*,
DOI 10.1007/978-1-4419-0531-4_1, © Springer Science+Business Media, LLC 2010

the missile are "oxygen-imbalanced" leading to incomplete combustion upon explosion and produces soot. In 1963 a brilliant Ukrainian physicist named Vladimir Danilenko found that soot from the explosion of well-known military explosive called Composition B contained nanodiamond in high concentration [1]. His invaluable observation marked the early discovery of *detonation nanodiamond* (DN), the central theme of this Chapter and this book [2, 3]. Composition B is a 1:1 mixture of TNT ($C_7H_5N_3O_6$)-hexogen ($C_3H_6N_6O_6$), and still used in military.

Crude DN is being produced on industrial scale mainly in Russia and China (Table 1.1) since about 1990 using the expired or overproduced stocks of Composition B as the raw material. Transformation of war explosives into artificial nanodiamond is certainly the most attractive chemical conversion we have ever heard of and actually our motivation to join the development of DN around 1999. At that time, the primary particles have been well known by TEM and X-ray diffraction studies on the crude product: extremely small single crystal of cubic diamond, 4–5 nm in diameter [4, 5]. However, we soon realized that the primary particles have never been isolated in dispersed form. Apparently, the importance of aggregation among nanoparticles has not been fully recognized and researchers in this field did not pay due attention to destroy the aggregation [6]. To our knowledge, the presence and hazards of especially persistent form of aggregation in DN was first pointed out by Shenderova [7], but, here again, her seminal review published in a local journal did not attract enough attention.

The manufacturers failed to recognize the stubborn aggregation in their raw product, and believed until recently that they were handling single-nano diamond particles. We named the special form of aggregation in the primary particles of DN as *agglutination* [8]. However, the agglutinate product had been given a

Table 1.1 World production of detonation nanodiamond in tons (estimated by Prof. A. Vul')

<table>
<tr><th>Country</th><th>2006</th><th>2007</th><th>2007</th></tr>
<tr><td>Russia</td><td>0.8</td><td>2.0</td><td rowspan="4">8.0</td></tr>
<tr><td>Ukraine</td><td>1.0</td><td>1.2</td></tr>
<tr><td>Byelorussia</td><td>0.05</td><td>0.06*</td></tr>
<tr><td>China</td><td>1.0</td><td>1.2</td></tr>
<tr><td>Japan</td><td>-</td><td>0.2#</td><td>0.3#</td></tr>
<tr><td>Total</td><td>2.8</td><td>4.7</td><td>8.3</td></tr>
</table>

* Agglutinates are decomposed by chemical etching.

\# Primary particles are produced by beads milling (NCRI procedure).

\- Estimates unavailable.

wrong and misleading general name of *ultradispersed diamond (UDD)*. To complicate the confusion, the agglutinates which consist of core particles with 60–200 nm in size were somewhat useful. We later found that the conglomerates of core agglutinates, which had grown to fine powders with several ten microns in diameter during drying process, can be readily dispersed into water by vigorous stirring into rather stable and thick slurry. These crude dispersions proved applicable for a few low-precision purposes like lubrication additives to engine oil, and polishing. The most successful application was found in creating fine ditch structures of about 1 nm in depth on the surface of harddisk made from glass and aluminum plates. An exception was successful reinforcing in metal plating, which should appear only in dispersed form. This point will be discussed again below in Sect. 5.2.

Having diameters less than 100 nm, the smaller DN agglutinates certainly fit the definition of nanoparticles and will continue to find small markets in the low-technology domain, where secretive policy prevails and only very few scientists work. However, the *UDD* was never accepted in science arena as a decent scientific object of research. UDD particles are in the intermediate stage of purification between the raw products of detonation to singlenanodiamond crystalline particles. There is no reliable scientific publication on UDD and virtually no mention is given in any textbook of diamond. Out of respect to the pioneers in DN research and certain usefulness for industrial purposes, we would like to define the agglutinates as the first generation DN.

We fortunately succeeded in disintegrating the agglutinates into the primary particles by stirred-media milling (more often called beads-milling, *vide infra*) [8]. In order to unambiguously distinguish these primary particles from their agglutinates, we call our final product as *single-nano buckydiamond* (SNBD) in this Chapter. This new naming is a bit too long, but more illustrative of its unique structure than calling them simply as dispersed nanodiamond. SNBD particles, having a remarkably small average size and narrow distribution, 4.8 ± 0.7 nm, represent complex core-shell structure with thin graphitic surface layer(s) around core diamond as will be mentioned in more detail soon. SNBD particle is not a simple diamond crystal but a novel nanocarbon with mixed structure, very likely the most stable among nanocarbons of this size range. Figure 1.1 illustrates various aspects of an idealized structure of a SNBD particle based more on theoretical calculations of Barnard and Sternberg [9] rather than experimental evidence which is hard to obtain.

Considering the increasing attention towards small and invisible nanoparticles as potential health risk, we close this section with a few words on the safety of SNBD. As can be readily anticipated from the well-known lack of chemical reactivity in diamond, screening of cytotoxicity against a variety of cell types so far did not show any sign of measurable toxicity. Genotoxicity and the rate of metabolism are still under investigation [10–12]. For safety, however, utmost care should be taken not to release the SNBD in its dispersed form to our environments under any circumstances.

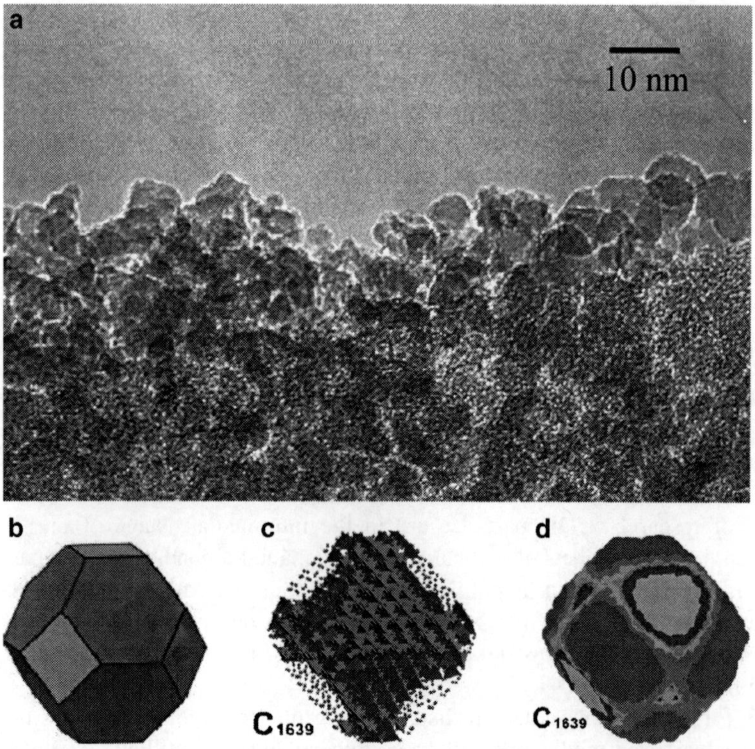

Fig. 1.1 **a**: TEM image of single-nano bucky diamond (SNBD) particles in solid, which are not dispersed but aggregated by incoherent interfacial Coulombic interactions [7]. Taken by Dr. M. Ozawa. **b**: Truncated octahedron used here as a model of SNBD primary particle. {100} facets in green, and {111} facets in red. **c**: An optimum structure of truncated octahedral C1639 model of SNBD particle after SCC-DFTB calculations by Barnard [6]. Carbon atoms on the {111} surface and some below have rearranged into sp^2 and sp^{2+x} (0<x<1) hybridization, which are represented by grey spots for clarity. The diamond carbon atoms that maintained sp3 hybridization are depicted by black tetrahedral. **d**: An example of electrostatic potential distribution over the surface of the optimized structure given in C. Red-orange-yellow colors correspond to positive potential up to 25eV, and the potential decreases in this order of coloring. Grey-purple-blue-green colors to negative potential up to -25eV, decreasing in this color order

1.2 Formation and Isolation

1.2.1 Detonation and Agglutination

The most critical step in the production of SNBD is undoubtedly the detonation process. After explosion of Composition B in inert medium, invisibly small diamond crystals will start to grow from deposition of unoxidized carbon atoms from the explosive molecules in the thin high-pressure high-temperature zone formed behind the rapidly propagating front of shock wave. It seems that a very large number of crystallization nuclei simultaneously start the diamond growth process under abundant supply of carbon atoms generated from incomplete combustion of Composition B. However, the crystal growth is suddenly and all at once suspended as the shock wave passes at supersonic speed from the diamond growth area.

At present, only little information is available on the incipient crystallization and diamond growth processes in the formation of DN, and their effects on the properties of raw product remain almost unknown for two obvious reasons. First, analysis of detonation of one of the most powerful explosives is inherently difficult. Nevertheless, we already know quite a few stable nanocarbon products from carbon plasma generated by high-energy processes involving irradiation of intense laser and shock waves from detonation, spark discharge and meteorite impacts. The known products include highly crystalline diamond particles, C_{60} and higher fullerenes, multishell fullerenes, and carbon nanotubes in addition to low-crystalline and amorphous carbons. Systematic mechanistic analysis of all these recent results from high-energy carbon processes should be a rewarding project. The detonation synthesis of nanodiamond will provide valuable information on the initial processes in the diamond crystallization. The second reason for the difficulty for us to follow the detonation process is the fact that most of the past detonation works have been exclusively carried out in the Soviet military regime and remain unpublished in western languages.

It should be noted that the lack of incentive to study the nanodiamond growth process in the detonation environments is at least partly due to the past misconception on the structure of agglutinates. Until recently, many DN researchers believed that the raw product is straightforward *van der Waals* aggregates of primary nanodiamond crystals and its most striking property of high polarity stems from functional groups like carboxyl and hydroxyl on the particle surface, in spite of the fact that these functional groups have never been confirmed nor analyzed. Today, our image of primary particle has changed into more complex structure.

We noticed the phenomenon of agglutination not by systematic efforts but by following our chemical instinct to go down to the primary constituent of matter with the first priority. We were fortunate to have begun the investigation at the time when nanoparticles were becoming the foci of attention by scientists and convenient tools of measuring the size and distribution of nanoparticles by dynamic light scattering (DLS) method available. However, when we were given in 1999 a sample of DN agglutinates for the first time from its importer, who was perplexed by its slow sales in spite of the high recommendation by manufacturers, we were the perplexed ones: why they sell DNs in powder form but not in colloidal solution? Does this mean that visible aggregates can be easily disintegrated? This is, however, unlikely because quick aggregation usually means difficulty in disintegration, hence even if disintegrated, the dispersion should reaggregate quickly. There was no instruction from the manufacturer on the conditions of dispersion.

In 2001, we began serious attempts to disintegrate the commercial DN powder into its primary particles, namely single nanocrystals of diamond particles while following the particle-size distribution by DLS measurements. We soon found out that the aqueous slurry of conglomerates in the micron-ranges could be quickly destructed by light sonication into a broad distribution centered at 200 nm in diameter. From here on, more and more intense sonication, first with a 400W laboratory supersonic processor, then eventually with a 2kW industrial ultrasonic rod, shifted the center of distribution to 100 nm and then to 60 nm, but never beyond. At this point, the slurry had turned into an almost transparent and gray-colored colloidal solution containing single but broad particle-size distribution. These colloidal particles are the core agglutinates that disguised the naïve engineers in small industries like plating houses for many decades that they have believed to be handling ultradispersed diamond particles.

By the fall of 2002, it was clear that we were encountering the unknown but extremely tight binding situation but we could think only of covalent chemical bond formation from our knowledge of the chemical bonds. It did not seem totally unlikely that, when the diamond crystal growth process in the center of detonation stopped, temperature and pressure are still high, and interparticle recombination of carbon radicals in the diamond growth centers will take place to bind diamond particles by C–C bonds together to form tight assembly. We have once mentioned this mechanism of C–C bond formation on the surface of primary ND particles in a publication [8]. Actually this supposition led us to apply brute force methods to break up interparticle C–C covalent bonds in the core agglutinates to isolate primary particles.

1.2.2 Isolation of Primary Particles

After testing a few known powerful pulverization methods including dry atomization technique, we encountered in 2002 with the wet beads milling method, for which a few commercial apparatuses of diverse types were also already on market, claiming to produce nanoparticles by top-down destruction of micron-sized particles with spherical ceramic microbeads which were also becoming available at that time. Luckily the beads milling with zirconia beads with 30 μm in diameter worked. Even now this is still the only method that produces primary DN particles from its agglutinates in large quantities with enough efficiency suitable for industrial production. With the smallest available mill having a capacity of 150 ml (Figs. 1.2 and 1.3), we can routinely produce 50–100 g of primary particles per day. By running larger beads-mills in parallel, it would be straightforward to increase the production capacity to a few kg per day.

Considering the novelty of this milling method for materials-related R&D laboratories, it may be appropriate to give a brief explanation on the mechanism of beads milling here (Fig. 1.4). Like ball milling, beads milling crushes the objects by concentrating the momentum of moving beads into a small area of collision involving spherical beads and the object. In order to give enough momentum to small and light beads, the following strategies are used:

1. High density material like zirconia (specific gravity 6.0 g/cm^3) is used to give beads higher momentum (to increase m term of momentum mv).

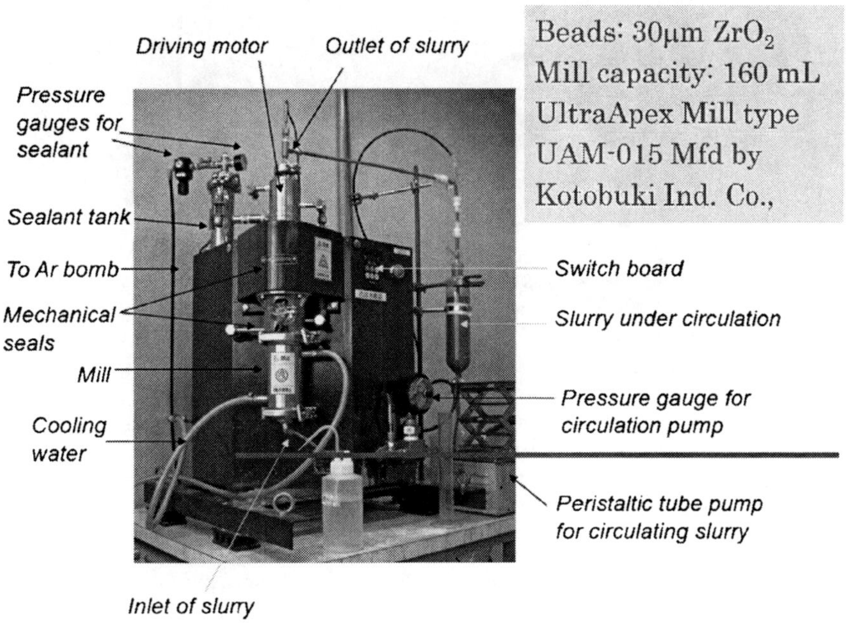

Fig. 1.2 A vertical-type beads-milling machine being used in author's laboratory to disintegrate detonation nanodiamond into its primary particles with 4-5 nm in diameter for test production.

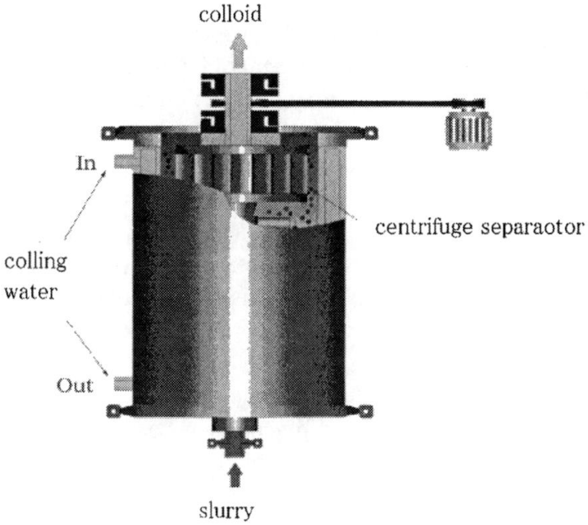

Fig. 1.3 Expanded illustration of vertical mill with a partial view of centrifuge separator, which can be readily decomposed into parts, cleaned, and reassembled. The separator, inner mill-wall and rotor blades (only partially visible) are made of zirconia

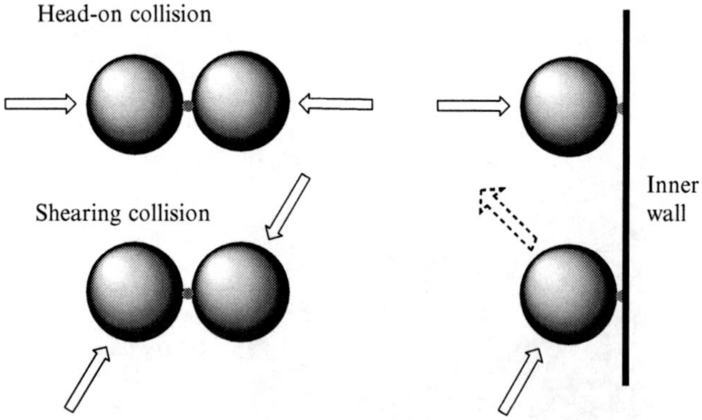

Fig. 1.4 Imaginative crushing mechanism of beads-milling. Pinching of an agglutinate (small red sphere) at a small contact area between a pair of zirconia beads (larg grey balls) in a head-on collision on at high speed is the most naïve image, but deemed unlikely in a rotating flow [top left]. More likely is the pinching between a pair of beads under shearing contact (bottom left) in turbulent flow or that between the inner wall and a bouncing bead (right). Small area of collision involving near-perfect spherical microbeads moving at high-speed, high density of zirconia beads, high-packing ratio (70-80% of mill space) of beads contribute to high probability of pinching as illustrated. Pinching produces large destructive impact upon the agglutinates to cause cleavage of the tight coherent interfacial Coulombic bonding between charged facets [18]. Simple calculations indicate that the local temperature at the point of pinching collision may reach 1800°C [Eidelman, E.D.; Siklitsky, V.I.; Sharonova, L.V.; Yagovkina, M.A.; Vul', A. Ya; Takahashi, M.; Inakuma, M.; Ōsawa, E. Diamond & Rel. Mater. 2005, 14, 1765-1769]. Note that the relative size of a bead and an agglutinates is much larger than illustrated (150~3000).

2. Beads are suspended in liquid medium (usually water) in order to remove heat evolved, and the suspension rotated at high speed in order to increase v term of momentum. However, the rotation rate cannot be increased beyond certain limit due to cavitation of water that occurs under the effect of gravity and consumes much of the rotational energy input. In the commercial beads mills, the rotational speed is generally of the order of 4,000 rpm which requires up to 10 m/s of periphery speed of the rotor.
3. Beads are prepared to precisely spherical shape in order to reduce the area of collision to as small as possible and to increase the impact of collision.
4. In order to increase successful crushing collision to produce astronomical number of nanoparticles (micron beads produces nanoparticles), the number of collision is increased by loading the mill space with as much beads as possible. Usually beads are loaded to 70–80% of the mill-space.
5. Not only the head-on collision-and-pinching of beads with an object particle (which is rather rare in the agitation by a blade-rotor as in the usual beads-mill), but also the shearing collision-and-pinching is utilized to crush the object. This effect is provided simply by adopting axial rotation, which involves large gradient in the rotational speed in radial direction, which in turn creates countless number of shearing collisions among the beads rotating at different speeds in the same direction.

6. Conventional beads-mills are equipped with circulation capability of slurry through the mill to repeat the crushing until the desired dispersity is reached and with external cooling device to remove large heats evolved inside the mill. At the exit of circulation beads are separated from the slurry of object being crushed by means of centrifugal or filtration devices. According to our experience, the centrifugal separation is much more efficient than filtration which tends to produce clogging of meshes by aggregation-prone nanoparticles.

1.2.2.1 Operation of Beads-Milling

Note that, in the beads milling of DN agglutinates, we are not crushing the diamond nanocrystals top-down but destroying between the agglutinated primary particles. It seems that cleavage of single nanocrystals of diamond does not occur under the milling conditions as the final particle size obtained by beads milling and that of primary particles in the agglutinate as seen under TEM agree. Nor any serious destruction of zirconia occurs during milling according to visual inspection of the used beads under optical microscopes as long as excessive contact with exposed diamond is avoided by suspending the operation as soon as the particle size reached the desired range. Nevertheless, surface ablation of zirconia beads by contact with diamond nanoparticles occurs as shown by semi-quantitative ICP analysis of the milled DN product to give contamination with zirconia up to a few % if milling operation was continued too long. Contamination of SNBD product with zirconia may pose a serious obstacle depending on the final uses as zirconia is hardly soluble in acid or alkali and the latter reagents lead to serious re-aggregation of primary nanodiamond particles. We are now considering to use yttria (Y_2O_3) beads, which is readily soluble in concentrated hydrochloric acid, instead of zirconia. At the moment, we still use zirconia and suppress the contamination level to below 0.2% by adopting optimum condition obtained by a multifactorial analysis based on experimental design [14, 15].

Operation Under the Optimum Condition

After about every ten milling operations, the vertical mill portion of a beads milling equipment UltraApex Mill Type UAM-150 (manufactured by Kotobuki Industries Ltd., Tokyo) is taken apart to wash the rotation blades and centrifugal device, all made of zirconia (Fig. 1.3). The cleaned parts are dried, reassembled and tightened carefully. Then, 112 ml of zirconia beads (30 μm in average diameter, manufactured by Netsuren Co., Tokyo, by plasma sintering procedure, well ground in a blank mill, washed well with DI water to remove sub-nano pieces of zirconia detached from the surface and dried in advance) are loaded in the mill through the funnel from the top of mill in small portions (Fig. 1.2). Separately, a 50 g portion of DN agglutinate that had passed our preliminary beads milling test (see Sect. 1.2.2.3 below) was mixed with DI water in a 500 ml tall beaker to make up 400 ml of suspension, which was subjected to vigorous stirring in a Robomix emulsifier (type F, manufactured by Primix Company, Osaka) for half an hour at 5,000 rpm. The gray, thick and homogeneous slurry was then immersed in a Bakusen supersonic washing bath (equipped with 24 + 32 kW supersonic generators, manufactured by Honda Electronic Co., Toyohashi) for 99 min to ensure thorough decomposition of large aggregates.

After starting the flow of cooling water through the jacket of mill and applying 0.4 MPa of Ar pressure to a pair of mechanical sealing in the inlet and outlet of mill, the external reservoir of a 500 ml separation cylinder is added with 100 ml of water, which was then slowly pumped in from the bottom of mill to wet the beads. Thereupon the rotating motor was turned on. While the rotor periphery speed increases, the reservoir is filled with the slurry of DN agglutinates. When the rotor speed reached the projected value of 7 m/s, the peristaltic tube pump is turned on to begin the circulation of slurry through the mill at a flow rate of 200 ml/min. As the total volume of slurry is 520 ml after combining washing water, it takes 2.60 min for the whole slurry to complete one circulation pass. According to experience, 30–40 passes are required to attain a complete dispersion. Hence, the circulation was continued for 117 min or about 2 h.

After the milling has been continued for about half an hour, the gray slurry slowly begins to acquire dark tone, which becomes darker and darker and finally pitch-black (see the content of slurry reservoir at the right of Fig. 1.2, which is already black). At the end of milling time, the initially thick and gray slurry containing fine and visible particles had changed into black and much less viscous colloidal solution of single nano buckydiamond particles. At this point the apparent concentration of SNBD exceeds 8%, and the colloid is not completely transparent but slightly turbid. At this concentration range, it is likely that partial reaggregation of primary particles already is taking place. When the solution was diluted later for particle size determination to 1–2%, the color changed into brownish and the solution became transparent. In the course of first 30 min of circulation, the volume of slurry decreased by about ten ml. A part of volume contraction may be caused by the release of air that was kept within nanopores in the agglutinates.

At the end of milling period, the agitation is maintained at the top speed while the colloid solution was taken out from the top of mill. Beads were washed by circulating five 50 ml portions of fresh DI water. Combined colloidal solution and washings (770–800 ml) are then irradiated with powerful supersonic wave in a circulating setup consisting of a 400W laboratory ultrasonic processor (type UP400S manufactured by Hielscher Ultrasonic Co., Teltow, Germany) attached with a titanium sonotrode (type H22), a vertical flow cell (type H22LD) and a peristaltic tube pump. Colloidal solution was circulated at a speed of 600 ml/min for 1 h. The sonication setup is similar to the beads milling circuit except that no mechanical sealing is provided; hence the pressure inside the flow cell is only slightly above atmospheric pressure. This feature imposes us to adjust the rotation speed of tube pump carefully so that no overflow of colloidal solution occurs from the loose top of flow cell. In this way, we noticed that the viscosity of colloidal solution decreased considerably during the supersonic irradiation. This observation suggests that the beads milling did not completely destroy all the interparticle bonding in the agglutinates but left considerable number of partially broken bonds (Coulombic, *vide infra*) to give somewhat viscous colloid under the operation condition of minimum passes. This situation seems to have been smoothed out by irradiation of powerful supersonic wave to complete the disintegration and reduce the viscosity. Remarkable effects of powerful sonication for certain classes of nanoparticles are well documented [13]. Actually DLS measurements before the sonication processing have often given erratic results. We preferred this way to over milling in order to suppress contamination of zirconia as low as possible.

Thereupon, the sonicated colloidal solution was subjected to centrifugal separation of still remaining large particles and other contaminants using a tabletop centrifuge (type 5200 manufactured by KUBOTA Corporation) equipped with a swing bucket rotor (type ST-720, 200 ml × 4) at 3,500 rpm for 1 h. Small amounts of gray solids precipitated at the bottom of buckets and thin layer of oily film floated on the surface. These were removed by decanting the centrifuged colloidal solutions into a 1 l separately funnel. After leaving the funnel standing overnight, about 30 ml of bottom liquid and a few ml of the supernatant oily layer were discarded and the rest was transferred into a stock bottle.

The solid contents in the final colloids were determined by pipetting out three 1 ml portions of the colloid into three 2.2 ml vials, whose tare weight with marked plastic caps had been measured, evaporating the water first under atmospheric pressure in a desiccator containing molecular sieve for a few days followed by heating at 200°C under a vacuum of 15 hPa in an electric vacuum oven overnight, leaving the oven to cool down to ambient temperature under vacuum, resuming normal pressure, quickly closing the vials with the plastic cap, and measuring the weight of capped vials containing completely dried solids. Particle size that falls within our provisional standard range of 4.8 ± 0.7 nm is accepted. This standard range is an average of three independent determinations (Scherer's method using X-ray intensity of an agglutinate sample, image analysis of a TEM picture of well-dispersed SNBD, and DLS determination over a wide concentration range of a standard SNBD sample). The stock black colloid solution was kept in refrigerator after substituting dead space in the storage bottle with Argon. The solutions show no sign of sedimentation at least for a year.

1.2.2.2 Other Methods of Disintegrating Agglutinates

High-power supersonic irradiation alone is not capable of destroying the DN agglutinates but effective when performed after the beads-milling as shown above. Ozawa combined the two methods into one by irradiating a slurry of DN agglutinates mixed with zirconia beads with Hielscher's laboratory ultrasonicator equipped with a horn-type sonotrode H3 [16]. Agitation is provided by the shock of irradiating focused sonic wave from the sonotrode. This setup offers a convenient method for disintegrating 50–500 mg of agglutinates in laboratory, but suffers from contamination with titanium particles (material of sonotrode), and requires fine manual tuning of conditions and is difficult to expand to larger scales.

1.2.2.3 Miniature Beads-Mill for Laboratory Testing

We must mention here that not all the agglutinates being manufactured in several small detonation factories in previous Soviet countries and in China are disintegratable. Sometimes the slurry of agglutinates became abnormally viscous soon after the milling operation was started, to make circulation impossible and the operation had to be suspended to protect the machine. It is therefore desirable to perform a small milling test for every batch of agglutinates received to see if it is crushable into primary particles or not. Hence, we designed a miniature beads mill capable of performing the initial stages of disintegration of DN agglutinates using less than 1 g of the sample.

In order to achieve effective destruction of DN agglutinates with small and light microbeads, the beads must be moved at a linear speed higher than 7 m/s. However, the speed of agitation by rotation is limited by cavitation as mentioned above, therefore achieving this limiting periphery speed becomes difficult as the diameter of agitator in a vertical mill is decreased. Actually the beads mill we used above (mill diameter 10 cm and capacity 150 ml) is the smallest available. For this reason, here we use so-called film mixer, wherein cavitation is avoided by pressing the rotating liquid medium against the inner wall of mixer to form a vertical thin liquid

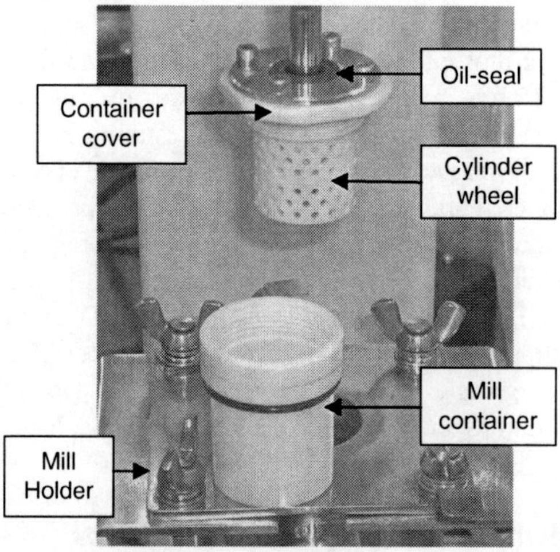

Fig. 1.5 Locally designed miniature beads mill made entirely of zirconia for the purpose of testing a new batch of agglutinate for suitability of beads-milling. Dimensions in mm: mill container innerφ30 h32, cylinder wheel outerφ26 h22. Cylinder wheel is fixed at the end of rotating axis through container cover holding an oil seal at its center hole. A narrow space of 2 mm is maintained between the inner wall of container and outer wall of wheel throughout the test milling. The bottom end of wheel is suspended a few mm above the inner bottom of container. Standard composition of test miture: 0.7 g of agglutinate, 5.8 g of 30 micron zirconia beads, and 7 ml of water. Wheel is rotated at 7,000 rpm for 15 min under external cooling with running water, rotation suspended and the milled mixture let stand for a few minutes to settle down beads at the bottom. Aliquots of supernatant colloidal solution are pipetted out, diluted to 2.2 to 1.8 w/v % and subjected to particle-size distribution analysis by dynamic light scattering

film moving upwards against gravity along the wall. The pressing is realized by using a cylindrical wheel having slightly smaller diameter than the wall. In the liquid film is generated a large gradient in the rotational speed along radial direction, thus the microbeads will collide with the large shearing force and destroy the pinched agglutinate particles.

This concept was realized by our in-house designed miniature beads mill shown in Fig. 1.5. In this miniature mill, the generated heat is removed by cooling with a jacket of running water and the pressure that is built up by the heat of disintegration is released through an oil seal attached at the top of the mill and holding the axis connecting a high-speed motor and the cylindrical rotation wheel. See the Figure caption for detailed operation conditions. Disintegration of DN agglutinates could be successfully performed to give almost comparable particle sizes as obtained with the Apex mill.

1.3 Characteristics

1.3.1 Spontaneous Polarization

To our great frustration, we have had difficult time for a long time after successfully isolating the primary particles of DN because of their unpredictable behaviors and persistent black color of the colloidal solution. The nature of force responsible for the formation of stubborn agglutinates remained unsolved for some years. Guiding theoretical perspective continued missing, primarily because the primary particle was too large to perform reliable calculations. Surface structure was not clear. Conflict between TEM observation (clean diamond) and Raman feature (high proportion of sp^2 carbon) continued [17]. A few interesting leads to potential applications were found, but we could not understand its mechanism in most cases. For example, SNBD was found to act as a promising platform for a new type of drug carrier, but again the nature of bonding between diamond and drug molecules was unclear. It was at such difficult time when Barnard and Sternberg published the results of their systematic SCC-DFTB (selfconsistent- charge density-functional tight-binding) calculations on the smaller models in late 2007 and early 2008 [9, 18].

In short, they discovered migration of electron from inside to outside of a diamond nanocrystal, which not only led to the first recognition of interfacial bonding by Coulombic force in the face-to-face orientation of neighboring nanocrystal particles, but also offered a reasonable explanation for the outstanding polar nature of SNBD particles. Extending their computational results on the smaller models to the much larger SNBD particles should be done carefully as the properties of smaller nanoparticles should be size-dependent in this size-range and the level of theory used is still quite low. Nevertheless, the self-polarization in the complex structure of DN primary particle, which consist of highly electropositive diamond core and highly electronegative fullerenic shell, seems quite likely [19].

1.3.2 Hydration

The importance of surprisingly high stability of aqueous colloidal solution of SNBD cannot be overemphasized [19]. Our early success in the isolation of primary particles owes much to the high affinity of water to the polar facets of SNBD crystals. If it hadn't been for this tight hydration onto the SNBD surface, the isolation work would have been much delayed. The presence of strong hydration shell is supported by experimental observations. We first noticed ready gel formation from colloid; colloidal solution forms soft gel upon evaporation of water in a rotary evaporator from the 5–10% colloid concentration obtained by beads-milling to above 12%. The gel becomes thicker and viscous and its black color more intense as more and more water is evaporated and finally gave dry mass that can be pulverized by grinding in mortar to fine and light brown powder. However, we were much astonished when we found 2–6% water still in such a powder that had been dried under a vacuum of 50 hPa at 60°C for several hours (Fig. 1.6).

Shortly before Barnard and Sternberg published their seminal theoretical results, we have observed by DSC analysis of the SNBD gel a fraction of water freezing only at −8°C [20, 21]. Detailed analysis led us to propose a model of hydration shell structure surrounding a dispersed colloidal SNBD particle in water (Fig. 1.7). According to this analysis, the hydration shell consists of two layers. The outer melting/freezing layer is in direct contact with bulk water, but water molecules in this layer are well-absorbed and -oriented to the particle surface that their reorganization to ice structure needs extra energy to freeze only at −8°C. However, there will be an inner layer, wherein water molecules are so strongly bound with the surface (charges) that these will never freeze nor exchange with outer layer. Namely the inner layer will always stay on the surface and move together with the particle. Then, it is attractive to interpret the persistent water contents after deep drying of soft gel as the result of strong Coulombic bonding between polar water molecule and electrostatic potential field found on the facet in the Barnard model of SNBD [18]. As water is bipolar, it can bind with positive or negative charged facets.

When drying of soft gel is stopped short and the slightly wet powder is taken out and dried slowly in nitrogen atmosphere until the water content dropped to 22–24%, we expect that the resulting particle have lost the outer hydration layer, but still carries the inner layer intact according to Fig. 1.7. This is an interesting gel state, wherein primary particles do not touch directly each other, therefore still perfectly dispersible, but solid state. Following this strategy, we actually obtained fine and dry powder retaining large amounts of water corresponding to about one fifth of its dry mass, but could be easily dissolved in water by immersing in a supersonic washing bath for 3 h to give perfectly dispersed colloidal solution. We call the solid-looking gel as *hard hydrogel*, and use it as a convenient substitute of a colloidal solution for long transportation. However, hard hydrogel easily loses water when left exposed to open air; hence, it should be always kept in a tightly closed bottle.

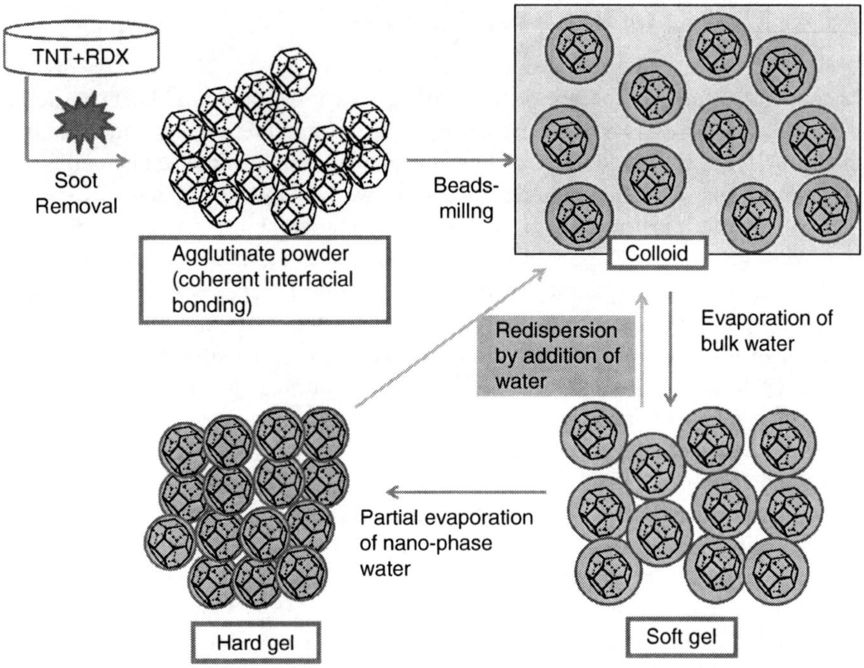

Fig. 1.6 Illustration of the dispersed states of SNBD primary particles of detonation nanodiamond in water. Decomposition of the agglutinates in 5-8% aqueous suspension by beads-milling first produces surprisingly stable, black and clear colloidal solution. Note that the formation of perfectly dispersed SNBD primary particles in concentrated colloid is quite an unprecedented event. Evaporation of water give soft and then hard gels, but the gels can be readily re-dispersed in colloidal solution upon addition of water and irradiation of ultrasonic wave. Note also that agglutinates cannot be regenerated by simple removal of water. Circles surrounding SNBD particles signify strongly bound layers of water molecules on the SNBD surface. Typically soft gel contains 12-20% of SNBD where hard gel contains 20-25% of water

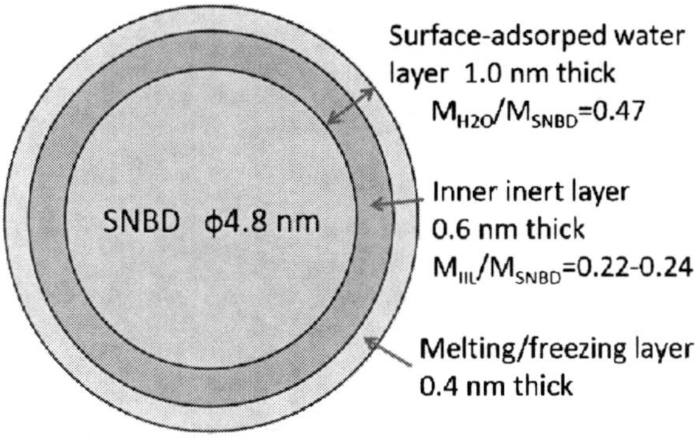

Fig. 1.7 Illustration of a double-layer model for hydration shell over the surface of SNBD. Relative thicknesses of shells and diameter of core are approximately proportional to those obtained by DSC experiments [20]

1.3.3 Mechanism of Ligand Exchange

Encouraged by successful interpretation of persistent hydration in SNBD particles, we will briefly discuss here the new perspective of interpreting and understanding the behaviors of SNBD in terms of ligand exchange reactions on the charged facets [19]. Hydration of SNBD particles can be viewed as a manifestation of new and general interaction principle which involves nanocrystal surfaces as the reaction sites. In chemistry, the reaction sites are always atoms in the molecule, but in the case of SNBD, it is more convenient to regard facets as the reaction site, because the electrostatic potentials extend over the facet and atoms within a facet have more or less similar charges and indistinguishable in this respect. Interparticle interaction can be treated as surface/surface interactions, but we will also encounter with surface/molecule, surface/ion, surface/electron interactions involving nanoparticles and environments. Here, the term surface includes facets, edges and apexes of crystal.

One characteristic of Coulombic bonding involving charged facet is the lack of directionality, in contrast to strong directional nature of covalent bonding involving *p*-atomic orbitals. This means that the ligand bound to diamond crystal surface is free to move over a facet. The edge between the two facets and apexes connecting more than two facets are sterically more open and vulnerable to the attack of incoming reagent than within facet. The exchange reaction will preferably take place at the edge or apexes involving S_N2-like transition state rather than in the center of facet (Fig. 1.8).

General expression of ligand exchange reactions on the surface of nanocrystals will be as follows:

$$F^- /(HOH)_n + E \text{ Æ } F^- /E(HOH)_{n-1} + H_2O \qquad (1.1)$$

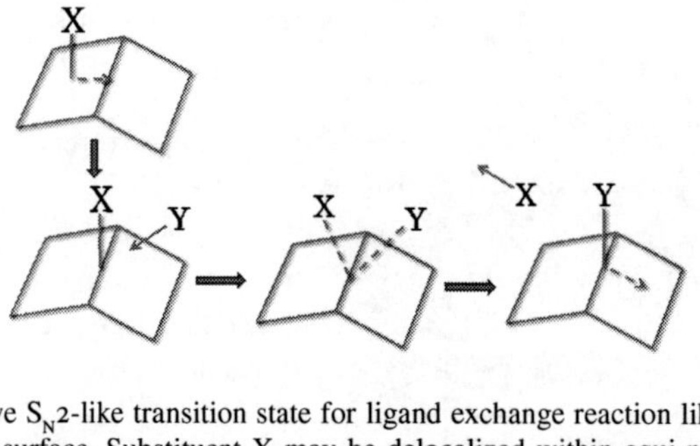

Fig. 1.8 Speculative S_N2-like transition state for ligand exchange reaction likely taking place at the edge of crystal surface. Substituent X may be delocalized within equi-potential region in a facet but sterically accessible to the attack of incoming reagent Y. S_N2-like transition state for ligand exchange reaction likely taking place at the edge of crystal surface

$$F^{+} / (OH_2)_n + N \text{Æ} F^{+} / N\,(OH_2)_{n-1} + H_2O \tag{1.2}$$

where F^- denotes a crystal facet having negative electrostatic potential, F^+ a facet having positive potential, E means an electrophile like electrophilic chemical species, positive cation, positive pole of a dipolar molecule or a polar functional group in molecule, and N a nucleophile like nucleophilic chemical species, negative cation, negative pole of a dipolar molecule or a polar functional group in molecule, and a sign "/" signifies Coulombic attractive bonding between the facet and ligand. At the moment, basic information and knowledge on the equilibria in, e.g., (1.1) and (1.2), are still totally missing.

Barnard suggested coherent interfacial Coulombic interactions between the neighboring particles of SNBD as the reason for the unusually tight agglutination observed [18]. This seems to be the most likely explanation to the long-lasting problem which remained unsolved since the discovery of detonation nanodiamond for 45 years. More specifically, two kinds of facet-pairs, $\{100\}/\{111\}_b$ and $\{111\}_a/\{111\}_b$, are charged in opposite signs, but the change distribution not necessarily simple. When these facets are aligned parallel in close distances and rotated around the common normal passing through these facet pairs, strongly attractive interactions of about 2.8 eV appeared at a distance of about 0.18 nm, which are comparable with a C–C covalent bond in strength. A number of these interfacial interactions are likely to have been formed among the pristine DN particles during a considerably long time of high-temperature period after the shock wave of detonation passed. Requirement of the long high-temperature period for the "coherent" bonding configuration to appear excludes occurrence of such strong agglutination under ordinary conditions like conventional drying SNBD powders. Therefore, the reaggregation of once beads-milled agglutinates is in general weak because only the "incoherent" interfacial Coulombic interactions are involved.

The hydration of crystal surface in SNBD cannot be very strong as the nature of interaction is a *charge-dipole* type. Interactions involving ions like proton, metallic ion, ammonium ion are of *charge-charge* type, and should be much stronger than hydration. As expected, the high sensitivity of colloidal stability of SNBD particles to pH values has been repeatedly observed. Most recently, Huang and Dai demonstrated large scale assembly of SNBD particles when its aqueous colloid is acidified to below *pH* 4 to produce whiskers, fibers, and films of SNBD [22]. Water-soluble ammonium salt of anti-cancer drug doxorubicin forms very strong bonding with SNBD particles in water, to precipitate as a complex gel. This gel is now being developed as a potential drug carrier. The drug carrier gel stays undissociated in the course of transportation through blood stream to cancer cell due to strong *charge-charge* interaction of the drug molecule with the crystal surface. After penetrating through the wall of cancer cell, the drug can be released by increasing salt concentration or lowering *pH*. All these changes should fit to the reaction schemes like (1.1) and (1.2), hence precise control over the rate of drug release should be possible by knowing the physicochemical features of the equilibria.

1.3.4 Core-Shell Structure

No less important to the understanding of the behaviors of SNBD particles is its internal and surface structure. The SCC-DFTB calculations of Barnard/Sternberg mentioned above [9] clarified several crucial aspects that have been unsettled for a long time. First thing to be mentioned is the confirmation of the significance of graphitic shell that have been thought to exist on the surface, but recently denied based on TEM observations [17, 23]. As shown in Fig. 1.1c, truncated octahedral all-diamond C_{1639}, the largest model used in their calculations, was geometry-optimized to produce extensively graphitized outer shell. According to their result, sp^2-hybridized carbon atoms comprised 34%, sp^{2+x} carbon 22%, and sp^3 carbon 44% of the total number of carbon atoms, respectively. Namely, more than half of the starting carbon atoms had rearranged from *sp^3 to sp^2 or its intermediate state of sp^{2+x}* hybridization state, respectively, in its energy minimum structure. The proportions of combined non-sp^3 carbon atoms are clearly overestimated in view of the results of the other, smaller models in which most of diamond carbons have been rehybridized into graphitic carbons. Notwithstanding the overestimation, we take these results as the strong evidence for the surface carbon atoms in the small nanodiamond to preferentially rearrange into more stable sp^2 carbon. For our SNBD particles, which is supposed to contain about 8,000 carbon atoms, it would be reasonable to estimate that the total non-sp^3 carbon atoms should be between 10 and 20% at the largest.

There are a couple of strong pieces of evidence as well for the existence of graphitic layer(s) on the surface (and below) of SNBD. One is the development of intense black color when agglutinates are disintegrated into its primary particles by beads milling (Figs. 1.2 and 1.9), and the other is intense Raman G-band at 1,600 cm^{-1}. Furthermore, the surface graphitic layers of SNBD are extremely difficult to remove. In the past few years we have been attempting almost all of the known oxidation reactions of graphitic carbon atoms to produce transparent and colorless single nano-diamond particles, but the best result so far obtained was heating with molten NaOH at 800°C for 5 h under nitrogen atmosphere, when careful Raman analysis revealed decrease in the intensity of broad G band to 55% of the starting value. Combining all these results, we conclude that the most likely and the most stable structure for the primary particle of DN is the novel triple-decked core-shell structure as originally suggested by Barnard and Sternberg [9] but modified slightly to fit a few experimental observations. Namely, the diamond core carries on its uppermost portions a few layers of strained structures in between diamond and graphitic shell, and one or two layers of graphitic shell covering the core (Fig. 1.10a). The diamond and graphitic shell are considered highly crystalline. Any attempts to remove the surface shell will destabilize the novel composite structure and also make the overall shape irregular.

The second important geometrical information that the calculations by Barnard and Sternberg provide is the exposure of diamond structure in the {100} facets (Fig. 1.10b). This is a natural consequence of the diamond-graphite rearrangement mechanism, but so far neglected in the structural discussion of DN, probably because of the very strong

Fig. 1.9 Photographs of various forms of SNBD products. Dried SNBD flakes are intense black in large particles (upper left) but becomes brown when pulverized well in mortar (lower left). Likewise aqueous colloid solutions above 1% in SNBD concentration are colored in intense black (right) but becomes brown in lower concentrations

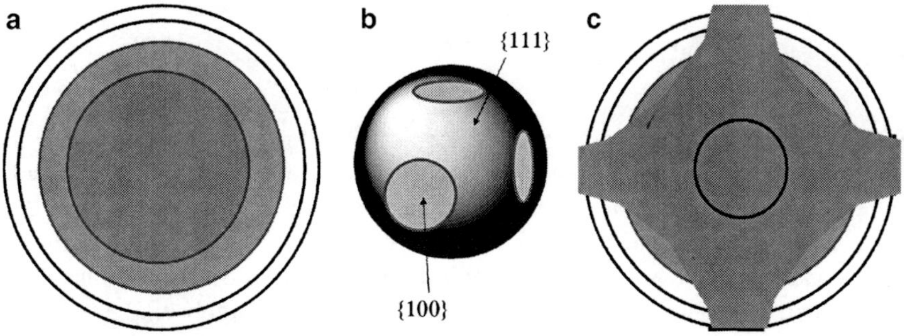

Fig. 1.10 Geometrical structure of SNBD primary particle. (**a**) The surface of diamond core (blue) is covered by and bonded to a layer of sp^{2+x} carbon atoms (vermillion). Then comes a thin empty space surrounded with graphitic shell, which in turn are layered with spacing. (**b**) However, the outermost surface has large voids, theoretically six, originating from {100} facets which remain ungraphitized (orange yellow). (**c**) More realistic cross-sectional view. Note that these pictures are based on a truncated octahedron model, which has not yet been fully justified

impression of TEM pictures on Banhart diamond synthesis, or onion-to-diamond transformation, where perfect one or two layers of spherical onion shells remain unconverted and surround the spherical diamond core [24, 25]. This is a diamond-to-onion transformation reaction and different from what we are discussing here.

Therefore, the idealized surface structure of our composite nanocarbon is a holey giant fullerene with holes of six {100} facets (Fig. 1.10c).

This recognition of partial surface exposure of diamond internal structure in SNBD rationalizes our earlier observation that SNBD particles provide very efficient seeds for the homoepitaxial growth of CVD polycrystalline diamond film. It is these patches of diamond {100} facets that served as the nucleation seeding (*vide infra*).

Much of the statements made above, including the presence of graphitic shells in SNBD, have been confirmed by a recent elegant research of fitting high-energy X-ray diffraction patterns of DN to real space by utilizing geometry optimization through molecular dynamics by Hawelek and his group [26]. The only difference is that Hawelek did not detect sp^{2+x}, but this is a natural consequence of missing X-ray diffraction from low-crystalline portions.

Finally, we must note that our previous belief that the highly polar nature of DN primary particles was caused by the presence of polar functional groups like COOH and OH on the surface was probably wrong. Infrared evidence for the existence of COOH and OH groups are weak and there is no chemical proof like alkaline titration for the existence of COOH groups. Ji *et al.* have long proved, by means of in situ infrared spectra of heated agglutinates powder, that OH groups originate from the absorbed water [27]. We are now very sure that there is no C–OH group on the surface of SNBD: all OH vibrations stem from the hydration water. The interpretation of ς-potential in SNBD (positive when washed with acid and negative when washed with alkali or ammonia) should be reconsidered.

1.3.5 Black Color

Still, the origin and removal of black color in the colloidal solution of SNBD are controversial and remain unsolved. In order to keep stability of SNBD, it may be helpful to keep the graphitic layers on the surface. Hence, if the origin of intense black color is mainly due to the graphitic layers, the color of SNBD will remain intense black, a bad news for optical applications of SNBD. On the other hand, the size of SNBD belongs to the range of Rayleigh scattering. Highly effective scattering of diamond nanocrystals in high concentration will certainly produce black color from the loss of emitting light (Fig. 1.8a).

1.4 Applications

1.4.1 Three Major Uses Under Developments

Having decided to leave black color as it is, we are exploring applications of the novel carbon nanoparticles of SNBD as they are now, namely black in color and core-shell in structure. From among more than a dozen of promising leads that have been found so far, we briefly introduce here three major developments that are going particularly well, namely drug carrier platform, CVD seeding, and lubrication.

1.4.1.1 Inactivated Drug Carrier/Slow-Releaser

Probably the most well-known application of SNBD, especially in the US, is its novel drug carrier capability, discovered by H. Huang and D. Ho of Northwestern University in 2007 [28]. The drug carrier actions of SNBD–drug complex gel have been developed well in collaboration with University of California, San Francisco [29–31]. As many chapters of this book are concerned with this project, we will mention only on the nature of binding between the crystal surface of SNBD and the drug molecules. Our approach has been mentioned under Sects. 1.3.1–1.3.3, in some detail.

1. There are several characteristic advantages in the Huang/Ho drug carrier: SNBD is a multi-pole, with ten positive and four negative facets in an idealized truncated octahedral model [9,18]. Therefore, one nanocrystal can carry several drug molecules. If we think doxorubicin hydrochloride (I) as the typical target to bind and carry, each drug molecule will cover almost one facet. In that case, one facet–drug interaction will have a binding energy comparable to that of a C–C bond energy.
2. Actually, the SNBD–drug complex forms tight gel, which forms a net work of incoherently aggregated SNBD particles. Nanopores having diameters of 8–10 nm are formed within the net work of aggregates [21] and drug molecules are considered trapped within the nanopores. This means that the drug molecules are strongly bonded with its ionic functional group not only to a charged facet, but also enclosed within the pore space surrounded by a number of SNBD particles and is likely to find multiple bonding sites within the pore. This is the reason for the secure inactivation of drug, and also for the absence of leakage of drug during transportation through the bloodstream.
3. Gross stability constants of SNBD–drug complex, saturated concentration of drug in the complex, the effect of salt concentrations and pH values of blood, and other physicochemical characteristics of the drug carrier may be obtained beforehand in vitro using model molecules for the drugs to be delivered.
4. Once settled in the target sick cell, we should be able to precisely control the release of drug, hopefully very slowly, by adjusting pH and/or salt concentration in patient's blood. Especially powerful will be proton concentration in the blood.

In short, the Huang/Ho method of drug delivery seems to turn the art of drug delivery from encapsulation-dominated mechanical/chemical manipulation to a more readily controllable physicochemical method. It is hoped that the method will be brought to use in the nearest future, especially in cancer chemotherapy.

1.4.1.2 CVD Seeding

In our opinion, it is not yet possible for the art of producing CVD polycrystalline diamond film to offer dense films without pinhole and with flat surface in affordable cost, simply because of the lack of appropriate seeding technique. A simple solution to this problem would be to use SNBD particles for the favorable homoepitaxial growth. At first, we were not enthusiastic about this idea due to our own misconception on the structure of graphitic shell, which seemed to cover entire surface of nanodiamond particle to make access of the growth species to nucleation sites hardly possible. The first attempt to use pristine SNBD particles for CVD seeding was made only in 2007 by O. Williams and he made an instant success [32]. The success reminds us of the well-known experiment on the formation of C_{60} by Kroto and his coworkers who made it in the first laser-ablation run, although in their case C_{60} was not the final objective.

The advantages of using SNBD particles as the CVD seeding are numerous and overwhelming:

1. We need only very dilute colloidal solution to coat an object with a layer of single particles, hopefully to the highest possible density. Water may be used as the dispersing medium without causing any problem to dry the seeded substrate [32].
2. By simply immersing the object into a dilute colloidal solution under supersonic irradiation followed by rinsing, anyone can achieve considerably high seeding density of the order of 10^{11} particles/cm^2. It is recommended to use polished, smooth surface for higher seeding (*vide infra*). Nevertheless, even at this seeding level, the grown diamond film is dense enough and has no hole from the beginning. If the growth can be stopped at the early stage, surface is smooth.
3. Because of easy and fast growth by homoepitaxial mechanism, we can choose substrates material at our will. Si wafer is convenient and also practical because the diamond films are well bonded to this substrate by forming Si–C bonds [33].

An immersion seeding on Si wafer under typical condition is mentioned below, which gave a seeding density of 1.1×10^{11} particles/cm^2 (Fig. 1.11). Immersion method is particularly advantageous for wrapping complex 3D objects like MEMS products with thin diamond film. Attempts to improve the seeding density towards the highest attainable calculated density of 4.0×10^{12} p/ cm^2 for SNBD are in progress in our laboratories. Our champion record attained so far is 2.4×10^{11} p/cm^2, which was obtained by using an ink-jet patterning machine (Fig. 1.11).

There is a lot of room for further elaboration and developments in this area [34]. A challenging subject will be to align SNBD particles on the seeding substrate. Suitable alignment would lead not only to higher seeding density but also to the possibility of oriented growth of CVD diamond film and eventually to the growth of single-crystalline diamond film.

Experimental Procedure Involving Si-Wafer as the Substrate of Seeding

Cleaning of Si-wafer. The surface of commercial Si-wafer may be contaminated or even oxidized for various reasons, hence subjected to RCA cleaning as follows. In a 300 ml beaker were added 20 ml of 25% ammonia water (Guaranteed Grade

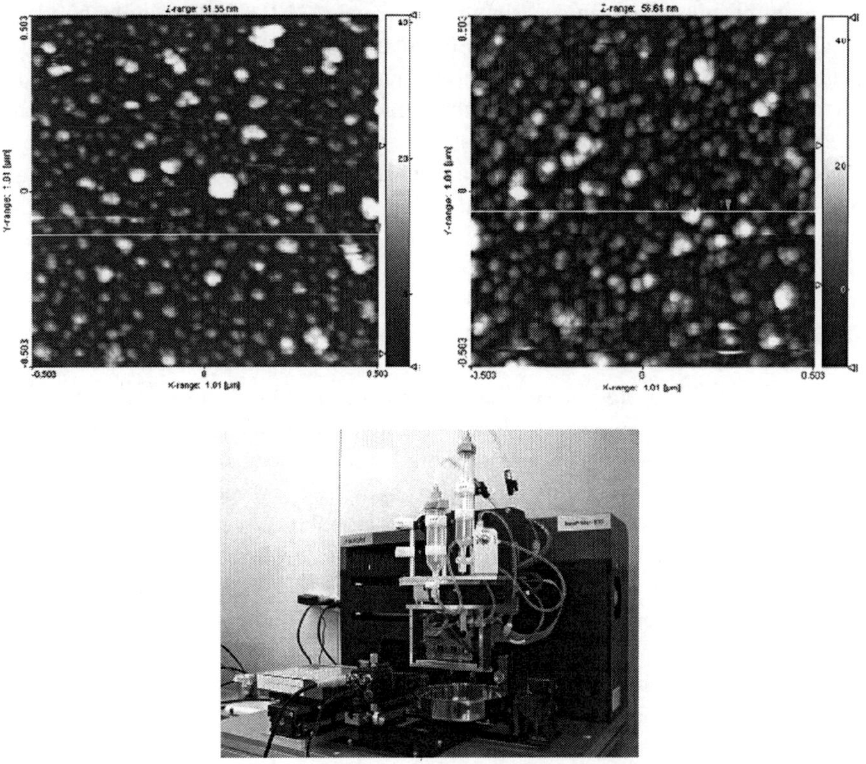

Fig. 1.11 AFM images of Si-wafer seeded by immersion method (top) and by ink-jet patterning method (middle) from 0.05% aqueous colloidal solution of SNBD particles dispersed to a size-distribution of 3.9±0.3 nm (98.6wt%). Seeding densities: 1.1×10^{11}/cm^2 by immersion, 2.4×10^{11}/cm^2 by ink-jet patterning. Below: Ink-jet patterning machine, NanoPrinter 500, MicroJet Co

from Wako Chemical Co.), 20 ml of 30% hydrogen peroxide solution, and 100 ml of DI water (NH_4OH:H_2O_2:H_2O = 1:1:5). To this solution was dipped a sheet of Silicon wafer ('50 mm in diameter, P<100>, 1–10Ω, 280±20μm, manufactured by Toyo AdTech Co.), and heated to 80°C for 10 min. The Si-wafer was taken out, rinsed with DI water, and dipped in the second 300 ml beaker containing 20 ml of 35% hydrochloric acid, 20 ml of 30% hydrogen peroxide solution, and 100 ml of DI water, and heated at 80°C for 10 min. The wafer thus treated was rinsed with DI water.

Immersion seeding of Si wafer. An aqueous colloidal solution (No. SI-22) of 4.91 w/v% SNBD was freshly prepared by beads-milling of agglutinate of detonation nanodiamond (Diamond Powder Co., Gansu, China, Lot No. XP1/056) according to the procedure mentioned above and determined to have a representative particle-size distribution of 3.9±0.3 nm (98.6 wt%), 51.9±21.2 nm (1.4 wt%). An aliquot of 1 ml was taken out from the stock solution by using Eppendorf pipette, diluted 100 times with DI water and subjected to ultrasonic treatment by immersing in a washing bath (Type 3210J-DTH, manufactured by Branson Ultrasonics Co.) for 60 min to obtain 100 ml of 0.05 wt% of colloidal solution. The Si wafer washed above was immersed, without drying, in this colloidal solution and irradiated with ultrasonic wave for 30 min. Thereupon the seeded wafer was immersed in DI water, sonicated for 10 min, and immediately placed in a small kitchen microwave oven (500W) for 20 s for quick drying. Seeding density was obtained by analyzing the AFM images according to the usual procedure (Fig. 1.11).

1.4.1.3 Nanospacer Colloid Lubrication

When C_{60} was discovered in 1985 and its macroscopic production began in 1990, we often heard talks among people about using it as an additive to engine oil for reducing the friction between moving parts, and later saw small packages of C_{60} sold in junk shops with claims for longer engine life and reduced CO_2 emission. Apparently this easy invention did not gain popularity, because these amateur inventors probably did not grasp the real size of C_{60} molecule relative to the roughness of polished surfaces, nor recognized high tendency of C_{60} molecules to coagulate in the solid state. However, the conception of using hard and ball-like small particles as sort of ball bearings in reducing friction is novel and worthwhile reconsideration. The basic idea is to use suitably sized nanoparticles as the rollers or ball-bearings between real contact points of interacting surfaces and prevents plastic deformation and flow from occurring in these points under boundary friction conditions. The effective size of a dispersed SNBD particles when used as nano ball bearings is about 8 nm (*vide infra*), hence appropriate for lubrication between conventionally polished surfaces. As the low nano-particles like SNBD quickly aggregate among themselves and become too large for boundary lubrication, it is absolutely necessary that these particles are dispersed in a liquid medium like oil or water to be used for this purpose. Finally, let us adopt the common strategy of wrapping the hard ball with soft shell from soft metal, high viscosity liquid and even oriented water cluster to absorb concentrated load at the real contact points under the boundary condition by plastic deformation. Thus, a new lubrication system is defined herewith. We temporarily name this system as nanospacer *lubrication*. Clearly there is no precedent for such a system, because SNBD is at the moment the only known dispersible and nearly spherical nanoparticle having suitable size.

It may be interesting to note that the nanodiamond colloid lubrication had been discovered in around 2006 in four places of Japan (AIST, Iwate University [35], RIKEN [36] and Nissan Motors Co. [37]), independently but all using dispersed SNBD particles supplied by us. However, none of us realized its novelty until recently [38]. For example, unequivocal decrease in the friction between sapphire slider and poly (acrylamide) hydrogel substrate was observed in aqueous colloid of SNBD (Fig. 1.12). Separate experiments confirmed that the effect is long lasting at least up to 9 h. The mechanism may be explained by invoking tight hydration shell, believed to move with the particle as mentioned above, which absorbs frictional force from the gel by plastic deformation (Fig. 1.13) [35]. We can readily envisage quick developments of *Lubrication Water*, the next-generation and environment friendly lubrication system based on the nanodiamond colloid lubrication system.

Somewhat higher effect was observed for the contact friction between cum and bulb lifter of automobile engine using synthetic oil as the medium and applied for patent [37]. In this case, glycerin mono-oleate (GMO), the well-known additive for ultralow friction between metallic surfaces [39], was used to form soft lubrication film over the surface of SNBD particle.

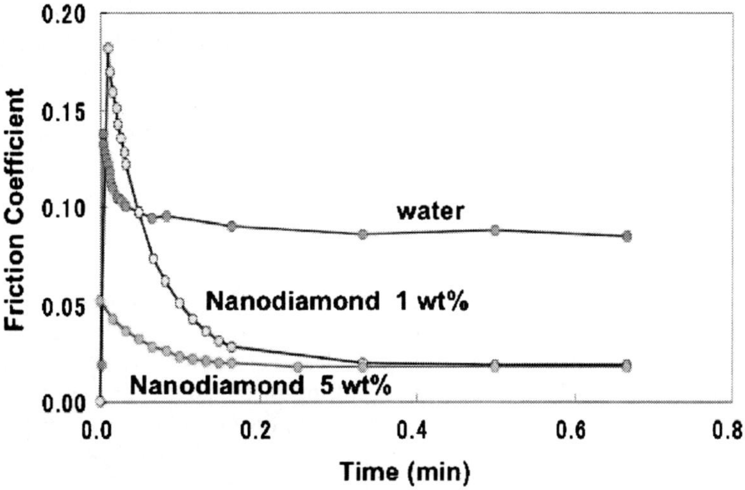

Fig. 1.12 Significant reduction of friction coefficient in sapphire slider /poly(acrylamide) hydrogel system in the nanodiamond aqueous colloid lubrication method. Experimental conditions: sapphire slider=φ2mm, load=10mN, contact distance =1.5mm, speed=10Hz

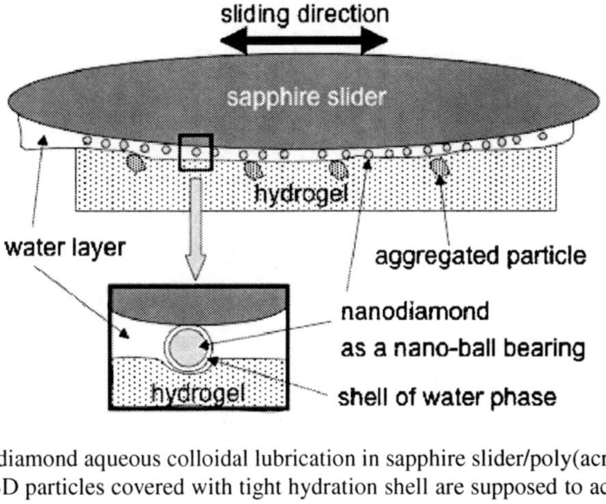

Fig. 1.13 Nanodiamond aqueous colloidal lubrication in sapphire slider/poly(acrylamide) hydrogel system. SNBD particles covered with tight hydration shell are supposed to act as a hard core/soft shell roller at the interface between ball and gel

1.4.2 Other Favorable Assets of SNBD as Industrial Material

Nanocomposites of SNBD are highly attractive as this is one certain way to escape from the inevitable trap that plagued industrial application of diamond: *the first shape is the final shape*. In other words, there was no way in the past to fabricate or mold diamond into desired final form because diamond cannot be processed nor deformed. However, if we could disperse SNBD particles into solid matrix, we could overcome the processability problem. In addition there seems to be almost unlimited option for the solid matrices, which include polymers, plastics, resins, metals, alloys, ceramics, glasses, etc. However, there is no general method known to disperse SNBD in solid matrices. We have seen a number of mediocre results that are obtained by simply mixing SNBD, either in the form of dispersed colloid or powder aggregates, using conventional methods of mixing. In general properties such as Young's modulus and hardness are improved appreciably by the addition of SNBD in considerably high proportions. Composites prepared in this way usually have poor dispersion regarding the distribution of SNBD particles and do not differ from the microcomposites. There is even a theoretical work on the invalidation of Hall–Petch theory in nanocomposites [40].

Recently, the first example of nanocomposite was reported wherein primary particles of SNBD are perfectly dispersed by means of *graft polymerization* [41]. Heating m-phenoxybenzoic acid in polyphosphoric acid in the presence of 1–30% of SNBD particles began with grafting of benzoyl cation at sp^2C–H of SNBD followed by Friedel–Crafts polymerization at the terminal m-phenoxy group to give a linear hydrocarbon polymer with a rigid SNBD particle in each polymer chain:

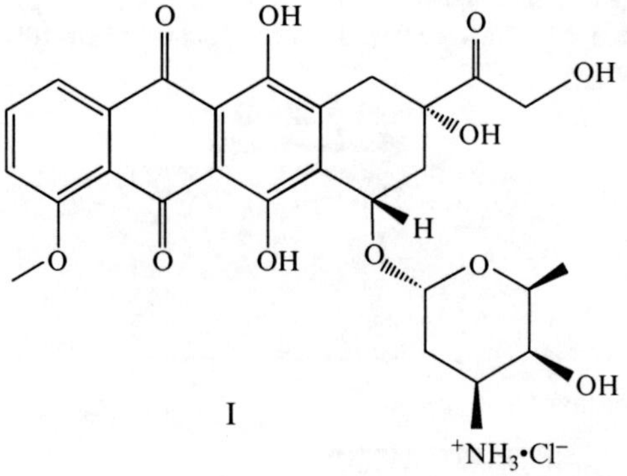

I

The meta-poly(etherketone)-grafted SNBD composites, mPEK-g-SNBD, thus prepared displayed homogeneous properties regarding color, viscosity, SEM, IR and other spectra. The most interesting feature of this polymer series is their thermal properties (Fig. 1.14). In the first DSC run (a), only the pristine mPEK polymer displayed endothermic *Tg* transition at 136°C, but all the other graft polymers show

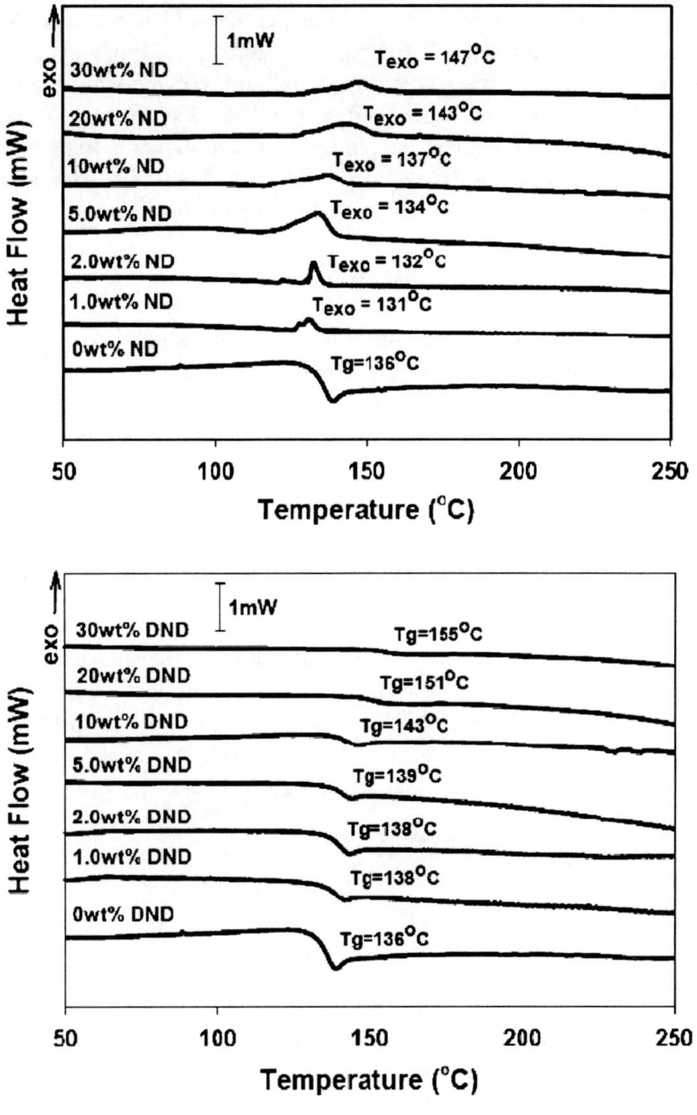

Fig. 1.14 DSC thermogram of mPEK-g-SNBD samples run twice (a, then b) with a heating rate of 10°C/min

only one exothermic peak at 131–147°C, increasing smoothly with increasing SNBD contents. In the second run (b), only *Tg* transition peak appeared increasing smoothly up to 155°C with increasing contents of SNBD. The exothermic peaks correspond to the release of shearing strain built up during polymerization with continued stirring in the increasing viscosity. Whereas this strain could be released well during cooling and storage of the flexible pristine mPEK polymer, graft polymers containing rigid and large SNBD particles stored the strain and could release it only when heated at temperatures higher than the polymerization temperature during DSC. Glass transition in the graft polymers took place only in the second run, after all the shearing strain had been released, but the transition energy decreased with increased SNBD contents. All these behaviors demonstrate hindered chain movements when grafted with SNBD particles, which were dispersed homogeneously in the polymers.

In the past few years, a highly successful *reinforced plating* technique has been developing in Russia. The thin metallic films obtained by reinforced plating can be regarded as a sort of nanocomposites of nanodiamond and metal. For example, Cr plating under universal condition (240 g/l of CrO_3, 2.5 g/l of 98% H_2SO_4 and 3.5 g/l of Cr^{3+}) containing 0.5–1 g/l of DN agglutinates at 45°C and cathode current density of 50 A/cm^2 gave Cr coating with Vickers's hardness about 25% higher when compared with the coating done without agglutinates. Use of SNBD instead of the agglutinates gave even higher increase (about 30%) in Vickers's hardness of the coating [42]. Initially, it was very hard for us to understand why such remarkable reinforcement effects appeared by the addition of large particles of DN agglutinate or SNBD aggregate in the acidic and electrolytic, aggregation promoting solution.

One likely explanation seems to involve adsorption of the charged facets of DN or SNBD conglomerates onto the cathode by Coulombic force followed by their disintegration on the surface of cathode. The first adsorption step (1.3) may be interpreted as a ligand exchange between adsorbed water and cathode on the facet of crystalline nanodiamond according to our basic scheme:

$$F^-F^+ \cdots F^-F^+ / aq + CAT^- \text{ Æ } F^-F^+ \cdots F^-F^+ / (CAT^-) + (H_2O)_x \qquad (1.3)$$

$$F^-F^+ \cdots F^-F^+ / (CAT^-) + H_2O \; \mathit{leftrightarrows} (F^- + F^+ + \cdots + F^- + F^+) / (CAT^-) \qquad (1.4)$$

where $F^-F^+ \cdots F^-F^+$ represents conglomerates of nanodiamond primary particles bonded together by interfacial Coulombic attraction (in DN agglomerates the interfacial bonding is coherent and strong, but in SNBD aggregates the bonding is incoherent and weak), aq denotes water molecules in hydration shell, and CAT means cathode. The second step (1.4) represents disintegration of dielectric particles attached on the surface of electrode of high-density direct current by electric polarization. Certain proportion of Coulombic facet–facet bonding in the absorbed aggregates or agglutinates will be loosened and destroyed by electric polarization with opposite direction. Once destroyed the separated facets will be immediately covered and inactivated for recombination by water or electrolyte ions. Therefore, there will be no recombination once the primary particles were generated.

According to this two-step mechanism, dispersion of ND particles on the electrode is only partial, in agreement with significant but hardly outstanding reinforcing effects of nanodiamond particles. The hypothesis of partial disintegration by electric polarization of dielectric material like DN agglutinates and aggregates should be confirmed. It is highly desirable to find a method of suppressing re-aggregation in SNBD colloid under plating condition. These will be highly rewarding efforts for the future improvements in the plating technique.

We would like to add two more particularly intriguing applications of SNBD. One is an ingenious *hard masking* technique with isolate SNBD particles, which was devised as the first step in a long preparative sequence of vertically aligned diamond nanowires functionalized with DNA [43, 44]. They first grew by MWPCVD B-doped polycrystalline diamond film on a commercial single diamond crystal surface, polished the surface of CVD fim to root-mean-square roughness of 0.8 Å, and then coated the polished surface with SNBD particles by immersing the assembly into dilute colloidal solution of SNBD as mentioned above (Sect. 1.4.1.2). In this way, they obtained a novel all-diamond structure consisting of poly- and single crystalline double-layered diamond topped with a single-particle layer of SNBD. Reactive ion etching of the surface by an oxygen plasma until the SNBD particles are burnt out, for 10 s under their experimental condition, resulted in a forest of vertical horn-shaped conductive diamond wires, 3–10 nm long and spaced about 11 nm.

While such an assembly should be useful also as cold electron emitting cathodes in addition to fabricating a DNA sensor, an interesting aspect of this experiment is the fact that the density of immersion coated SNBD particles back-estimated from the average separation of diamond wires was 8×10^{11} particles/cm^2, much higher than those obtained on Si-wafers under the same coating method (1.1×10^{11}/cm^2). Actually, this density exceeds our best value of 2.4×10^{11}/cm^2 recorded on Si-wafers by using a commercial ink-jet patterning machine and corresponds to 20% of the theoretically highest possible density of 4×10^{12}/cm^2 for our SNBD particles. Advantage of the polished diamond surface as the substrate for high density seeding of SNBD over the unpolished non-diamond surfaces is clearly demonstrated.

The last example of promising application of SNBD to be mentioned in this article is the wide open area of *nanocatalyst supports*, as can be readily anticipated from its multi-pole electronic structure and very large surface area in dispersed state. Various bonding patterns between the polar or ionic nanocatalysts and the charged facets of SNBD particles may be realized. Note that the truncated octahedron model used by Barnard and others as an ideal shape of SNBD is identical with the basic structural element of zeolite catalysts. Finally, an interesting geometrical analogy between the truncated octahedron model of SNBD and Mackay crystal type P, recently proposed as the most realistic approach to the hypothetical Schwarzites, a group of infinite and continuous curved surfaces having negative Gaussian curvature at any point of the surface. In SNBD, among its 14 facets, eight {111} facets consist of net work of hexagonal sp^2-carbon rings. In Mackay crystal P, its continuous surface consists of net work of hexagonal and octagonal sp^2-carbon rings, and its asymmetric unit is a curved variation of polyhedral truncated octahedron (Fig. 1.10c). Both forms belong to the same point group, O_h. Mackay crystal P is predicted to be ferromagnetic and harder than diamond [45, 46].

1.5 Perspectives and Conclusions

1.5.1 Cost Perspectives

Three-step production scheme of SNBD from Composition B is illustrated in Table 1.2 together with the cost and production levels at present and projection for the year 2012. In view of the slow developments in the agglutinates production in the last few years, it is clear that the agglutinates have only limited applications and cannot remain as the final product for too long, although certain special markets like texturing of hard disk will remain. Transition of major product from agglutinates to SNBD will be a critical condition for the growth of nanodiamond industry, but for this transition to occur, we need to create at least several secondary products from the SNBD having very large markets.

As to the cost reduction, the most effective will be the continuous detonation process, but at the moment we do not see any perspective. A milder method of growing nanodiamond like combustion as in the C_{60} production is desirable because the chance of converting a milder batch process into continuous one should be higher. Without the continuous process, the best attainable level of production cost will remain somewhere near $3,000/kg.

Table 1.2 Scale and cost estimations in the three-step production of single-nano bucky-diamond SNBD

Material	Operation	Cost, $/kg		Consumption, ton	
		2008	2012	2008	2012
Composition B		-	-	320	960
	Detonation/H_2O				
Detonation soot		100	30	16	48
	Oxidation				
Agglutinates		1,000	300	8	24
	Beads-milling				
SNBD		20,000	5,000	0.001	12

1.5.2 Conclusions

Practically every aspect of SNBD particles has been discussed above. Having completed the analysis, we cannot refrain from concluding that we have finally encountered with the right material to study nanoscience at the right time. Unexpected behaviors are unfolding one after the other, like high stability of its aqueous colloid, high hygroscopicity, unprecedented difficulty in removing the surface graphitic layers, mysterious agglutination, etc. Almost every day we find new problems.

At the moment, we have more problems than answers. We began to understand these behaviors, but only slowly and always only partly. Prevailing ineffectiveness of our work stems from our total inability to purify this material. We still do not have "authentic standard sample," nor determined any of the fundamental physical constant of SNBD. We are still trying to extract truth from "impure" material. For this reason, our reliance on the theoretical tools including atomistic simulation is high, even though the number of atoms in a typical nanoparticle like SNBD is still too large for the level of theory we wish to use. Arrival of ever bigger and faster computer is earnestly waited for.

If we look at the situation from different angle, we feel blessed because everything is almost entirely new. For example, Barnard and Sternberg in their two short papers published in 2007 and 2008 [9, 18] described such remarkable and diverse discoveries as a new type of interparticle interaction (interfacial Coulombic), a new type of carbon hybridization (sp^{2+x}), spontaneous polarization of electrons in diamond, and novel multipole particles. They also solved the cause of agglutination phenomenon that embarrassed scientists for more than 40 years. How can we expect so rich harvest in such a short time with simple tool (tight binding theory!) in chemistry? Chemical fields have been exploited so thoroughly for the past 200 years and not much is left. Why do chemists stay in the wasted land even now? The grass of your neighbor is deep green.

Acknowledgments Professor Dr. A. Krüger performed the first experiments on nanodiamond in my laboratory in Toyohashi University of Technolgy. In NanoCarbon Research Institute, Mr. F. Kataoka played a vital role in introducing beads milling method, and Mr. M. Takahashi optimized the milling conditions to enable industrial production of SNBD. Virtually all applications were found by friends outside my own group: drug carrier action by Dr. H. Huang and Prof. D. Ho, CVD seeding by Dr. O. Williams, novel lubrication system by four Japanese research groups (*vide supra*), hard masking by a group led by Dr. Ch. Nebel. Special thanks are due to Dr. K. Uemura for untiring moral and technical supports. Generous funding was provided by MEXT, NEDO, JST, Futaba Corporation, and Nippon Kayaku Co.

References

1. Danilenko VV (2004) Phys Solid State 46:595–599
2. Ōsawa E (2008) Pure Appl Chem 80:1365–1379
3. Ōsawa E (2007) Diam Relat Mater 16:2018–2022
4. Aleksenski AE, Baidakova M, Vul' A Ya, Siklitski VI (1999) Phys Solid State 41:668–671

5. Baidakova MV, Siklitsky VI, Vul' A Ya (1999) Chaos Solitons Fractals 10:2153–2163
6. Donnet J-B, Lemoigne C, Wang TK, Peng C-M, Samirant M, Eckhardt A (1997) Bull Soc Chim Fr 134:875–890
7. Shenderov OA, Zhirmov VV, Brenner DW (2002) Crit Rev Solid State Mater Sci 27:227–356
8. Krüger A, Kataoka F, Ozawa M, Aksenskii A, Vul' YA, Fjino Y, Suzuki A, Ōsawa E (2005) Carbon 43:1722–1730
9. Barnard AS, Sternberg M (2007) J Mater Chem 17:4811–4819
10. Schrand AM, Dai L, Schlager JJ, Hussain SM, Ōsawa E (2007) Diam Relat Mater 16:2118–2123
11. Schrand AM, Huang H, Carlson C, Schlager JJ, Ōsawa E, Hussain SM, Dai L (2007) J Phys Chem B 111:2–7
12. Schrand AM, Johnson J, Dai L, Hussain SM, Schlager JJ, Zhu L, Hong Y, Ōsawa E (2008) In: Webster TJ (ed), Safety of nanoparticles: from manufacturing to clinical applications, Chapter 8. Springer, New York, pp 159–188
13. Doktycz SJ, Suslick KS (1990) Science 247:1067–1069
14. Box GE, Hunter WG, Hunter JS (2005) Statistics for experimenters: design, innovation, and discovery, 2nd edn. Wiley, Hoboken, NJ
15. Takahashi M, Ōsawa E Unpublished results.
16. Ozawa M, Inaguma M, Takahashi M, Kataoka F, Krüger A, Ōsawa E (2007) Adv Mater 19:1201–1206
17. Iakoubouvskii K, Mitsuishi K, Furuya K (2008) Nanotechnology 19:155705(5pp)
18. Barnard AS (2008) J Mater Chem 18:4038–4041
19. Ōsawa E, Ho D, Huang H, Korobov MV, Rozhkova NN (2009) Diam Relat Mater doi.org/10.1016/j.diamond.2009.01.025
20. Korobov MV, Avramenko NV, Bogachev AG, Rozhkova NV, Ōsawa E (2007) J Phys Chem C 111:7330–7334
21. Korobov MV, Efremova MM, Avramenko NV, Ivanova NI, Rozhkova NN, Ōsawa E (submitted) Langmuir
22. Huang H, Dai L, Wang DH, Tan L-S, Ōsawa E (2008) J Mater Chem 18:1347–1352
23. Hu S, Sun J, Du X, Tian F, Jiang L (2008) Diam Relat Mater 17:142–146
24. Banhart F, Ajayan PM (1996) Nature 382:433–435
25. Huang JY (2007) Nano Lett 10.1021/nl0709975
26. Hawelek L, Brodka A, Dore JC, Honkimaki V, Tomita S, Burian A (2008) Diam Relat Mater 17:1186–1193
27. Ji S, Jiang T, Xu K, Li S (1998) Appl Surf Sci 133:231–238
28. Huang H, Pierstorff E, Ōsawa E, Ho D (2007) Nano Lett 7:3305–3314
29. Lam R, Chen M, Pierstorff E, Huang H, Ōsawa E, Ho D (2008) ACS Nano 2: 10.1021/nn800465x
30. Chen M, Pierstorff ED, Lam R, Li S, Huang H, Ōsawa E, Ho D. (2009) ACS Nano 3:10.1021/nn90048m
31. Huang H, Pierstorff E, Ōsawa E, Ho D (2008) ACS Nano 2:203–212
32. Williams OA, Douheret O, Daenen M, Haenen K, Ōsawa E, Takahashi M (2007) Chem Phys Lett 445:255–258
33. Arnault JC, Saada S, Nesladek M, Williams OA, Haenen K, Bergonzo P, Ōsawa E (2008) Diam Relat Mater 17: doi:10.1016/j.diamond.2008.01.008
34. Williams OA, Nesladek M, Daenen M, Michaelson S, Ternyak O, Hoffman A, Ōsawa E, Haenen K, Jackman RB, Gruen DM (2008) Diam Relat Mater 17:1080–1088
35. Mori S, Kanno A, Nanao H, Minami I, Ōsawa E (2008) In: Vul' A, Baidakova M (eds) Proceedings of the 3rd International Symposium on Detonation Nanodiamonds: Technology, Properties and Applications, July 1–4, 2008, St. Petersburg, Russia, Ioffe Physico-Technical Institute, St. Petersburg, Russia, pp 21–28
36. Kato T, Omori H, Lin I, Ōsawa E (2009) Tribologists (In Japanese) 54:122–129
37. Mabuchi Y, Nakagawa A (Nissan Motors Co.) (2006) Japanese Open Patent No. 2006-241443

38. Ōsawa E, Mori S (2009) Mon Tribol (in Japanese) 258:39–43
39. Kano M et al (2005) Tribol Lett 18:245–251
40. Schiotz J, Jacobsen KW (2003) Science 301:1357–1359
41. Wang DH, Tan L-S, Huang H, Dai L, Ōsawa E (2009) Macromolecules 42:114–124
42. Burkat A, Dolmatov VV, Ōsawa E (in preparation)
43. Yang N, Uetsuka H, Ōsawa E, Nebel CE (2008) Angew Chem Int Ed 47:5183–5185
44. Yang N, Uetsuka H, Ōsawa E, Nebel CE (2008) Nano Lett 10.1021/nl801136h
45. Makino K, Tejima S, Minami I, Nakamura H, Ōsawa E (2009) RIST News (in Japanese) 46:28–37
46. Ōsawa E (2002) Petrotech (in Japanese) 25:746–752

Chapter 2
Molecular Dynamics Simulations of Nanodiamond Graphitization

Shashishekar P. Adiga, Larry A. Curtiss, and Dieter M. Gruen

Abstract Nanocarbons have attracted great interest due to their potential applications in nanoscale devices, medicine, lubrication and composite materials. Recently, nanocarbons with a variety of morphologies are reported to have been obtained after annealing nanodiamonds above 1,200 K. Here, we have investigated the transformation of 2–5 nm nanodiamond particles upon annealing using molecular dynamics simulations. The simulations show that nanodiamonds undergo annealing-induced graphitization by a progressive sp^3 to sp^2 conversion of carbon atoms that begins at the surface. The extent of this conversion depends on the size and morphology of the nanodiamond. It is found that while graphitization proceeds easily from {111} surfaces towards the core, the presence of {100} surfaces leads to residual sp^3 carbon atoms. We will also discuss different steps involved in nanodiamond graphitization, the formation of onion-like carbon and vibrational spectra of these structures.

2.1 Introduction

The science of carbon materials was immeasurably enriched some 20 years ago with the discovery of the nanocarbon family comprising fullerenes [1], nanotubes [2], ultrananocrystalline diamond [3] and more exotic forms of carbon such as nanohorns [4] and nano-onions [5]. More recently, a renewed interest in graphene has emerged. Understanding the properties of the nanocarbons including the conditions under which one form transforms into another has become an important enterprise. This is due to the existence of many basic science issues such as the close energetic relationships between all the forms of carbon as well as the tremendous potential for important applications of these materials in a wide variety of fields of application.

S.P. Adiga (✉), L.A. Curtiss, and D.M. Gruen
Materials Science Division, Argonne National Laboratory, 9700 S. Cass Avenue, Argonne, IL, 60439, USA
e-mail: shashi.adiga@gmail.com

D. Ho (ed.), *Nanodiamonds: Applications in Biology and Nanoscale Medicine*,
DOI 10.1007/978-1-4419-0531-4_2, © Springer Science+Business Media, LLC 2010

Among the variety of nanocarbon materials explored, nanodiamonds have spurred great excitement among researchers due to their potential applications in many engineering fields. Nanodiamonds are diamond molecules of 2–5 nm size and are found in meteorites [6] and interstellar dust [7]. Recently, nanodiamonds have been produced synthetically in the form of films [3] and powders [8–10]. Of particular success in producing large quantities of nanodiamond powder has been the detonation synthesis from explosive carbonaceous mixture. Potential applications for nanodiamonds include abrasive pastes and suspensions for high precision polishing [11], nanodiamond-polymer composites [12], wear-resistant surface coatings, cooling fluids, lubricants, and electroplating baths [13]. Another promising area for nanodiamonds is in biomedical applications [14]. A large number of potential applications, including biocompatible composites [15], drug delivery [16, 17], have been identified thanks to their superior physical and chemical properties and biocompatibility [14]. They have also been considered as potential medical agents due to high adsorption capacity, high specific surface area, and chemical inertness. They have also been explored as fluorescent biomarkers [18, 19].

Nanodiamonds are also important to the formation of nanocarbon ensembles which are being studied because of their favorable configurational entropies to improve thermoelectric performance particularly at high temperatures. Because of their high Debye temperatures, nanocarbons are among very few nanomaterials that preserve their nanocrystallinity and therefore their desirable "quantum" properties even at temperatures of 2,500 K. Despite being one of the actively researched nanomaterials, the understanding of the structure of nanodiamonds remains open. Much of the attention has been focused on the coexistence of sp^3 and sp^2 hybridized atoms, how it is affected by particle size and shape, and the ability to alter their relative amounts through the application of heat and pressure. In fact, carbon atoms in nanodiamond do not have a purely diamond structure, rather they have an intermediary structure, with a diamond-like core covered by an outer shell of graphitic/amorphous carbon [20, 21]. It has also been demonstrated that nanodiamond could transform into carbon onions [22, 23] and vice versa [24]. The structure of nanodiamond particles synthesized by detonation and their annealing induced transformation into onion-like carbons has been studied by various experimental techniques [25]. Notwithstanding many experimental studies and attempts by many theoretical efforts [26–29], the graphitization mechanism of nanodiamond is still not sufficiently understood, and so far there has been no systematic study of the graphitization of nanodiamond crystals of different shapes, sizes and hydrogen termination. There is clearly a need for a thorough study of nanodiamond graphitization and how it is likely to depend on temperature, surface termination, particle size and shape. As new applications involving surface functionalization and surface reactivity are developed, a detailed understanding of the interaction of nanodiamond/carbon-onion with its environment, as a function of the inherent changes in surface hybridization, reactivity and electrostatic potential becomes important. Particularly, in biological applications, such fundamental knowledge will help predict the potential sites for protein/DNA binding and design novel biofunctionalized nanomaterials.

In understanding the transformation between different forms of carbon at the nanoscale, it is of the essence to capture information at the atomic level. To this end, molecular modeling has played an important role in understanding the structure, stability, and phase transitions in nanocarbon materials. For example, Barnard et al. have performed extensive work on the relative stability of fullerenes, nanodiamonds and bucky diamond particles using quantum chemical calculations [30–32]. Similarly, the surface structure of nanodiamond has been studied by Hu et al. [33] using molecular dynamics (MD) simulations and Raty et al. [34, 35] using density functional theory (DFT) calculations. More recently, Barnard and Sternberg, based on the analysis of surface electrostatic potential of nanodiamond particles derived from tight-binding DFT calculations, have pointed out preferred orientations for interaction between nanodiamond particles [36].

In this chapter, we present results from classical MD simulations of graphitization of nanodiamond particles. To obtain detailed and quantitative information about the graphitization of nanodiamond, we have performed annealing simulations on several model nanodiamond structures. The emphasis is on the early stages in the heat induced sp^3–sp^2 transformation of nanodiamond particles of different size and shape. Section 2.2 contains a brief discussion of the method used including the interatomic potential. Section 2.3 describes the structure of nanodiamond particles with emphasis on surface termination and morphology. The different morphologies considered in this study are also discussed. In Sect. 2.4, results from our simulations are presented and the graphitization process is discussed with respect to the effect of temperature, shape, size and hydrogen termination. The structure of onion-like carbon (OLC) that results from graphitization and the vibrational spectra of nanodiamond particles before and after annealing is also presented.

2.2 Computational Method

The second generation reactive empirical bond order (REBO) potential [37] was used to describe the short range interaction between carbon atoms. This potential function has been widely used for hydrocarbon systems and it successfully models both sp^2 and sp^3 bonding, depending on local coordination and the degree of conjugation. Since the original REBO potential lacked both repulsive and dispersive van der Waals interactions for interatomic distances beyond which REBO is zero, later implementations included Lennard–Jones (LJ) long-range interactions. In this study, the LJ interactions were implemented with the use of cubic spline functions and parameters as described in the work by Mao et al. [38] Classical equations of motion for each atom were numerically integrated using a third-order Nordsieck predictor-corrector algorithm and a time step size 0.5 fs. The system temperature was constrained using the Langevin method [39]. The appropriateness of the interatomic potential was tested by simulating the annealing of a diamond (111) slab (a total of 21 (111) bilayers parallel to the surface) at 1,800 K. After about 600 ps annealing at 1,800 K, the graphitization of all the layers was complete and the process resulted in a hexagonal pattern of sp^2 carbon atoms corresponding to (0001) plane of graphite with an average interlayer spacing of 3.4 Å.

2.3 Nanodiamond Structures

The nanodiamonds also referred to as ultradisperse diamonds are particles in the 2–5 nm size range. While diamond is metastable relative to graphite under atmospheric pressure, in the nanoscale regime, diamond particles less than 5 nm are more stable than graphite. For example, Barnard [30–32] compared the energies of graphite, diamond, and fullerene on the basis of density functional theory to conclude that diamond could be the stable phase of carbon clusters in the size range of 1.9–5.2 nm. Nanodiamond is often described as a crystalline diamond core with a perfect diamond lattice surrounded by an amorphous shell with a combination of sp^2/sp^3 bonds or onion-like graphite shell. Even though theoretical studies to date have considered both spherical and polyhedral nanodiamond particles, high resolution transmission electron microscope (HRTEM) images of nanodiamond have confirmed the latter [40]. Barnard et al. [30–32], using the first principles computer simulations, have shown how the polyhedral shape affects the stability of nanodiamond particles. In particular, they considered nanodiamond particles of octahedral, cuboctahedral, and cubic morphologies up to 2 nm in diameter and discussed the shape dependent stability of nanodiamond particles.

We have considered different sizes of nanodiamond particles in three shapes: octahedral (OC), truncated octahedral (TO) and cuboctahedral (CO) as shown in Fig. 2.1. The OC particle consisted of eight {111} facets (Fig. 2.1a). The TO particle consists of eight {111} facets whose two opposite vertices were truncated to produce (100) surfaces (Fig. 2.1b). The CO particle consists of eight {111} and six {100} facets (Fig. 2.1c). The models were cut from an ideal diamond crystal with the lattice constant 3.56 Å. Each of these facets in a cluster is at about the same distance from the center of the cluster.

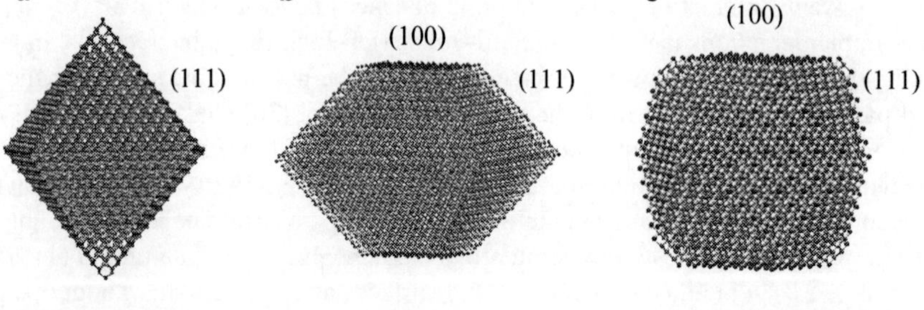

Fig. 2.1 Nanodiamond morphologies studied in this work: models of (**a**) octahedral (**b**) truncated octahedron and (**c**) cuboctahedron nanoparticles

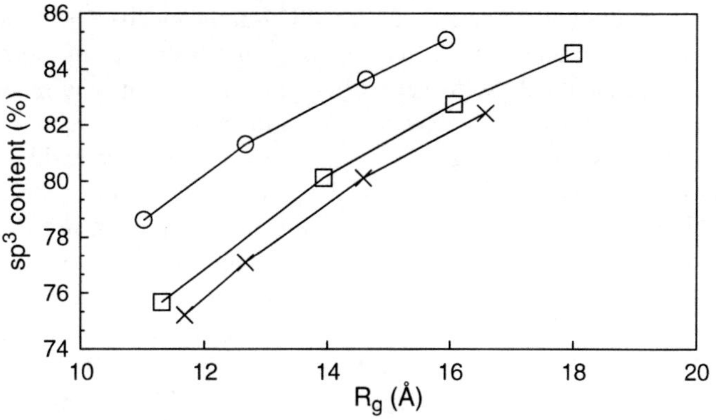

Fig. 2.2 Fraction of sp^3 hybridized atoms as a function of nanodiamond size for Octahedron (*crosses*), truncated octahedron (*squares*), and cuboctahedron (*circles*) shapes. The nanoparticles have unreconstructed surfaces. R_g is the radius of gyration of the particle

As generated, the surfaces of these nanodiamonds are unreconstructed. We first carried out MD simulations at 300 K to follow surface relaxation/reconstruction. The {111} surfaces are commonly found in the relaxed Pandey chain 2×1 reconstruction [41]. That reconstruction would yield a considerable energy gain of 0.85 eV/surface atom in the binding energy. Similarly, {100} planes usually undergo a 2×1 dimer reconstruction [42]. However, these reconstructions require substantial reorganization of the surface bonds that can only be achieved by overcoming a rather high energy barrier. Owing to this high barrier, we did not observe these reconstructions during our MD simulations at carried out at 300 K. However, as will be mentioned in the next section, we did observe these reconstructions at higher annealing temperatures before they became unstable or full of defects at temperatures above 1,800 K. After relaxation for 1 ns at 300 K, we obtained the fraction of sp^3 atoms in these nanodiamond particles. The fraction of sp^3 atoms indicates how close the nanoparticle is to the ideal diamond structure which is widely believed to determine interactions with other molecules. The variation of sp^3 content as a function of size of the nanodiamonds of OC, TO and CO morphologies is illustrated in Fig. 2.2. The size of the particle is calculated by means of radius of gyration (Rg), which is the root mean square distance of all atoms from the center of mass of the nanoparticle. It is to be noted that a higher sp^3 fraction in CO nanoparticle is due to a lower surface to volume ratio as compared to the TO and OC nanoparticles.

2.4 Results

2.4.1 Graphitization Mechanism

Our first goal is to understand the initial stages of graphitization and how the transformation proceeds in nanodiamonds. While there are several experiments reporting that nanodiamond particles can be transformed into spherical and polyhedron carbon onions at high temperatures [22, 23, 43, 44], the mechanism of such a transformation is not well understood. In contrast, graphitization of the surface of bulk diamond has been studied extensively using both experiments and modeling. Both MD [45–47] and molecular mechanics [48] simulations of graphitization of the diamond (111) surface have suggested that the formation of fullerene-like or graphitic surface structure is one of the most effective ways of minimizing the surface free energy at high temperatures. In addition to being a dominant face in the morphology of diamond, the reason for {111} planes playing a prominent role in the graphitization process stems from the fact that there is a close geometrical relationship between them and the individual graphene sheets. The diamond (111) plane contains a pair of atomic layers and resembles "puckered" graphite (0001) layer with a hexagonal pattern. This geometrical relationship together with the lengthening of carbon–carbon bonds from 1.42 Å in graphite to 1.54 Å in diamond results in the near epitaxial relations.

Vita et al. using ab initio molecular dynamics (MD) methods, showed that at flat diamond (111) surfaces graphitic features can be formed at 2,500 K by a direct transformation [45]. They showed that the decisive step in initiating the transformation is the breaking of some of the bonds between the first and second bilayers at the surface. To explain why a (111) surface readily undergoes graphitization, Kuznetsov et al. [48] compared graphitization on diamond (111) and (110) surfaces using the cluster models. Their calculations showed that graphitization of (111) surface is more favorable by 0.24 eV/surface atom than graphitization of a (110) surface. Also, the barrier for transformation was found to be higher for the (110) surface by 0.93 eV/surface atom. These calculations demonstrated that graphitization of a (111) surface is clearly preferred over that of a (110) surface. In addition to the direct transformation, they have also suggested a "zipper-like" migration mechanism in which two graphene sheets are formed from three diamond (111) planes with the carbon atoms of the middle diamond layer being distributed equally between the two growing graphitic sheets as the graphite diamond interface advances. However, the relative importance of these to mechanisms or any other mechanism in nanodiamond graphitizaton is yet to be determined.

Recently, there have been a few modeling studies concerning the heat induced graphitization of nanodiamond particles. Fugaciu et al. investigated graphitization of nanodiamonds with a diameter in the 1.2–1.4 nm range using the tight-binding DFT [26]. Lee et al. [27] employed tight-binding MD simulations and found that nanodiamonds of ~1.4 nm diameter transformed into a tube-shaped fullerene via annealing. These simulations provided important information on the nature of transformation of nanodiamond particles which consist of several hundred carbon atoms. Modeling graphitization of experimentally relevant nanodiamonds, i.e., in the size range 3–5 nm, is currently beyond the reach of ab initio methods. On the other hand, Bro'dka et al. investigated the graphitization of a 3 nm spherical nanodiamond annealed at temperatures 1,200, 1,500 and 1,800 K [28]. However, it was later demonstrated by Leyssale and Vignoles [29] that the Berendsen thermostat used by these authors is not suitable to the study of isolated nanoparticles and suggested using a more suitable thermostat. These studies also did not incorporate repulsive and dispersive van der Waals interactions,

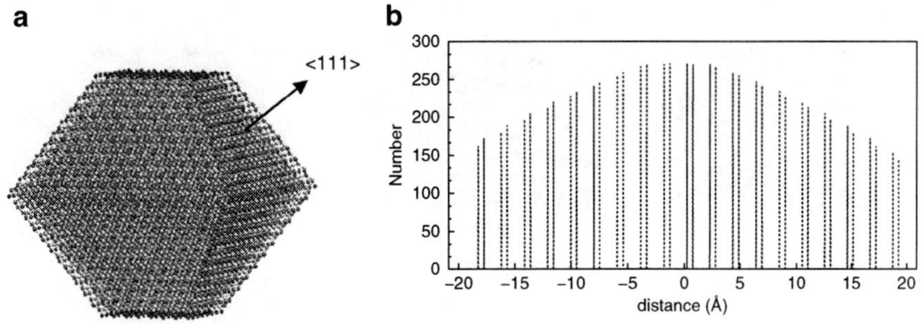

Fig. 2.3 Structure of a TO nanodiamond consisting of 8,459 atoms prior to annealing: (**a**) atomic model with <111> direction marked and (**b**) number of atoms as a function of distance from the origin along <111> direction, the diamond (111) double layers are clearly seen

which are thought to be important in modeling the graphitic phase. We have addressed the shortcomings of the previous classical MD studies by using both a more suitable Langevin thermostat and a more appropriate interatomic potential that includes LJ non-bonded interactions as described in Sect. 2.2.

Before we discuss the dependence of the extent of graphitization on temperature, particle size and shape, and hydrogen termination, we will demonstrate the progressive transformation of predominantly sp^3 atoms into sp^2 atoms with the example of a TO nanodiamond particle (8,459 atoms) annealed at 1,800 K. The initial structure of the nanodiamond particle is shown in Fig. 2.3a. Also, the ideal diamond structure of the nanodiamond particle is further characterized by plotting the number of atoms as a function of distance from the origin along the <111> direction (Fig. 2.3b). The plot shows the characteristic double peaks corresponding to diamond (111) bilayers and that the distance between two adjacent bilayers is 2.02 Å (interplanar distance).

In Fig. 2.4, the time evolution of graphitization is illustrated with a series of snapshots of the nanodiamond particle after t=0, 5, 10, 20, 200, and 800 ps of annealing at T=1,800 K. The 2-, 3-, and 4-coordinated carbon atoms are colored black, gray and silver, respectively. The two (100) surfaces undergo 2×1 reconstruction within the first 5 ps of annealing. Graphitization begins at {111} facets on the surface and progresses towards the center of the nanodiamond particle. Figure 2.5 further illustrates the initial stages of the graphitization process that involves progressive disappearance of the layered structure characteristic of diamond lattice. The plots represent number of atoms as a function of distance from the center of mass (COM) of the particle along the <111> direction. The plots are given for annealing times t=10, 200, and 400 ps at T=1,200, 1,500, and 1,800 K. The graphitization process is initiated at the surface (111) layer with thermal displacements of surface atoms from their equilibrium positions. During the initial period of about 10 ps, the diamond particle graphitizes at the surface forming graphitic layers enclosing a diamond core. Then, the diamond short-range order in the core of the particle gets more and more disordered as indicated by straightening of

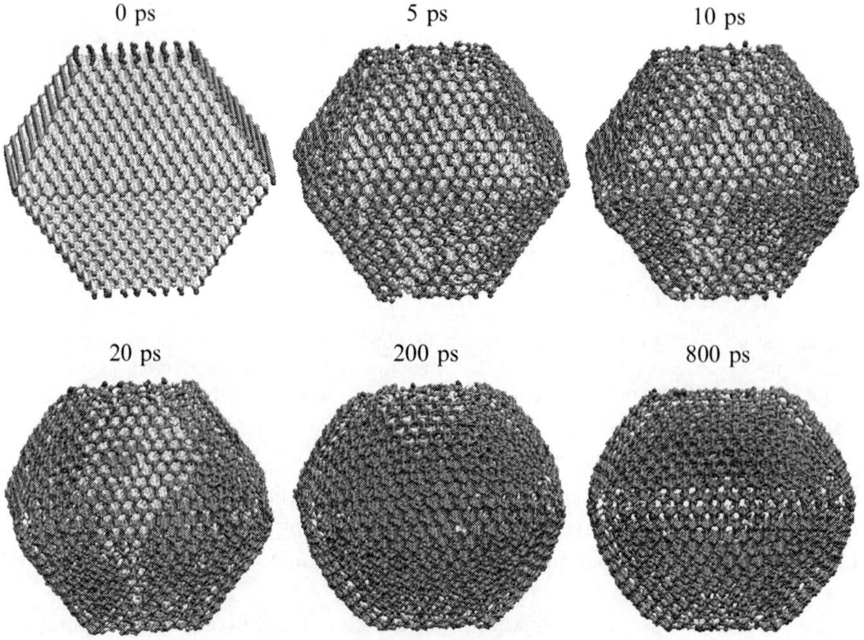

Fig. 2.4 Initial stages of graphitization. Snapshots from MD simulations of a truncated octahedral cluster (8,459 atoms) after $t = 0$, 5, 10, 20, 200, and 800 ps of annealing at 1,800 K. The 2-, 3-, and 4-coordinated carbon atoms are colored black, gray and silver, respectively. The TO nanoparticle relaxed at 300 K for 1 ns was used as the starting structure ($t = 0$ ps). The two (100) surfaces (*top* and *bottom*) undergo 2×1 reconstruction within the first 5 ps of annealing. Graphitization begins at {111} facets on the surface and progresses towards the center of the nanodiamond particle. The regions underneath surface (100) facets (top and bottom) undergo a slower transformation and contain residual sp^3 atoms. The (111) facets become curved as a result of annealing

the puckered (111) bilayers into a single layer and broadening of the peaks. Once a graphitic region is formed within the surface layer, the graphitization starts to proceed towards the center in a direction perpendicular to the (111) surface layer. A similar mechanism for the progress of graphitization has previously been observed in MD simulations of planar diamond (111) surfaces [45].

A key feature of the graphitization process is the formation of graphitic "blister" on the surface. As the outermost layer is delaminated it bulges out and assumes a convex shape that is under tensile stress. This is evident from the larger interplanar distance between the top two layers as illustrated in the density profiles (Fig. 2.5). Evidence for the formation of such curved graphitic features has been provided by HRTEM images of annealed nanodiamond particles [25].

Further information on the evolution of the graphitization during the early stage of annealing is analyzed in terms of the fraction of carbon atoms with trivalent coordination (sp^2) as a function of time at annealing temperatures $T = 1{,}200$, 1,500,

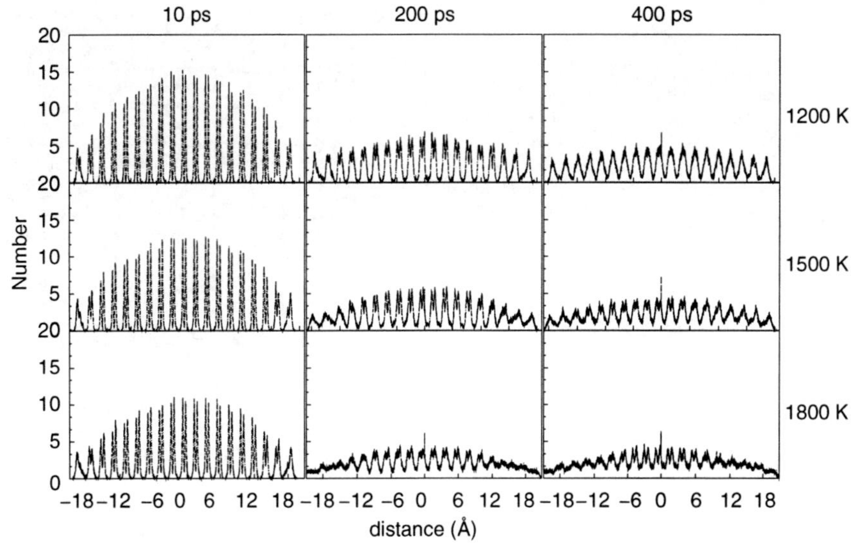

Fig. 2.5 The evolution of graphitization at different temperatures. The number of atoms along <111> as a function of distance from the center of mass of the particle is plotted at 10 (*left*), 200 (*center*) and 400 (*right*) ps after annealing at 1,200 (*top*), 1,500 (*middle*) and 1,800 (*bottom*) K

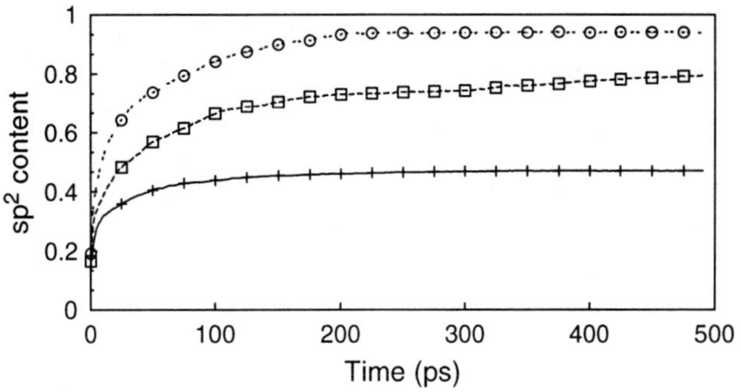

Fig. 2.6 Progression of sp^3–sp^2 conversion during the initial stages of annealing. The fraction of three coordinated carbon atoms in the TO nanodiamond particle (8,459 atoms) is plotted as function of annealing time at $T = 1{,}200$ (*crosses*), 1,500 (*squares*), and 1,800 (*circles*) K

and 1,800 K (Fig. 2.6). Two carbon atoms are considered bonded if the distance between them is less than 1.85 Å, corresponding to the first minimum in the radial distribution function of graphite. At 1,200 K, about 47% of all atoms have trivalent coordination. Increasing the temperature to 1,500 K accelerated the sp^3 to sp^2 conversion, and eventually about 80% of carbon atoms had trivalent coordination. The increase of the annealing temperature to 1,800 K increased the fraction of three coordinated atoms to 95%. For all temperatures studied, the fraction of sp^2 atoms did not change beyond about 500 ps.

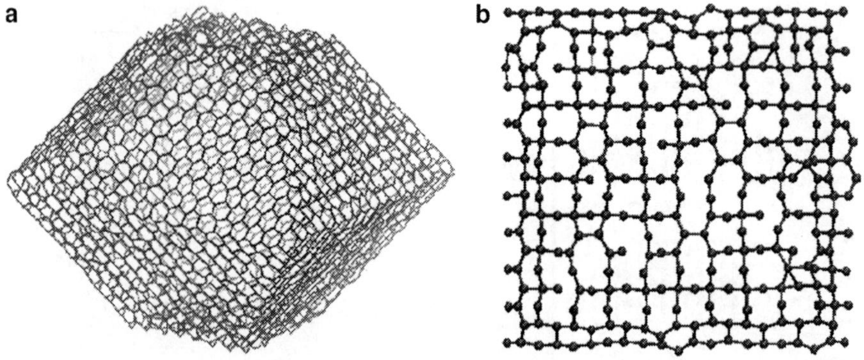

Fig. 2.7 The final structure of the TO nanodiamond cluster after 2 ns of annealing at 1,800 K. The cluster is predominantly sp^2 (95%). (**a**) A wire frame representation showing graphitic hexagonal pattern formed on the original nanodiamond (111) facets. (**b**) Surface layer corresponding to the (100) facet of the original nanodiamond

The final structure after 2 ns of annealing at 1,800 K shown in Fig. 2.7a retained the original TO shape to a large extent. It consisted of nested cages of mostly three coordinated atoms characterized by graphene-like hexagonal network structure parallel to the original nanodiamond {111} facets. Parallel to the two {100} facets of the nanodiamond, the resulting layers contained carbon atoms with mixed coordination. The surface layer corresponding to the (100) facet of the original nanodiamond is shown in Fig. 2.7b. It has a mix of 3, 5, 6, 7 and 8 sided polygons.

2.4.2 Factors Affecting Graphitization

2.4.2.1 Annealing Temperature

Many experimental studies have suggested that nanodiamond particles transform into graphitic structures after annealing at temperatures ranging from 1,373 to 2,173 K [49, 50]. The extent of graphitization, however, depended on the temperature. For example, it has been reported that annealing at 1,373 K results in partial graphitization and complete transformation happens only for temperatures above 1,673 K [50]. Incomplete graphitization has been indicated by sp^3 signal from electron energy loss spectroscopy, X-ray diffraction or Raman spectroscopy. The residual sp^3 signal has been variously attributed to the presence of dangling bonds, or atoms connecting small sp^2 domains, or the presence of a diamond core surrounded by graphitic shells [20, 21]. To explore the effect of temperature on the extent of transformation, the cuboctahedron nanoparticle with 3,130 atoms and the octahedron nanoparticle with 4,495 atoms were annealed at temperatures $T = 1{,}200$, 1,500, 1,800 and 2,500 K. In Fig. 2.8, a plot of the fraction of sp^2 atoms vs annealing temperature for an annealing time of 2 ns is given. For the octahedron, the fraction

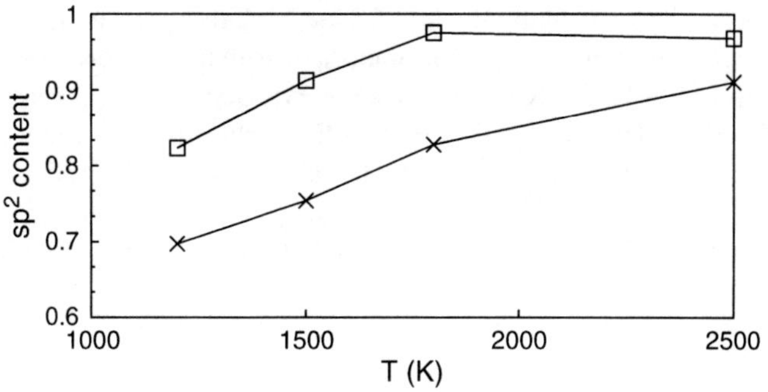

Fig. 2.8 The effect of annealing temperature on the extent of graphitization. The fraction of sp^2 atoms in a CO with 3,130 atoms (*crosses*) and an OC with 4,495 atoms (*squares*) nanodiamonds as a function of annealing temperature is plotted. The annealing was performed for 2 ns

of 3-coordinated sp^2 carbon atoms increased from 0.82 at 1,200 K to 0.97 at 1,800 K and slowly plateaued off. The CO particle, on the other hand, increased its sp^2 fraction from 0.69 at 1,200 K to 0.91 at 2,500 K. As discussed in the next paragraph, the residual sp^3 atoms belong mainly to the untransformed regions below {100} facets of the CO nanoparticle and hence, particle shape plays a big role in the extent of transformation, besides temperature. Whereas in the OC particle, the small amount of sp^3 atoms are mostly isolated and often connect two adjacent graphitic shells.

In contrast to our simulation results, experimental studies have reported a complete transformation to spherical onion-like carbons at 2,000 K. For example, a nanodiamond particle of diameter ~5 nm was transformed into spherical carbon onions at 2,000 K and into polyhedron carbon onions at 2,300 K [44]. The final structures obtained from our 2 ns long simulations contain residual sp^3 atoms. However, it is possible that, if the simulations can be performed for a sufficiently long time to allow for large scale rearrangement of atoms between the graphene layers, a spherical carbon onion shall be obtained. It is also possible that the presence of defects and surfaces and contact with other nanodiamond particles may facilitate the complete transformation observed in the experiments.

2.4.2.2 Nanodiamond Shape

As previously mentioned, the transformation of {111} planes occurs more easily as compared to {110} or {100} planes. The fact that transformation rate of the {111} diamond planes to graphite-like sheets is higher than those of the other planes, which has been confirmed by both experimental and theoretical investigations. For example, Pantea et al. [51] while studying the graphitization of single crystal diamond at different pressures and temperatures, observed oriented growth of graphite on {111} faces when compared with the disoriented graphitic layers on {100} faces, and suggested dissimilar graphitization mechanisms for different diamond faces.

Since the surface of nanodiamond particles, depending on the morphology, can be terminated with different facets, it is important to explore how graphitization is affected by their shape. To investigate this, we considered OC, CO and TO nanodiamond particles of the same size in terms of the surface to nanodiamond center of mass (COM) distance. These particles with 4,495, 3,864, and 3,130 atoms, respectively, were annealed at 1,800 K.

Because of the difference in the barrier and the energy of transformation between different facets, it is interesting to probe how morphology plays a role in the extent of graphitization. In Fig. 2.9, we illustrate this with OC, TO and CO nanoparticles of roughly the same size (i. e. distance from the center of mass at which the nanoparticle facets lie is ~15 Å). These particles consisted of 4,495, 3,864, and 3,130 atoms, respectively. In the bottom panel of Fig. 2.9, a plot of the fraction of sp^2 atoms as a function of annealing time is given for these three nanoparticles. The OC, TO and CO nanoparticles rapidly reach sp^2 fractions of 0.97, 0.95, and 0.84, respectively, within 100 ps. In addition to the extent of transformation, it is apparent

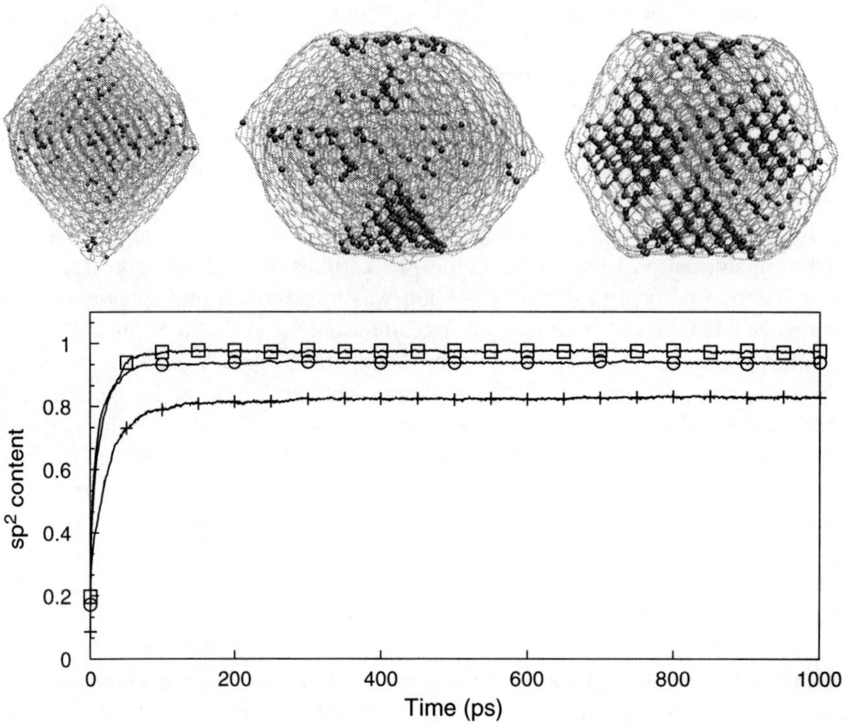

Fig. 2.9 The effect of particle shape on nanodiamond graphitization. *Top*: Snapshots of the nanoparticles octahedron (OC) with 4,495 atoms (*left*), truncated octahedron (TO) with 3,864 atoms (*center*), and cuboctahedron (CO) with 3,130 atoms (*right*) after 2 ns of annealing at 1,800 K. The 4-coordinated atoms are represented by ball-stick model to illustrate the presence of residual diamond-like regions mostly beneath {100} surfaces. Bottom: The evolution of sp^2 fraction for OC (*squares*), TO (*circles*) and CO (*crosses*) nanoparticles with annealing time

that the rate of transformation in the initial 100 ps is much higher for OC which has 100% {111} termination. The TO and CO nanoparticles with larger {100} surface coverage graphitize to a smaller extent is further illustrated with MD snapshots in the top panel of Fig. 2.9. These snapshots of the three nanoparticles are taken after 2 ns annealing at 1,800 K. The 3- and lower coordinated atoms are shown in with wireframe representation, whereas the 4-coordinated atoms are represented by ball-stick model to illustrate the location of residual sp^3 atoms. In the case of OC, the residual sp^3 atoms are sparse and isolated and they connect two neighboring graphitic shells. In contrast, the residual sp^3 atoms in TO and CO nanoparticles are often part of larger diamond domains and they represent untransformed regions below {100} surfaces. In the case of CO, for example, six such sp^3 regions are present (two on the back side), that lie underneath the six {100} surfaces of the original nanodiamond particle. Hence, the presence of {100} facets along with {111} facets has the effect of arresting the otherwise fast transformation of nanodiamond particles leading to the residual diamond like regions.

2.4.2.3 Nanodiamond Size

Graphitization of diamond also strongly depends on the diamond particle size. Crystals with smaller sizes graphitize faster [52]. In annealing the experiments conducted by Xu et al., nanodiamonds were annealed in different gas atmospheres [53]. They found that the graphitization of nanodiamond in argon begins at 940 K, which is significantly lower than the value 1,800 K for bulk diamond. The difference was attributed to the large surface-to-volume ratio and high thermal conductivity of nanodiamond. In Fig. 2.10, the evolution of graphitization is analyzed for CO nanoparticles consisting of 2,067 (CO1), 3,130 (CO2) and 6,232 (CO3) atoms at 1,800 K. The values obtained for the fraction of sp^2 atoms for the three

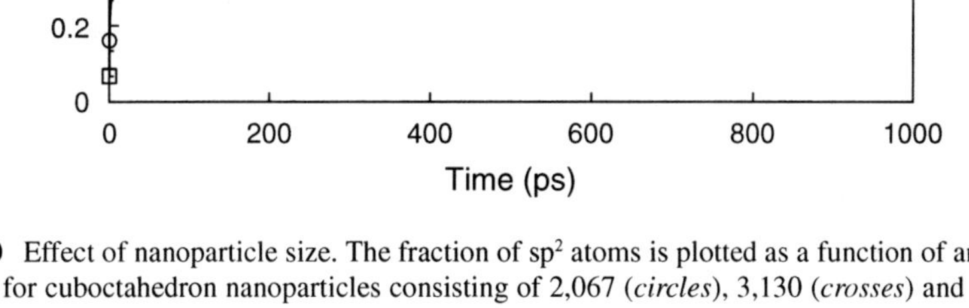

Fig. 2.10 Effect of nanoparticle size. The fraction of sp^2 atoms is plotted as a function of annealing time for cuboctahedron nanoparticles consisting of 2,067 (*circles*), 3,130 (*crosses*) and 6,232 (*squares*) atoms at $T = 1{,}800$ K

nanoparticles are 0.86, 0.84 and 0.83 for CO1, CO2, and CO3, respectively. Given the relatively smaller size range we have considered, the small change in the extent of graphitization for the three sizes is reasonable.

2.4.2.4 Hydrogen Termination

The discussion so far considered graphitization of bare nanodiamond particles. Hydrogen termination is expected to stabilize the nanodiamond structure. For example, Russo et al. [54] performed DFT calculations of hydrogenated nanodiamonds of both the octahedral and cuboctahedral morphologies and found that in addition to passivating the surface hydrogenation stabilizes diamond structure against graphitization. To probe this, we considered a TO nanoparticle with 8,459 carbon atoms terminated with 1,454 hydrogen atoms. The annealing was performed at 1,800 K for 2 ns. The resulting atom density profile along the <111> direction is shown in Fig. 2.11. The {111} bilayers of the nanodiamond particle remain intact as opposed to the unpassivated nanodiamond particle (Fig. 2.5).

2.4.3 Onion-Like Carbons

Carbon onions are multilayered spherical nanostructures made up of nested fullerene cages. A variety of methods have been used to make carbon onions including, for example, high-energy electron irradiation, arc discharge and thermal treatment of carbonaceous materials. This class of nanocarbons has been referred to with multiple

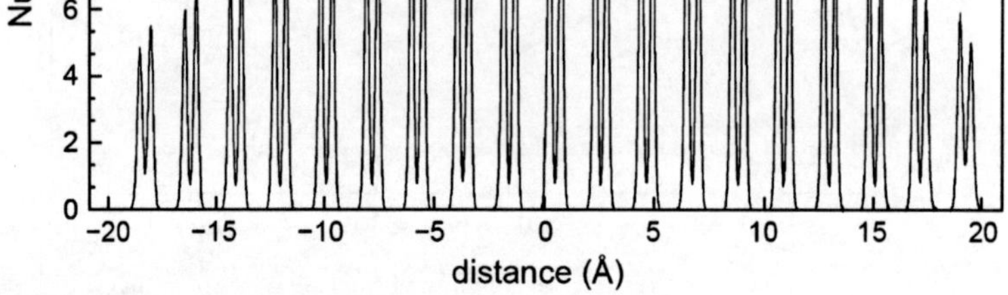

Fig. 2.11 Effect of hydrogen termination. The number of atoms along <111> as a function of distance from the center of mass of the particle is plotted for the hydrogen terminated TO particle with 8,459 carbon atoms annealed at 1,800 K for 2 ns

names including carbon onions, carbon nano-onions (CNO), multilayer fullerenes, onion-like fullerenes and onion-like carbon (OLC). Recently, the transformation of detonation nanodiamond (DND) particles into the nested fullerene-like carbon structures at temperatures above 800 K has been explored as a way to make the nanographitic structures. In the literature, these multilayered cages obtained by annealing of DNDs has been referred to as OLC, mainly to distinguish them from the ideal carbon onions consisting of layers of enclosed fullerene molecules of different sizes [22]. The predominantly sp^2 hybridized structures obtained in our simulations as a result of annealing nanodiamonds closely represent the OLC structures described in the literature. In this context, we will discuss the structure obtained after annealing the OC1 cluster at 1,800 K for 2 ns.

The final structure shown in Fig. 2.12, consists of seven concentric graphitic shells around a few central atoms which can be better described as a small carbon cluster. The structure has ~3% residual sp^3 atoms, a majority of which link two adjacent graphitic shells. The OLC structure more or less retains the original octahedral shape of the nanodiamond. In Fig. 2.12, the average distance $D_{i,\,i+1}$, between layers i and i+1, is plotted as a function of layer index i. The layers are numbered from the center. The interlayer distance between the two outermost layers is about 2.9 Å, close to the equilibrium distance between two graphene sheets, whereas the inter layer distance towards the center is ~2.35 Å , slightly more than that for diamond (111) planes (=2.06 Å). Based on the interlayer distances, one can expect a relatively high pressure region at the center of the onion. It is interesting to point out that Banhart and Ajayan [55] demonstrated the reverse transformation of the cores of carbon onions to diamond via electron irradiation at temperatures above 900 K. Unlike the nanodiamond-OLC conversion that starts at the surface of the nanoparticle, the nucleation of sp^3 diamond region occurs at, the center of the carbon onion. Banhart [56] proposed that the high pressure in the concentric-shell fullerene is a prerequisite for the nucleation of diamond. Additionally, he also speculated about the presence of sp^3 cross links between the adjacent graphitic shells.

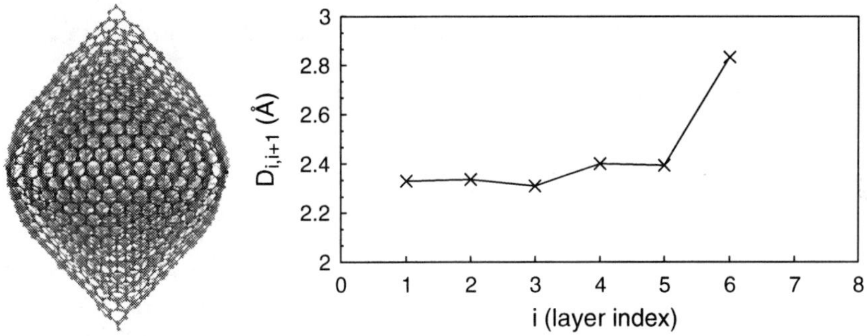

Fig. 2.12 The structure of onio-like carbon obtained after annealing an OC nanodiamond particle. *Left*: A ball-stick representation of the cluster shows nested graphitic cages. *Right*: Plot of interlayer distance as a function of layer index

Perhaps it is worthwhile to point out a key aspect of nanodiamond to OLC transformation to better understand why the interlayer distance in the multishell graphitic structure resulting from nanodiamond annealing is less than 3.37 Å. Diamond to graphite transformation is characterized by a substantial increase in volume, as diamond is ~1.53 times denser than graphite. Let us consider the transformation of an octahedral nanodiamond particle with all {111} termination such as the one discussed in this section. Such a nanodiamond cluster will essentially consist of nested cages made up of diamond (111) bilayers. The first step in graphitization of these individual cages involves straightening of the buckled (111) layer that results in merging of the bilayers into a single layer, structurally similar to a polyhedral graphene cage. The next step involves an increase in the distance between the two consecutive shells from 2.06 Å, corresponding to the diamond (111) interplanar distance, to 3.37 Å of the distance between (0001) planes in graphite. This will require a substantial volume expansion that leads to a decrease in the number of atoms per unit area in the expanding layer. The deficit in atom density created because of the expansion has to be compensated for in order to relieve the surface tension. These "extra" atoms needed in the outer layers can be hypothesized to be supplied through a flow of atoms from the inner layers to the outer. We can intuitively see how such a process can lead to OLCs with hollow cores. However, it can also be speculated that such a process has a large energy barrier due to substantial rearrangement of atoms. Perhaps, the presence of the other particles/ surface in the vicinity can relieve the surface tension of the graphitizing nanodiamond particle and assist in the completion of the transformation.

Another feature we have not considered in this study is the presence of defects. It has been reported that graphitization can be initiated at surface defects. For example, stepped {111} surfaces are graphitized spontaneously [57]. Based on the MD simulations of bulk diamond surfaces it has been proposed that twin boundaries also can promote spontaneous graphitization [47]. In future studies, it will be interesting to study the effect of defects on the graphitization of nanodiamond particles.

Further information on the structure of onion-like carbon can be obtained by analyzing the radial distribution function, which gives the average number of atoms as a function of distance around any given atom. The radial distribution function $g(r)$ is defined as $g(r) = \dfrac{\langle n(r,\Delta r)\rangle}{4\pi r^2 \Delta r \rho}$ where $n(r, \Delta r)$ is the average number of atoms in a spherical shell of thickness Δr at a distance r from an atom and is an average over all atoms sampled over 100 ps. The number density of atoms ρ is taken as the value corresponding to the mass density of diamond (3.53 g/cm^3). A value of $\Delta r = 0.01$ Å is used. Figure 2.13 shows the radial distribution functions ($g(r)$) for the octahedral nanodiamond before (*solid line*) and after (*dotted line*) annealing at 1,800 K respectively. The $g(r)$ analysis was obtained for both structures from simulations performed at 300 K. The first four peaks in $g(r)$ for the cluster annealed at 1,800 K are at 1.41, 2.45, 2.85 and 3.76 Å. These peak positions correspond well with the first to fourth neighbor distances in a graphene sheet and may be seen as

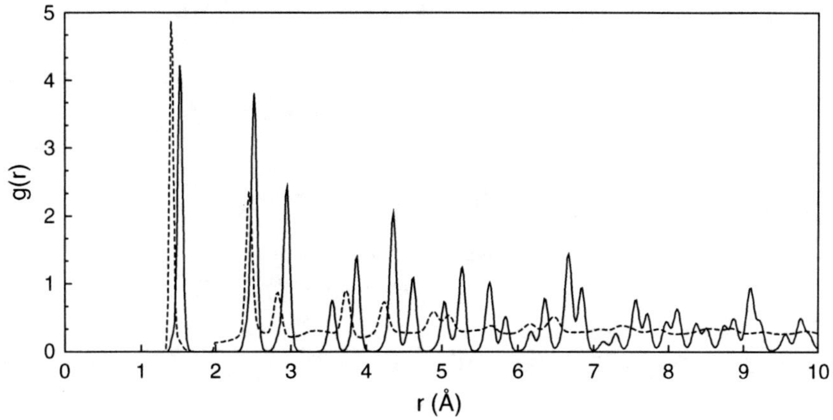

Fig. 2.13 The radial distribution function for 4,495 atom octahedral nanodiamond cluster at $T=300$ K before (*solid line*) and after (*dotted line*) annealing at 1,800 K

an indication that a phase transition to sp^2 carbon has occurred. It is important to note that, unlike periodic structures, the nanoparticle $g(r)$ calculated here does not assume a value of 1.0 at long distances.

2.4.4 Vibrational Spectra

Raman spectroscopy has been a key technique used for studying the nature of the c–c bond as it is able to distinguish between the sp^2 and sp^3 content quite effectively. To illustrate how annealing induced graphitization changes the vibrational properties of nanodiamond particles, we have calculated the vibrational power spectra (vibrational densities of states) by following the atomic trajectories during the MD simulation. The vibrational density of states (VDOS) was determined by calculating the Fourier transform of the velocity auto correlation function (VACF). The VACF gives the correlation between the velocity of an atom at time $t=0$ and the velocity of an atom at time $t=t$. It is important to point out that the computed VDOS and the experimental Raman vibrational spectra are not identical, as the former contains all vibrational atomic motions, and the latter have only Raman active modes with intensities that are different than the computed VDOS. However, the calculated band positions and experimental spectra can be compared. To calculate VACF, the velocities were sampled every time step for 10 ps.

In Fig. 2.14, VDOS at 300 K for the CO nanoparticle with 3,130 atoms before and after annealing at 1,800 K are compared. The VDOS for the nanodiamond particle is characterized by a strong peak at 1,380 cm^{-1}, corresponding to sp^3 bond stretching. The experimental number for this peak is lower, in the range 1,320–1,335 cm^{-1}. Another noteworthy feature is at around 1,620 cm^{-1}, which can be attributed to the

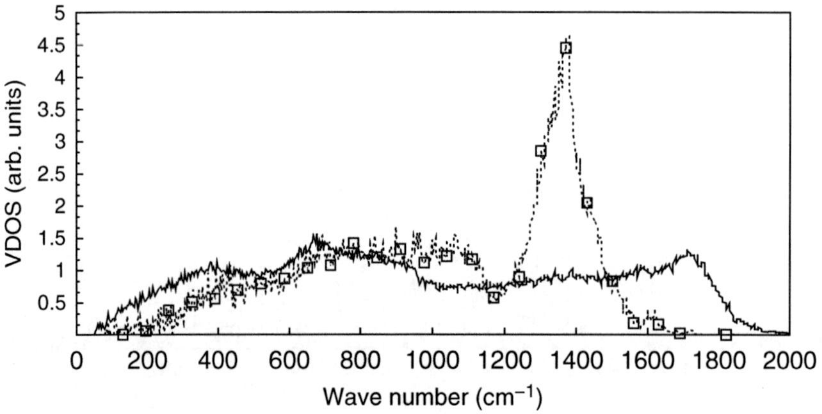

Fig. 2.14 Vibrational density of states obtained for the cuboctahedral nanodiamond cluster with 3,130 atoms: before (*dotted line* with *squares*) and after (*solid line*) annealing at 1,800 K

undercoordinated surface atoms. The key difference in the VDOS of the annealed structure is the diminishing and the broadening of the diamond peak (1,380 cm^{-1}) and appearance of a broad peak at 1,730 cm^{-1}. This peak from sp^2 carbon atoms reflects the effect of annealing. This signature for sp^3–sp^2 transformation provided by the VDOS of the nanodiamond particle is comparable with the previous experimental Raman studies used to monitor the nanodiamond graphitization [25].

2.5 Summary and Outlook

In summary, the classical MD simulations study discussed above provides insights into the structure of nanodiamonds and their annealing induced transformation into graphitic carbon. The simulations show, as has been indicated by previous experiments, that graphitization begins at nanodiamond surface and progresses to the center. The graphitization process preferentially advances perpendicular to {111} surfaces over {100} surfaces. In addition to {100} surfaces, hydrogen termination is found to stabilize sp^3 hybridization at high temperatures. The results show a strong correlation between shape, annealing temperature, surface hydrogenation and the extent of graphitization.

From the biomedical applications perspective, it is desirable to investigate the nature of bonding at the surface of nanodiamonds and their interaction with biomolecules and eventually design tailor made interfaces between them. In this respect, as a logical extension of this work, it is important to explore the interaction of proteins and nucleic acids with nanodiamond particles and investigate the effect of their relative sp^2/sp^3 content. MD simulations can provide valuable information on the nature of interaction between nanocarbon and how nanodiamond particles influence the structure and function of biomolecules. The future holds great promise for the development of multiscale methods that allow modeling an ensemble of large biomolecules interacting with nanocarbon particles and make the goal of complete understanding of the nature of nanodiamond-biomolecule interfaces feasible.

Acknowledgments This work was supported by the US Department of Energy's Office of Basic Energy Sciences, under contract no. DE-AC02-06CH11357. Use of computer resources from Argonne National Laboratory Computer Resource Center and US DOE National Energy Research Supercomputer Center is gratefully acknowledged.

References

1. Kroto HW, Heath JR, O'brien SC, Curl RF, Smalley RE (1985) Nature 318:162–163
2. Iijima S (1991) Nature 354:56–58
3. Gruen DM, Liu S, Krauss AR, Luo J, Pan X (1994) Appl Phys Lett 64:1502–1504
4. Iijima S, Yudasaka M, Yamada R, Bandow S, Suenaga K, Kokai F, Takahashi K (1999) Chem Phys Lett 309:165–170
5. Ugarte D (1992) Nature 359:707–709
6. Lewis RS, Tang M, Wacker JF, Anders E, Steel E (1987) Nature 326:160–162
7. Dai ZR, Bradley JP, Joswiak DJ, Brownlee DE, Hill HGM, Genge MJ (2002) Nature 418:157–159
8. Greiner NR, Philips DS, Johnson JD, Volk F (1988) Nature 333:440–442
9. Vereschagin AL, Sakovich GV, Komarov VF, Petrov EA (1993) Diam Relat Mater 3:160–162
10. Kuznetsov VL, Chuvilin AL, Moroz EM, Kolomiichuk VN, Shaikhutdinov ShK, Butenko YuV (1994) Carbon 32:873–882
11. Artemov AS (2004) Phys Solid State 46:687–695
12. Shenderova O, Tyler T, Cunningham G, Ray M, Walsh J, Casulli M, Hens S, McGuire G, Kuznetsov V, Lipa S (2007) Diam Relat Mater 16:1213–1217
13. Dolmatov VY (2001) Russ Chem Rev 70:607–626
14. Schrand AM, Huang H, Carlson C, Schlager JJ, Sawa EO, Hussain SM, Dai L (2007) J Phys Chem B 111:2–7
15. Khabashesku VN, Margrave JL, Barrera EV (2005) Diam Relat Mater 14:859–866
16. Kam NWS, Jessop TC, Wender PA, Dai HJ (2004) J Am Chem Soc 126:6850–6851
17. Huang H, Pierstorff E, Osawa E, Ho D (2007) Nano Lett 7:3305–3314
18. Narayan RJ, Wei W, Jin C, Andara M, Agarwal A, Gerhardt RA, Shih CC, Shih CM, Lin SJ, Su YY, Ramamurti Y, Singh RN (2006) Diam Relat Mater 15:1935–1940
19. Yu SJ, Kang MW, Chang HC, Chen KM, Yu YC (2005) J Am Chem Soc 127:17604–17605
20. Kuznetsov VL, Chuvilina AL, Butenkoa YV, Stankusb SV, Khairulinb RA, Gutakovskiic AK (1998) Chem Phys Lett 289:353–360
21. Tomita S, Burian A, Dore JC, LeBolloch D, Fujii M, Hayashi S (2002) Carbon 40:1469–1474
22. Kuznetsov VL, Chuvilin AL, Butenko YV, Malkov IY, Titov VM (1994) Chem Phys Lett 222:343–348
23. Aleksenskii AE, Baidakova MV, Vul' A Ya, Siklitskii VI (1999) Phys Solid State 41:668–671
24. Banhart F, Fuller T, Redlich Ph, Ajayan PM (1997) Chem Phys Lett 269:349–355
25. Mykhaylyk OO, Solonin YM, Batchelder DN, Brydson R (2005) J Appl Phys 97:074302
26. Fugaciu F, Herman H, Deifert G (1999) Phys Rev B 60:10711–10714
27. Lee GD, Wang CZ, Yu J, Yoon E, Ho KM (2003) Phys Rev Lett 91(26):265–701
28. Bro'Dka A, Zerda TW, Burian A (2006) Diam Relat Mater 15:1818–1821
29. Leyssale J-M, Vignoles GL (2008) Chem Phys Lett 454:299–304
30. Barnard AS, Russo SP, Snook IK (2003) Diam Relat Mater 12:1867–1872

31. Barnard AS, Russo SP, Snook IK (2003) Phys Rev B 68:073406
32. Barnard AS (2006) Stability of Nanodiamond In: Shenderova OA, Gruen DM (eds),Ultrananocrystalline diamond: synthesis, properties and applications. William Andrew Publishing, New York, pp 117–154
33. Hu Y, Shenderova OA, Hu Z, Padgett CW, Brenner DW (2006) Rep Prog Phys 6:1847–1895
34. Raty JY, Galli G (2003) Nat Mater 2:792–795
35. Raty JY, Galli G, Bostedt C, van Buuren TW, Terminello LJ (2003) Phys Rev Lett 90:037401
36. Barnard AS, Sternberg M (2007) J Mater Chem 17:4811–4819
37. Brenner DW, Shenderova OA, Harrison JA, Stuart SJ, Ni B, Sinnott SB (2002) J Phys: Condens Matter 14:783–802
38. Mao Z, Garg A, Sinnott SB (1999) Nanotechnology 3:273–277
39. Adelman SA, Doll JD (1976) J Chem Phys 64:2375–2388
40. Tyler T, Zhirnov VV, Kvit AV, Kang D, Hren JJ (2003) Appl Phys Lett 82:2904–2906
41. Pandey KC (1982) Phys Rev B 25:4338–4341
42. Zapol P, Curtiss LA, Tamura H, Gordon MS (2004) Theoretical studies of growth reactions on diamond surfaces In: Curtiss LA, Gordon MS (eds) Computational materials chemistry: methods and applications. Kluwer Academic Publishers, London, pp 266–307
43. Tomita S, Fuji M, Hayashi S (2002) Phys Rev B 66:245–424
44. Tomita SS, Fuji M, Hayashi S, Yamamoto K (1999) Chem Phys Lett 305:225–229
45. Vita AD, Galli G, Canning A, Car R (1996) Nature 379:523–526
46. Wang CZ, Ho KM, Shirk MD, Molian PA (2000) Phys Rev Lett 85:4092–4095
47. Jungnickel G, Porezag D, Frauenheim Th, Heggie MI, Lambrecht WRL, Segall B, Angus JC (1996) Phys Status Solidi A 154:109–125
48. Kuznetsov VL, Zilberberg IL, Butenko YuV, Chuvilin AL, Segall B (1999) J Appl Phys 86:863–870
49. Qian J, Pantea C, Huang J, Zerda TW, Zhao Y (2004) Carbon 42:2691–2697
50. Qiao Z, Li J, Zhao N, Shi C, Nash P (2006) Scr Mater 54:225–229
51. Pantea C, Qian J, Voronin GA, Zerda TW (2002) J Appl Phys 91:1957–1962
52. Chen PW, Ding YS, Chen Q, Huang FL, Yun SR (2000) Diam Relat Mater 9:1722–1725
53. Xu NS, Chen J, Deng SZ (2002) Diam Relat Mater 11:249–256
54. Russo SP, Barnard AS, Snook IK (2003) Surf Rev Lett 10:233–239
55. Banhart F, Ajayan PM (1996) Nature 382:433–435
56. Banhart F (1997) J Appl Phys 81:3440–3445
57. Davison BN, Picket W (1994) Phys Rev B 49:14770

Chapter 3
The Fundamental Properties and Characteristics of Nanodiamonds

Alexander Aleksenskiy, Marina Baidakova, Vladimir Osipov, and Alexander Vul'

Abstract The review is devoted to nanodiamond produced by detonation synthesis. The past results related to the main features of detonation technology for producing nanodiamond are highlighted. Effects of technology on the structure of nanodiamond particles as well as functionalization of nanodiamond surface to chemical properties are discussed. The real structure of single nanodiamond particles has been critically reviewed and its aggregation problem has been emphasized.

The review demonstrates that while retaining the merits inherent in diamond, nanodiamonds exhibit a number of essential features, both in structure and in physico-chemical characteristics. These features give one ground to consider nanodiamonds as a specific nanocarbon material.

3.1 Introduction

On the turn of the last century, the newly coined word "nanotechnologies" has become one of the most popular and widely used terms. This popularity was apparently rooted in the hope cherished by "nanotechnologists" to develop structures that would approach in complexity those of the protozoa. An analogy between the artificial and living organisms was pointed out in the very first monographs on fullerenes, a new allotropic modification of carbon [1]. Today this analogy is universally recognized, and structures of the members of the nanocarbon family discovered at the end of the last century, more specifically, the fullerenes, nanotubes, and onion form of carbon are imaged as symbols of nanotechnology conferences, and nanocarbon structures have won a prominent place among materials planned for use in nanobiotechnology [2].

This Chapter will address, necessarily briefly, the technology of production, the structure and properties of a member of the nanocarbon family, namely, detonation nanodiamonds (DND). While retaining the merits inherent in diamond, nanodiamonds

A. Aleksenskiy, M. Baidakova, V. Osipov, and A. Vul' (✉)
Ioffe Physico-Technical Institute, Russian Academy of Sciences, St. Petersburg, 194021, Russia
e-mail: Alexandervul@mail.ioffe.ru

D. Ho (ed.), *Nanodiamonds: Applications in Biology and Nanoscale Medicine*,
DOI 10.1007/978-1-4419-0531-4_3,

exhibit a number of essential features, both in structure and in physico-chemical characteristics. These features give one grounds to consider the nanodiamonds as a specific nanocarbon material.

The detonation-based synthesis of diamond particles had been discovered more than 40 years ago, but it is only in 1988 that readily accessible publications on the subject appeared. In 1988, in the Soviet magazine "Reports of the Academy of Sciences" [3] and the Nature magazine [4] published papers to which most of the authors of scientific papers on detonation nanodiamonds presently refer as landmarks. In the 1990s, studies of DND were naturally dealing with attempts at refining the technology of their production and chemical purification [5], and it is in this period that the first publication of researchers of the Ioffe Institute, among them the present authors, was made [6].

In our opinion, real progress in the technology of production of DND, combined with proper visualization and understanding of the vast area of their possible applications, besides the fairly obvious one as a means for grinding and polishing, has materialized at the turn of the last century. At that time, the programs of International "diamond" conferences began featuring reports on nanodiamonds produced by the detonation method. In 2003–2004, the first dedicated conferences on detonation nanodiamonds were held [7, 8], and the first collective monograph was published [9]. The explosive interest being witnessed in the recent years in DND can be traced both to progress in the technology of their purification and surface modification [10–12] and to the expansion of possible areas of their application [13, 14].

3.2 Detonation Nanodiamond Technology

3.2.1 Technology of Synthesis

It was universally known that it is graphite that represents a thermodynamically stable allotropic form of bulk carbon at room temperature and atmospheric pressure. In industrial synthesis of microcrystalline diamond from graphite, the graphite diamond phase transition in the presence of metal catalysts is run, as a rule, at temperatures of 1,500–1,800°C and pressures above 5 GPa [15].

The detonation synthesis of nanodiamonds draws essentially from two original ideas:

- To use the shock wave produced in detonation of an explosive as a source of high pressure and temperature for synthesis of crystalline diamond from a carbon material.
- To form a diamond crystal by assembling carbon atoms of the explosive itself.

The pressures and the temperature reached at the shock front ($P \approx 20$–30 GPa, $T \approx 3{,}000$–$4{,}000$ K) correspond to the region of thermodynamic stability of diamond in the phase diagram of carbon and are substantially higher than those employed in static synthesis of diamond from graphite, but the time of synthesis is considerably shorter and amounts to a few fractions of a microsecond.

DND were found to have an extremely narrow distribution in size. This remarkable feature is usually attributed to the theoretical prediction that it is diamond rather than graphite that should be thermodynamically stable for carbon particle dimensions less than 5 nm [16–19]. We are going to dwell in more detail on this point in the following session.

Figure 3.1 illustrates a general schematic of detonation – based synthesis. The process of production of DND can be broken up essentially into three stages, more specifically, nucleation of the diamond core, formation of the detonation carbon structure and chemical separation of the diamond phase from the detonation carbon (soot).

Consider briefly the mechanism of formation of the diamond core [3, 20–22].

There are two different viewpoints on this mechanism. The first of them is close in approach to formation of the fullerene molecule. Actually, this model assumes the assembly of tetrahedrally (sp^3) coordinated carbon atoms (the diamond core) in the gas phase from individual atoms at the instant of the explosion. In this case, the size of the forming diamond particle is obviously determined by the time of existence of the shock wave in the detonation chamber [23].

The second approach draws from the concept of growth of the diamond core from a liquid nanosized drop [24] or plasma [21, 22]; in other words, the diamond is assumed to start growing after the completion of the detonation process.

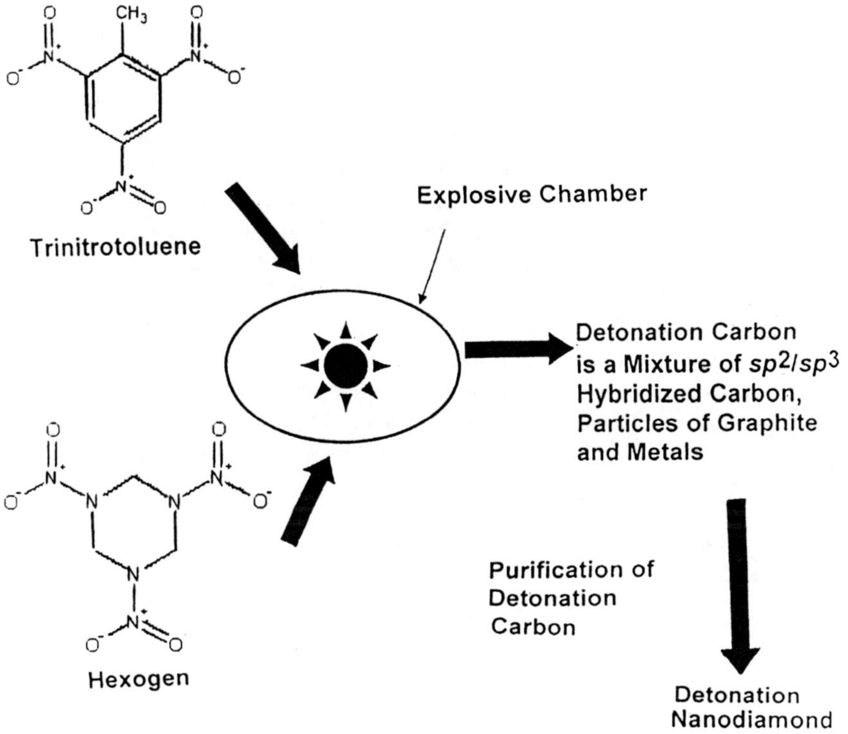

Fig. 3.1 Simplified scheme of detonation-based synthesis

In this case, the process should be mediated by thermodynamic conditions. As follows from theoretical calculations, for carbon particle sizes about 3 nm, it is diamond rather than graphite that is a thermodynamically stable form of carbon at detonation temperatures [16, 17]. One may therefore assume that as the particle grows in diameter, the strict tetrahedral structure corresponding to sp^3 hybridization will become distorted in the growing layer. The existence of such a partially disordered layer was experimentally corroborated [25–27]. Its thickness should naturally depend on the kinetics of cooling of the detonation products and on the rate of pressure drop and is, as a rule, about 1 nm. This is why most of the publications report about 4–5 nm for the size of the detonation diamond particle, and it is this size of the ordered region of the diamond lattice in a DND particle that was repeatedly revealed by XRD and Raman spectroscopy [20, 28]. We defer a comprehensive discussion of the structure of diamond particles thus formed to a later Section.

In a certain stage of the detonation synthesis, further drop of the pressure and temperature drives the *P–T* parameters to the region where the diamond is thermodynamically unstable, and it is in this region that detonation carbon as a mixture of sp^2/sp^3 hybridized carbon atoms forms.

If the temperature in this stage is still high enough to maintain a high mobility of carbon atoms (above the Debye temperature of diamond, $T_D \approx 1{,}800$ K), the diamond graphite phase transition will be more favorable than if the transition to the region of thermodynamic stability of graphite occurred at $T < T_D$. The higher the cooling rate, the shorter is the time the product of detonation synthesis stays in the region of the kinetic instability of diamond and, accordingly, the lower the probability of the reverse diamond graphite phase transition in the course of synthesis. Thus, it is the relative magnitude of the pressure drop and cooling rates (dP/dt and dT/dt) that determines the final sp^3/sp^2 ratio of the hybridized carbon shell coating the surface of a nanodiamond particle in detonation carbon [20].

Depending on the medium in which the process is run, the synthesis is called "dry" if the explosion chamber was filled by an inert gas, and "wet" or "icy", if the explosives were detonated in water or dry ice, respectively. Obviously enough, the rate of temperature drop in "wet" synthesis is higher than that in the "dry" process, so that, in accordance with the above consideration, the fraction of the diamond phase in detonation carbon before its chemical purification should grow in the order gas–water–ice, with the yields of not more than 40, 63 and 75 wt. %, respectively [29]. This is why the "wet" synthesis is characterized by a higher efficiency in the industrial scale production of DND.

3.2.2 *Isolation of Diamond from Detonation Carbon*

3.2.2.1 Removal of Impurities and of the sp^2 Phase

Isolation of DND from detonation carbon and their subsequent purification from contaminants is the key step in the synthesis technology.

As already mentioned, detonation carbon (detonation soot) is essentially a mixture of sp^2/sp^3 hybridized phases, with insignificant amounts of attendant inorganic impurities, primarily metals and their oxides present.

The technology of chemical separation and purification of DND can be divided into three stages, namely, dissolution of the metallic component, dissolution of non-diamond (sp^2) forms of carbon (NDC) and the removal of contaminants. Any method of oxidation (gasification by the reaction $C+2O_2 \rightarrow CO_2$) is based on the NDC having a higher reactivity than diamond, with this difference increasing with increasing dispersity, defect concentration and amorphism of NDC.

The two major NDC oxidation regimes employed are oxidation of detonation carbon in flowing or diffusing gas (gas phase) and in a liquid suspension (liquid phase) (see Table 3.1).

Liquid phase oxidation enjoys the widest application. This approach, apart from the purification itself by oxidation, involves also partial dissolution of the inorganic contaminants (with the exclusion of inert components like SiO_2). Liquid-phase purification of detonation carbon was performed with oxidizing systems of different types (see Table).

The technologically simplest approach is purification with chromic anhydride or chromates solution in concentrated sulfuric acid. The reaction is run in conventional glass equipment at atmospheric pressure and a temperature of 120–150°C [30].

Purification of DND with 50–75% chloric acid $HClO_4$ can also be carried out in a standard glass equipment at atmospheric pressure [4, 31]. The reaction temperature, 80 - 100°C, is lower than that employed in oxidation with chromium compounds. This method has an obvious advantage of a low consumption of the oxidizer and that the DND thus obtained is not loaded additionally by inorganic contaminants. On the other hand, chloric acid is an explosive compound, which makes a large-scale application of this method that is hardly recommendable.

In the initial stages of the development of industrial technology for detonation carbon purification, one employed oxidation with nitric acid in concentrated sulfuric acid [32, 33].

To achieve a high enough reaction rate, a higher temperature, above 200°C, is required, because nitric acid is a less active oxidizer than chromium salts or chloric acid. Among the merits of the method is that it lends itself readily to industrial realization. The main disadvantage is that it produces large amount of wastes requiring sophisticated equipment for their regeneration, which may be even more sophisticated in operation than that employed in the purification process itself.

The universally accepted method of industrial scale purification of DND involves oxidation of the NDC with a water solution of nitric acid (50–67 % concentration) [34–36].

The process is run at an elevated pressure (80–100 atm) in reactors made of titanium alloys. After thorough washing, the DND product thus obtained reveals only low-level contamination by inorganic compounds, because nitric acid forms easily soluble salts with practically all known metals.

A more complicated approach involves the use of a HNO_3–H_2O_2 mixture in place of pure HNO_3 [37, 38].

Table. 3.1 Comparison of various chemical methods of DND isolation from detonation soot and its purification

Method of purification	Oxidizer; conditions	Degree of purification; residual content of NDC	Content of inorganic impurities	References
Air, catalytic oxidation	O_2 + catalyst; 200–300°C	Average; 1–2%	5–0% without additional washing in acids	[39, 40]
Air, high-temperature oxidation	O_2; 420°C	No information	5–10%, washing of starting material with HF/HNO_3 mixture	[12, 42]
Ozone oxidation	O_3; 20°C	Very high level; <0.1%	5–10% without additional washing in acids	[41]
Chromic oxidation	CrO_3 in H_2SO_4 medium; 100°C	High; <0.2%	1–3%	[30]
Oxidation by perchloric acid	$HClO_4$ in water medium; 50–70%; 100°C	Average; 0.5–1.0%	0.5–1.0%	[4, 31]
Oxidation by a mixture of sulfuric and nitric acids	HNO_3 in H_2SO_4 medium; 200–250°C	Average; 0.5–2.0%	0.5–1.0%	[32, 33]
Oxidation by nitric acid	HNO_3 in water medium; 50–65%; 200–250°C	Average; 0.5–2.0%	0.5–1.0%	[34–36]
Oxidation by hydrogen peroxide	$HNO_3 + H_2O_2$ in water medium	Average; 0.5–2.0%	0.5–1.0%	[37, 38]

Oxygen of the air is the simplest and most widely used gaseous oxidizer. The temperature of the onset of the oxidation process is ordinarily lowered by adding an appropriate catalyst. Here, one may choose, manganese or iron among boric anhydride B_2O_3 [39] and chromium acetylacetonate [40].

The principal advantage of this purification process lies in its simplicity. On the other hand, the control of oxidation is anything but a simple procedure, and the technological conditions of the process are not the same for different constituents of the product. The process may be plagued by poor selectivity, because heating above 400°C brings about partial oxidation of the diamond yield as well.

Besides, oxidation does not remove any inorganic contaminants, and their content may become even enhanced by the presence of the catalyst. As a result, air (or any other gas phase) oxidation requires incorporation of an additional stage of acid processing to remove the soluble inorganic contaminants.

To sidestep this limitation, a process of high-temperature air oxidation, which does not involve catalysts was developed [12]. In this process, one first treats detonation carbon with a HNO_3–HF mixture, a step that reduces the level of inorganic contamination of detonation carbon down to ~1%. The product is then treated with air at 420°C for 100 h. The DND thus obtained has a 3.5% contamination, a figure, which exceeds nevertheless the standard level reached in the liquid phase processes.

Purification by ozone is one more variant of gas phase oxidation [41]. The oxidizer here is ozone produced in corona discharge. This method of oxidation stands out in its selectivity. The oxidation parameters are readily controllable and are practically the same for the product as a whole.

The principal merit of this method lies in the very high degree of purification from NDC, probably the highest reached thus far. Among its obvious shortcomings are the sophisticated equipment required, high power consumption and, as a consequence, the high cost of ozone-purified DND.

The graphitization oxidation two stage method of DND purification mentioned earlier [12] has been recently gaining popularity.

The first announcement of the results and efficiency of this method of DND purification has been apparently made in [42]. It was reported that graphitization of DND powder in nitrogen environment at 1,000°C, followed by oxidation in air at 450°C, results in a substantial removal of the sp^2 phase from the surface of the DND particles. The technique used in [42] to analyze the content of the sp^2 phase, infrared spectroscopy, cannot, however, be considered sensitive enough. This method of purification was subsequently analyzed experimentally in a onsiderable detail by thermogravimetric analysis, X-ray diffractometry, and Raman spectroscopy [12, 43], and optimum regimes of oxidation providing proper removal of the sp^2 phase were formulated. In our opinion, however, the interest in this method stems primarily from the fact that the graphitization – oxidation process brings about DND deaggregation, an aspect we are going to dwell briefly on in the next Section.

3.2.2.2 Deaggregation of Detonation Nanodiamond Particles

The commercial powder of diamonds produced by the explosive detonation and their water suspension consist, irrespective of the way they were purified, of particles about 200 nm in size, i.e., nearly two orders of magnitude larger than the abovementioned size of the ordered diamond lattice region of 4–5 nm.

Thus, researchers here actually do not deal with nanosized but rather with submicron diamond particles [44, 45]. DND powder is essentially a typically fractal structure which exhibits self-similarity under variation of scale [46]. There is nothing remarkable in the aggregation of nanosized particles into submicron and micron formations, if one considers their high surface energy; the only intriguing thing is the strength of the 200-nm aggregates coined [47] the core aggregates. It was found impossible to break up these aggregates into the 4-nm primary diamond particles by traditional ultrasonic treatment of the water suspension; partial success and comminution of aggregates down to 40–60 nm in size was reached only by subjecting the suspension to shock wave treatment [48].

It is only quite recently that one has succeeded in nearly complete (up to 90%) breakup of the core aggregates and obtaining a suspension consisting of single ultra-nanocrystalline diamond particles 4–6 nm in size by applying the so called stirred media milling [49], a method involving intense mixing of DND powder in a water medium together with micron sized zirconium dioxide balls. This method of comminution gives rise inevitably, however, to two new problems, namely, formation around each 4–6 nm DND particle of a sp^2 shell, which results from the diamond –graphite phase transition on the particle surface driven by the comminution-induced local heating [50], and contamination of the surface by zirconium dioxide.

This is why the experimentally realized possibility of reducing the size of the aggregates down to 50–60 nm by the graphitization–oxidation treatment has been attracting an ever increasing interest in the two recent years [12, 42, 51, 52].

Despite a certain progress demonstrated in deaggregation of commercial DND, the problem of obtaining stable hydrosols of 4-nm DND particles is still far from solution.

We are inclined to associate this failure with a lack of a clear model accounting for the strength of the DND aggregates. A recent model [20] treats the aggregate as a polycrystal made up of 4-nm misoriented diamond crystallites. As a result of this misorientation, both X-ray diffraction and Raman spectroscopy clearly reveal 4-nm features as the size of a perfect diamond crystal lattice. At the same time, the strength of an aggregate is determined by the same factor as in conventional polycrystals, i.e., by the chemical bonding coupling its grains, which in this case is apparently covalent.

It appears only natural to assume that the interfaces between the 4-nm crystallites, which represent actually disordered regions (a mixture of the sp^2/sp^3 phases), become graphitized and oxidize subsequently much more easily than the pure sp^3 phase. This is what can account for the deaggregation in the graphitization–oxidation process.

Summing up this Section, it has to be added that even after additional purification, commercial DND is presently still not a powder or a hydrosol of 4-nm diamond particles. In the best case, the results quoted in literature relate to small amounts of 60–100-nm aggregates, which contain a certain percentage of both inert contaminants and metallic salts and oxides, primarily of iron. While DND of this quality may be fully appropriate for industrial applications, to tailor it to wide use in biology and medicine, one would have both to reduce substantially the level of impurities to a level less than 0.1% and to obtain sols of 4-nm diamond particles.

3.3 Structure of the Detonation Soot and DND Particles

Let us address now the present ideas concerning the structure of a 4-nm DND particle.

The standpoint that was dominant in the late 1990s visualized DNDs as perfect diamond lattice crystallites characterized by a narrow distribution in size, with an average of about 4 nm. Detonation soot was considered to consist not only of the original DND aggregates of different sizes and amorphous graphite, but also of larger secondary aggregates, onion-shaped carbon, graphite ribbons and so on [53, 54].

Despite the very complicated structure of the detonation soot, analysis of experiments performed in different chemical purification conditions, which culminated in the isolation of DND of an acceptable purity level [55], combined with studies of the properties of DND under thermal annealing [56, 57], have brought several groups of researchers, independently and practically simultaneously, to the formulation of the so-called core-shell model of DND structure.

A major achievement reached with the core-shell models of the structure of detonation soot particles [55] and bucky nanodiamonds [58] was a successful explanation of many experimental observations (see [59] and references therein). Bucky diamonds are all carbon core-shell particles that are characterized by a crystalline diamond sp^3-bonded core, encapsulated by a single- or multi-layer sp^2-bonded fullerenic shell that either partially or fully covers the particle surface. The fullerene-like shell is formed in the partial graphitization of diamond in the course of detonation synthesis.

At the same time, the model does not consider the actual location of functional groups, whose presence on the DND surface is corroborated, for instance, by nuclear magnetic resonance and IR spectroscopy [53, 60, 61]. The model suggests, however, that such groups should be located primarily on the surface of the diamond core of the DND particle, with a part of them possibly bonded to the edges of the graphene sheets of the fullerene-like shell.

Indeed, it was demonstrated [60] that in chemically purified DND samples, in which the core surface is partially covered by a fullerene-like shell, the functional coating is bonded to the core. The actual ratio of different functional groups present depends primarily on the method employed to separate DND from the detonation soot and the techniques chosen for its modification [26, 29, 42].

The effect of the state of the surface and of the particle shape on stability of the bucky-diamond structure has been treated any length by theorists in the recent years [18, 19, 25, 58, 62, 63]. As follows from thermodynamic calculations, depending on the temperature and pressure of the hydrogen and carbon gases used in the diamond growth process, diamond would grow into nanoparticles of about 3 nm in size, with a non-hydrogenated, fullerene-like surface [18]. While first principles computer simulation experiments suggest that the surfaces of dehydrogenated nanodiamonds are structurally unstable, surface hydrogenation may induce some stability [19]. This is true for the particles of spherical shape. Polyhedral particles show a tendency to surface reconstruction with formation of fullerenic bubbles on the surface (see [62] and references therein).

Interestingly, transmission electron microscopy images show DND particles to be of polyhedral, primarily octahedral or cuboctahedral, shape and twinned crystals [64, 65]. The truncated octahedral shape, which is best suited for mathematical description in this case, is bounded to 76% by the {111} faces, and to 24%, by the {100} surfaces. If graphitization extends over less than 76% of the surface, the particle will remain stable with the diamond structure coated by a thin (one to two monolayers thick) shell [63].

Significantly, the core-shell model assumes the core to consist of sp^3-hybridized carbon atoms, i.e., to have a perfect diamond lattice.

Regrettably, far from all experimental data are amenable to description in terms of this model. It was shown [25–27, 59, 66] that one has to introduce a transition layer between the perfect diamond core and the sp^2 shell, which means transition to a three-layer model of the DND particle. The diamond core size was estimated [26, 27] as 3 nm, with the transition region consisting of 2–4 layers (0.8–1.2 nm) for a particle 3.5–4.5 nm in size.

There are more than one approach to the description of this transition layer.

The interatomic distance in the transition region was pointed out [26, 66] to be smaller than that in the DND core, a characteristic feature of the mixed sp^3/sp^2 bonding of carbon atoms. Also, the sp^3/sp^2 ratio for hybridized carbon in the transition region depends on the detonation product cooling kinetics and the method chosen for the isolation of DND from this product of synthesis [26, 29]. Both the fullerenic shell and the heavily defected, primarily sp^2-bonded part of the transition region can be removed by the corresponding oxidative technique [29]. The fullerenic shell grows, however, immediately again on top because the pristine {111} surfaces undergo graphitization as soon as they become exposed [67].

Thus, the structure of the transition layer depends essentially on the method and stage of purification of the detonation soot.

Another group of researchers treats the absence of the sp^2 phase in the transition layer from a different angle [25]. A comparison of a molecular dynamics simulation with X-ray diffraction experiments performed with synchrotron radiation (wavelength $\lambda = 0.13\,\text{Å}$) brought them to the conclusion that tetrahedral packing transforms gradually to a fullerenic-like shell with a disordered sp^2-hybridized transition layer. This process leaves graphite-like fragments on the {111} facets of DND particles, whereas the {100} surfaces retain their diamond-lattice ordered state.

The above consideration can be summed up as follows (see Fig.3.2.). DND particles are distinctly faceted polyhedra consisting of three parts. The central part is a diamond core built by sp^3 carbons. Directly adjoining the core is a transition layer formed primarily by disordered sp^3 carbon. This near surface transition region is partially coated by a fullerene-like shell of sp^2 carbons. Carbon atoms residing on free areas of the outer surface become bound to hydrogen and oxygen atoms to produce a variety of functional groups that terminate the dangling bonds. It is the diamond core that accounts for the main "diamond" properties of DND, more

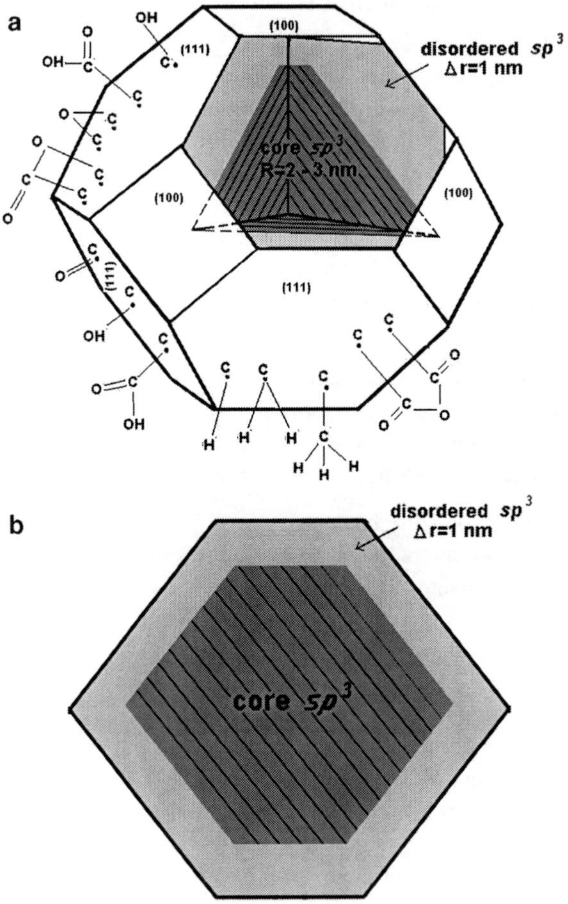

Fig. 3.2 (**a**) Schematic model of DND structure in the form of a truncated octahedron. The (100) faces are hatched, (111) faces are not hatched. Shown on the (111) faces are various functional groups, namely, the hydroxyl, carboxyl, epoxy, cetone, anhydride, ester, methyl, methilene, methine as most likely ones to become attached to DND isolated from detonation carbon in the course of chemical purification (see Chap. 4.2). The (100) faces have reconstructed surface, and the (111) ones can be either covered by the above functional groups or feature a fullerene-like form. (**b**) Cross-section of the DND structure, the functional groups are not displayed

specifically, its thermal and chemical stability, thermal conductivity, low electrical conductivity, quasi abrasiveness and quasi hardness; by contrast, the transition layer, together with the outer functional coating, is responsible for such not less essential characteristics of the material as absorption, adsorption, chemisorption, chemical composition of the functional groups, colloidal stability of DND in liquid media.

3.4 Methods for Characterization and Modification of Detonation Nanodiamonds

In the last Section of this Chapter, we are going to address the methods employed in characterization and the possibilities of surface modification of DND, because it is nanodiamonds with a modified surface that offer, as we believe, the most significant potential for the biological and medical applications [68].

The relevant studies were originally performed on poorly purified samples, in which DND particles coalesced, depending on the pH value used, into agglutinates a few hundred to a few thousand nanometers in size.

In selecting a proper method for DND investigation, one is guided by its potential to determine

- Average size of the DND cluster and size distribution of DND clusters
- sp^2/sp^3 ratio of hybridized carbon atoms in a cluster, or, in other words, the degree of chemical purification of diamond from the graphite phase
- Presence and type of impurities in the bulk and on the surface of a DND cluster

These requirements defined the choice of the methods adequate to the problem, with the most informative of them being X-ray diffractometry and small-angle X-ray scattering, Raman and IR spectroscopy, high resolution transmission electron microscopy and scanning electron microscopy, nuclear magnetic resonance and electron spin resonance (ESR).

Application of these methods made it possible to amass noncontradictory information on the "diamond" properties of DND particles and their structure and to develop the core-shell model. A detailed review of the above methods of investigation can be found in [20, 69].

In that stage of research, most of the published results related to cooperative properties of the composite material which consisted of sp^3-bonded diamond embedded in amorphous carbon with sp^2-bonded inclusions. It proved to be fairly difficult to isolate from these data specific information bearing on the properties of the 4-nm primary particles and their shells. Subsequent progress in DND purification and deaggregation provided a sound basis for the refinement of the core-shell model and for a dedicated study of the properties of single 4-nm particles [60, 70]. A proper approach to the problem of deaggregation made it necessary to invoke the other methods of investigation as well, primarily of dynamic light scattering [71], measurement of the electrokinetic potential (the ζ potential) [72].

We are going to dwell below, necessarily briefly, on the results obtained only in ESR and IR spectroscopy, as methods which, in our opinion, are capable of providing a major contribution to the characterization of samples intended for the application in biology and medicine.

3.4.1 Electron Spin Resonance and Magnetic Studies

It is known that ESR is a powerful tool for probing intrinsic paramagnetic defects and transition ferromagnetic metals present as impurities in DND [73].

The intrinsic paramagnetic signal of a DND particle is related intimately with the presence of lone orbitals (free radicals) produced in the fracture of the C–C sp^3 bonds in the diamond crystal lattice, both inside the core and on the nanodiamond particle surface.

The characteristics of the intrinsic ESR signal produced by a DND particle are as follows: signal width (0.80–0.87) mT, *g*-factor 2.00265, paramagnetic spin concentration $N_s = (6–7) \times 10^{19}$ spin/g. Interestingly, this value of the paramagnetic spin concentration yields 13–15 spins for an estimate of free radical concentration on the surface of a DND particle. This value is smaller by about two orders of magnitude than the number of surface atoms on a DND particle with a 5-nm diameter (≈2,600).

The behavior with temperature of the integrated ESR signal intensity of a DND follows the Curie–Weiss law within the temperature interval of 3–300 K; the Curie temperature here is $\Theta \approx -1.1$ K, and the width of the ESR signal grows only insignificantly with the decreasing temperature to 1.2–1.3 mT at $T = 3$ K. All this implies a localized point origin of the intrinsic defects in DND, which appear to be of the same nature, with a weak antiferromagnetic spin coupling.

Industrial scale DND material is usually contaminated by impurities of transition metals, such as iron, titanium and cobalt, which may reside either on the surface or in the bulk of the DND aggregates, both as single atoms and in the form of metallic nanoparticles. The latter form, however, only in high-temperature DND treatment at $T > 800°C$, when metallic impurities in DND aggregate to form a metallic nanophase [74]. This is why, the above metallic impurities of transition metals may be present in DND both in the paramagnetic and in the superparamagnetic and ferromagnetic forms.

Impurities of iron and of the other transition metals in DND produce a broad ($\Delta H_{pp} \sim 30–35$ mT) signal with a *g*-factor of 4.13–4.28 in the half-field region ($H \sim 155$ mT), which is detected reliably by X-band ESR at a frequency of 9–9.4 GHz [73].

Special chemical treatment of DND powder permits one to practically remove the trace amounts of iron and other transition metals completely [75]. This is evidenced reliably by the disappearance of the broad ESR peak ($\Delta H_{pp} \sim 30–35$ mT, $g = 4.13$) at $H \sim 155$ mT [73, 76].

Studies of the magnetic properties revealed that low-temperature ($T = 2$ K) magnetization curves of DND subjected to deep purification from transition metals should be fully assigned to a system of intrinsic $S = ½$ spins present in a concentration of $(6–7) \times 10^{19}$ spins/g. This concentration is reliably derived from the Curie constant in the temperature dependence of the reciprocal static magnetic susceptibility of DND powder within a broad temperature interval extending from 5 K to 380 K [77].

It should be stressed that while after this deep purification, the magnetization curve obtained at $T=2$ K does not reveal any contribution due to metals with spins $S \geq 1$ [76], in place of the now absent, broad ESR signal ($\Delta H_{pp} \sim 30$ mT, $g=4.13$) one finds a new doublet signal consisting of two narrow lines ($\Delta H_{pp}^{(1)} \sim 2.9$ mT, $g_1=4.29$; $\Delta H_{pp}^{(2)} \sim 1.3$ mT, $g_1=4.013$ at $T=300$ K) separated by a distance of 10.5 mT on the magnetic field scale and located in the 150–160 mT interval [77]. The total integrated intensity of the doublet signal is 10^4–10^5 times lower than that of the main ESR signal ($g=2.00265$) in DND.

Additional studies provided information on the nature of the doublet ESR signal in the region of $H \sim 155$ mT. It originates from exchange-coupled spin pairs of the intrinsic paramagnetic defects in DND and the so called "forbidden" ($\Delta M=2$) transitions between Zeeman-split states of the thermally populated triplet level with $S=1$ at the half-field (the so-called singlet – triplet model for exchange coupled spin pairs [77]).

Estimates based on a calculation of integrated intensity of the doublet ESR signal suggest that the concentration of observed exchange coupled spin pairs is about one dimer per (100–150) 5-nm DND particles. The small energy separation between the singlet and triplet states in an exchange-coupled spin pair ($\Theta \approx -1$ K) accounts for the practically equal population of both levels within a broad temperature interval, including the low-temperature domain, $T=4$ K, thus making possible observation of $\Delta M=2$ transitions in a magnetic field between the Zeeman-split states of the $S=1$ triplet level.

Studies of the behavior with temperature of the doublet ESR signal showed its integrated intensity to obey closely the Curie – Weiss law and to correlate with that of the main ESR signal in DND throughout the temperature interval covered, including the low-temperature one (down to 4 K). This provides additional evidence for the doublet ESR signal and the signal from the intrinsic paramagnetic defects in DND being of the same nature.

Significantly, the double ESR signal at half magnetic field is always present in the spectra of the DND powder samples contaminated by iron-containing complexes as well, but its detection is made difficult or even impossible by the superposition of the weak doublet signal on the broad ESR feature ($g \sim 4$) originating from the iron-containing complexes. The integrated intensity ratio of the two ESR signals is not less than 1000 : 1.

Observation of the doublet ESR signal combined with simultaneous disappearance of the broad ESR signal deriving from iron-containing complexes (see Fig. 3.3) may serve as a useful indication of high DND purification from iron and transition metal impurities [77].

3.4.2 Characterization of DND Surface by Infrared Absorption

It is well known that defect-free bulk diamond is transparent in the IR spectral region due to the high symmetry of its crystal lattice. Lowering of the C–C bond symmetry in a diamond crystal near its surface may give rise to the appearance of a band at 1,332 cm^{-1}. This effect was observed in IR spectra of synthetic diamond with a specific surface area of 22 m^2/g after surface treatment [78]. The powder was pretreated by a mixture of sulfuric and nitric acids, with subsequent annealing in hydrogen environment.

As a result of the high specific surface of DND (250–350 m^2/g), the pattern of the DND IR spectrum is determined by the surface adsorbed impurities or functional groups grafted to the surface. The presence of water adsorbed on the surface also becomes evident in varying degrees in the IR spectra of DND. Inorganic impurities in DND include both metal oxides and salts and traces of acids. The functional groups bound to the DND surface are predominantly oxygen-containing compounds which are formed on the surface in the course of DND synthesis, in the stage of removal of the non-diamond sp^2 component from detonation soot.

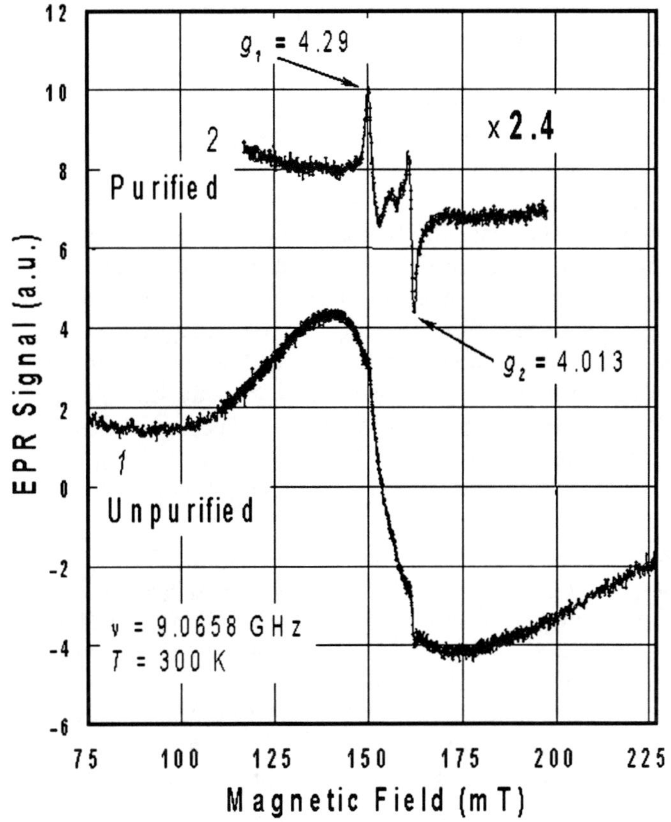

Fig. 3.3 ESR spectra of DND powder samples in the region of half magnetic field, $H \sim 155$ mT. *1*-non-purified DND powder revealing traces of transition metals and iron-containing compounds; *2*-DND powder after multiple special chemical purification treatments in boiling hydrochloric acid. Graph *2* shows a well-resolved doublet signal consisting of two narrow lines with a width $\Delta H_{pp}^{(1)} \sim 2.9$ mT; $\Delta H_{pp}^{(2)} \sim 1.3$ mT

Functional groups can be removed or modified by chemical treatment. This modification of DND surface is a useful tool permitting control of its properties. IR spectroscopy is a powerful method to follow the process of the modification.

Figure 3.4 presents the IR absorption spectra of DND. Curve *1* was obtained on a DND sample after the treatment of detonation soot with strong acids, a procedure most commonly accepted. The pattern of the spectrum (curve *1*) is in accord with available data measured on DND samples by different research groups in 1991–2006 [26, 43, 47, 52, 53, 61] The other spectra presented in Fig. 3.4 (curve *2*) relate to a chemically purified DND sample subjected to additional hydrothermal treatment and to the same sample which was fluorinated following hydrothermal treatment (curve *3*) [79]. As shown later on, hydrothermal treatment produces the same features in the IR spectrum as annealing in hydrogen, because both modification processes are run in a reducing medium [80].

Consider some typical features in the DND IR spectra.

The IR spectra of DND samples most frequently demonstrated in literature reveal a system of water absorption bands, which consists of a broad composite band close to 3,600 cm^{-1} and a somewhat weaker band, near 1,620 cm^{-1}. The first of them derives from the symmetric and asymmetric stretch vibrations of water molecules, and the other, from the OH bend vibrations [81, 82]. The disagreement with 1,645 cm^{-1}, the value usually quoted in the literature, is usually assigned to adsorption effects.

By heating DND embedded in KBr and AgCl pellets in vacuum (140°C, 4 h), one succeeded in obtaining an IR spectrum of practically dehydrated DND [81]. The stretch vibration is practically suppressed in the spectrum, thus evidencing a complete removal of water from the sample. Interestingly, the reverse process, i.e., adsorption of water from air, takes only a few minutes to come to the end.

The OH bend vibration band (1,620 cm^{-1}) does not vanish completely under dehydration, and one may therefore assume that this band masks the intrinsic IR spectrum of DND in this region.

The inert impurities usually occurring in DND (SiO_2, TiO_2) do not produce strong absorption lines in the 400–4,000 cm^{-1} region. Some spectra may reveal a weak absorption band at 820 cm^{-1} (curve *2*). Disappearance of this band after fluorination (curve *3*) permits assignment of this absorption to a TiO_2 impurity.

The bend vibration band system in the 1,600–1,500 cm^{-1} region characteristic of the aromatic carbon compounds [83] is not strong and can be masked by oxygen-containing functional groups. The CH stretch vibrations observed in aromatic systems near 3,000 cm^{-1} are not usually present in DND spectra. This suggests that the amount of aromatic carbon left over in DND after the standard industrial acidic purification is fairly small. On the other hand, in the Raman spectra of DND one observes also, besides the broadened main diamond line, slightly shifted by phonon confinement [28], an intense signal near 1,600 cm^{-1}, which originates from the absorption in aromatic rings. It is believed that the reason for this disagreement between the results obtained by two spectroscopic methods lies in the anomalously large absorption cross section for aromatic carbon in the Raman spectrum.

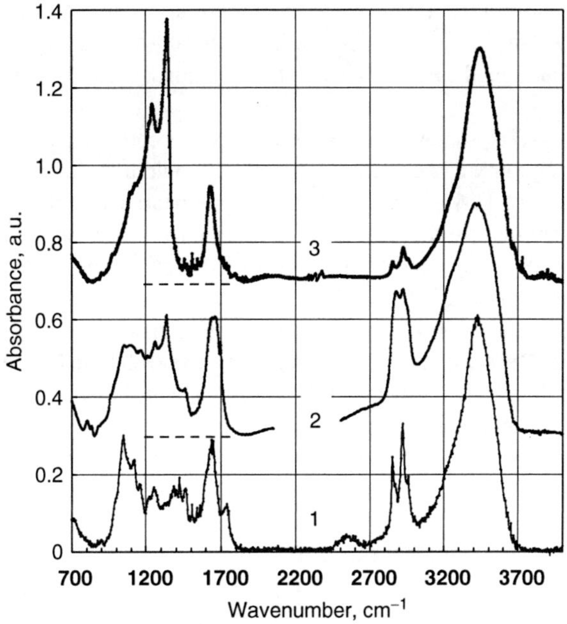

Fig. 3.4 Infrared absorption spectra of DND samples: *1*-oxidized DND obtained after additional purification of commercial DND with hydrochloric acid and subsequent drying in air at 110°C; *2*-hydrothermally treated DND obtained by subjecting standard DND to water under critical conditions (650°C, 100 MPa) for 36 h, followed by drying in vacuum at room temperature; *3*-fluorinated DND produced by treating sample *2* with gaseous fluorine (500°C, 1 atm) for 3 days. Curves *2* and *3* were translated upward to make the graphs more revealing. Dashed lines identify zero absorption levels for curves *2* and *3*. The curves are normalized against the amplitude of the absorption band at ~3,450 cm^{-1}. Curve *2* in Fig. 3.3 and curve *1* in Fig. 3.4 were obtained on the same sample

Thus, the bands that can be assigned to intrinsic absorption of conventional (non-modified) DND are: the group of maxima in the 2,850–2,950 cm^{-1} interval, the band at 1,720–1,750 cm^{-1}, the moderately strong band at 1,640 cm^{-1} masked by water absorption, and a system of bands with frequencies of 1,450, 1,340, 1,250, 1,160, 1,110 and 1,050 cm^{-1} [26, 43, 47, 52, 53, 61, 81, 84]. The exact positions and the relative intensities of these bands may vary somewhat from one publication to another.

The group of bands at 2,850–2,950 cm^{-1} is usually associated with CH stretch vibrations in the aliphatic carbon chains (methyl and methylene groups). This assignment is borne out by the strong enhancement of intensity and broadening of these bands in DND treated in a hydrogen environment [61, 81] (curve *2*). By contrast, when fluorine is substituted for hydrogen in the course of fluorination (curve *3*), the intensity of these lines drops noticeably, although not to zero. A similar behavior is observed with the bands at 1,450, 1,250 and 1,160 cm^{-1}, which are usually assigned to bending vibrations of CH_2 groups in a variety of environments.

Deep fluorination (14 at. %) of nanodiamond activates intense bands at 1,350, 1,245 and 1,120 cm^{-1} [79], which derive from vibrations in the CF_2 and CF chains, respectively. The spectrum in this region is demonstrated, however, to be more complex [85].

The band at 1,720–1,750 cm^{-1} is usually considered to belong to the ketone group with a secondary carbon atom. Reduction should transform it to the hydroxyl or methylene group. This is supported by the disappearance of this band in curve *2*. Some researchers believe that this group can transform to a carboxyl or anhydride group, to be accompanied by its shift by 20–40 cm^{-1} when treated by oxygen at temperatures above 400°C [26, 43, 52].

This band (1,720–1,750 cm^{-1}) could derive also from CO stretch vibrations in carboxyl groups. The presence of carboxyl groups is corroborated also by the presence of the complex bands at 1,420 and 1,320 cm^{-1}, which, however, could be masked by bending vibration lines of methyl groups.

The vibration bands at 1,110 and 1,340 cm^{-1} may suggest the presence of ternary alcohol groups [47], which become smeared in the course of reduction and disappear under fluorination.

A more accurate assignment of the absorption bands in DND IR spectra would require additional studies, including the preparation of dehydrated samples [86], performing purposeful surface modification and application of different techniques of sample characterization [87].

3.4.3 Functionalization of DND Surface

Purposeful surface modification by selective substitution or grafting of desired functional groups rests upon exact knowledge of the groups, which become bound to the surface in the course of a given method of DND isolation from the detonation soot.

The surface of DND was shown [10, 43, 52, 82, 88] to be partially terminated by carboxyl groups. Surface carboxyl groups dissociate in water suspension with formation of $-COO^-$ ionic derivatives bonded covalently to the surface and of the H^+ cations [10, 82]. As a result, the surface of DND particles in water suspension turns out charged negatively (as shown in Fig. 3.5). This makes DND in water suspension an ideal object for purposeful modification of the surface by various organic and inorganic atomic groups, metal ions, fluorescent organic complexes and biologically active compounds (biomolecules) [68, 89]. Complete or partial dissociation of surface carboxyl groups in DND water suspension suggests that mixing a DND suspension with water solutions of proper compounds may initiate ion exchange reactions on the DND particle surface with cations of the substances dissolved in water [10].

This idea initiated the development of a method [74] for obtaining DND powder with metal ions distributed uniformly over the particle surface. The method is based essentially on the substitution of metal ions for hydrogen in carboxyl groups. Because DND hydrosols used in [10] were composed of particles about 200 nm in

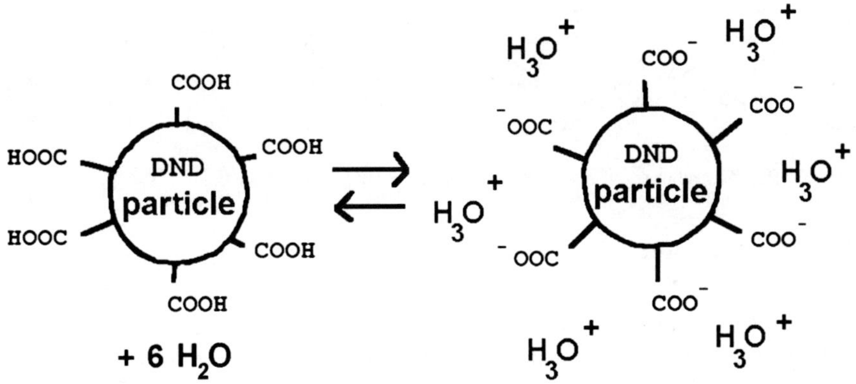

Fig. 3.5 Dissociation of carboxyl groups on the surface of a DND particle and formation of a double electric layer

size, the metal ions were distributed primarily on the surface of the aggregates rather than on the 4-nm DND particulates. After solving the problem of deaggregation, this method should reveal its potential in saturating the surface of a 4-nm DND particle directly with metal ions.

Detailed information on various methods of nanodiamond surface modification including the gas-phase treatment, "dry" and "wet" oxidation, plasma functionalization, UV-induced amination and chlorination can be found in [90].

A new approach has been developed and tested [91] to the purification and functionalization of nanodiamond surface, which consists essentially in its treatment in a gas phase containing chlorine, argon, and ammonia at high temperatures (400–1,100°C). The results obtained in a comparative analysis of the properties of ND powder following treatment in pure Ar and in a CCl_4/Ar mixture at 450°C are presented. It is pointed out that the DND powder treated in the gas mixture practically does not absorb water vapor, in other words, it becomes hydrophobic.

To produce stable hydrosols, it was proposed to modify DND by heat treatment in air [52]. The hydrosols thus obtained had particle sizes of down to 60 nm. The authors point out that while heat treatment improves hydrosol stability, on the whole it exerts a negative effect, because it initiates the formation of bridge bonding between the disparate groups on the surface of neighboring DND particles.

Indeed, besides the pure van der Waals interaction, DND particles may be linked by bridge bonding differing in nature (see Figs. 3.6a–c). Significantly, it is not only the anhydride groups that can affect the particle aggregation in DND hydrosols. This aggregation can be initiated by dimers of carboxyl groups and many-valent metal impurities as well [92]. Drying and thermal treatment of DND at the elevated temperatures may bring about irreversible aggregation.

The above methods of modification have not apparently succeeded in producing the monofunctional DND surface. It was demonstrated [87], however, that heat treatment in flowing hydrogen makes it possible to obtain a bifunctional DND surface, which hopefully can be transformed to monofunctional.

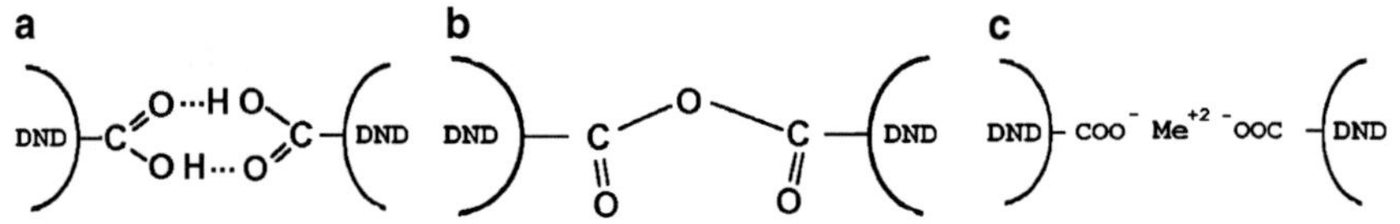

Fig.3.6 Possible schemes of bridging bonds linking nanodiamond particles

3.5 Conclusion

In conclusion, we should like to stress the problems related intimately with the structure and properties of detonation nanodiamonds, which are still awaiting solution.

The list of these problems, in our opinion, should include:

- Refinement of the core-shell model for a single DND particle
- Surface topology of a single primary DND particle
- Factors mediating the strength of primary 100–200-nm DND particle aggregates
- Mechanism accounting for the stability of DND hydrosols and suspensions

The above problems are connected in varying degrees with obtaining stable sols of primary DND particles, and it is their solution that has been pursued in studies during the recent 2 years [63, 70].

An aspect that appears to be of particular importance, both for basic science and for the field of applications, bears on the possibility of introducing desired impurity atoms into the bulk of a DND particle and development of purposeful functionalization of the surface of a DND particle, the subject of the last section of this Chapter.

In concluding the Chapter, we should like to point out that it is the recent years that have witnessed an explosive growth of interest in the promising areas of research and application of detonation nanodiamonds, an aspect which we feel reflects a general attitude to, and progress in the technology of carbon nanostructures.

Acknowledgment Studies of the present authors in the field of nanodiamonds have been supported by the Russian Foundation for Basic Research and by Programs of the Russian Academy of Sciences.

References

1. Dresselhaus MS, Dresselhaus G, Eklund PC (1995) In: Science of fullerenes and carbon nanotubes. Academic Press, San Diego
2. Piotrovskii LB and Kiselev ON (2006) In: Fullerenes in biology. Publisher house "Rostok", St. Petersburg, Russia, 336 pp
3. Lymkin AI, Petrov EA, Ershov AP, Sakovitch GV, Staver AM, Titov VM (1988) Dokl Akad Nauk USSR 302:611–613
4. Greiner NR, Philips DS, Johnson JD, Volk F (1988) Nature 333:440–442
5. Vul'A Ya, Dolmatov V Yu, Gruen DM, Shendorova O (eds) (2006) In: Detonation nanodiamonds and related materials. Bibliography index, 2nd issue. Ioffe Physico-Technical Institute, St Petersburg, Russia

6. Aleksensky AE, Baidakova MV, Boiko ME, Davydov V Yu, Vul' A Ya (1995) Application of diamond and related materials: 3 rd International Conference, Gaithersburg, MD, USA, 21–24 August 1995. NIST Special Publication Issue 885, pp 457–460
7. Detonation nanodiamonds: fabrication, properties and applications: Proceedings of the first International Symposium. St.-Petersburg, Russia, 7–9 July 2003 (2004) Phys Solid State 4: 595–769
8. Gruen D, Shenderova O, Vul' A Ya (eds) (2005) In: Synthesis, properties and applications of ultrananocrystalline diamond, vol. 192. Springer, Dordrecht
9. Shendorova OA and Gruen DM (eds) (2006) In: UltraNanocrystalline diamond. Synthesis, properties, and applications. William Andrew Publisher, Norwich, NY, USA
10. Aleksenski AE, Yagovkina MA, Vul' A Ya (2004) Phys Solid State 46:685–686
11. Petrov I, Shenderova O, Grishko V, Grichko V, Tyler T, Cunningham G, McGuire G (2007) Diam Rel Mater 16:2098–2103
12. Pichot V, Comet M, Fousson E, Baras C, Senger A, Le Normand F, Spitzer D (2008) Diam Rel Mater 17:13–22
13. Shi XQ, Jiang XH, Lu LD, Yang XJ, Wang X (2008) Mater Lett 62:1238–1241
14. Liu K-K, Cheng C-L, Chang C-C, Chao J-I (2007) Nanotechnology 18:325102 (10 pp)
15. Burchell TD (ed) (1999) In: Carbon materials for advanced technologies. Elsevier Science Ltd, Amsterdam
16. Gamarnik MY (1996) Phys Rev B 54:2150–2156
17. Badziag P, Verwoerd WS, Ellis WP, Greiner NR (1990) Nature 343:244–245
18. Raty J-Y, Galli G (2003) Nat Mater 2:792–795
19. Barnard AS, Russo SP, Snook IK (2003) Philos Mag Lett 83:39–45
20. Vul' A Ya (2006) In: Shenderova O, Gruen D (eds), Ultra-nanocrystalline diamond: syntheses, properties and applications. William Andrew Publisher, Norwich, NY, USA, pp 379–404
21. Titov VM, Tolochko BP, Ten KE, Lukyanchikov LA, Pruuel ER (2007) Diam Rel Mater 16:2009–2013
22. Tolochko BP, Titov VM, Chernyshev AP, Ten KA, Pruuel EP, Zhogin IL, Zubkov PI, Lyakhov NZ, Lukyanchikov LA, Sheromov MA (2007) Diam Rel Mater 16:2014–2017
23. Danilenko NV (2003) In: Synthesizing and sintering of diamond by explosion. Energoatomizdat, Moscow, Russia, 272 p
24. Fenglei H, Yi T, Shourong Y (2004) Phys Solid State 46:616–619
25. Hawelek L, Brodka A, Dore JC, Honkimaki V, Tomita S, Burian A (2008) Diam Rel Mater 17:1186–1193
26. Kulakova II (2004) Phys Solid State 46:636–643
27. Mykhaylyka OO, Solonin YM, Batchelder DN, Brydson RJ (2005) J Appl Phys 97:074302
28. Alexenskii AE, Baidakova MV, Vul' A Ya, Davydov V Yu, Pevtsova Yu A (1997) Phys Solid State 39:1007–1015
29. Dolmatov V Yu, Veretennikova MV, Marchukov VA, Sushchev VG (2004) Phys Solid State 46:611–615
30. Gubarevich TM, Gamanovich DN (2005) In: Gruen DM, Shenderova OA, Vul' A Ya (eds) Synthesis, properties and applications of ultrananocrystalline diamond proceedings of the nato advanced research workshop on synthesis, properties and applications of ultrananocrystalline diamond St. Petersburg, Russia June 7–10, 2004, NATO Science Series II: Mathematics, Physics and Chemistry, vol. 192. Springer, Netherlands, pp 337–344
31. Adadurov GA, Bavina TV, Breusov ON, Drobyshev VN, Messinev MJ, Rogacheva AI, Ananiin AV, Apollonov VN, Dremin AN, Doronin VN, Dubovitsky FI, Zemlyakova LG, Pershin SV, Tatsy VF (1984) Method of producing diamond and/or diamond-like modifications of boron nitride. US Patent N4483836: from 20.11.84
32. Filatov LI, Chukhaeva SI, Detkov P Ya (1997) A technology for nanodiamond purification. Patent RU N2077476 from 20.04.97
33. Gubarevich TM, Larionova IS, Kostyukova NM, Ryzhko GA, Turitsyna OF, Pleskach LI, Sataev PP A diamond purification technology. Avt Svid USSR N1770272 from 22.06.92

34. Gubarevich TM, Larionova IS, Sataev PP, Dolmatov V Yu, Pyaterikov VF A technology for purification of nanodiamonds from non-diamond carbon. Avt Svid USSR N1819851 from 12.10.92
35. Dolmatov V Yu, Suschev VG, Aleksandrov MM, Sakovich GV, Vishnevskij EN, Pyaterikov VF, Sataev PP, Komarov VF, Brylyakov PM, Shitenkov NV A way of separating synthetic nanodiamonds. Avt Svid USSR N1828067 from 13.10.92
36. Dolmatov V Yu, Suschev VG, Marchukov VA, Gubarevich TM, Korzhenevskii AP (1998) Method for recovering synthetic ultradispersed diamonds. Patent RU N2109683 from 27.04.98
37. Gubarevich TM, Gamanovich DN (2005) In: Gruen DM, Shenderova O A, Vul' A Ya (eds) Synthesis, properties and applications of ultrananocrystalline diamond proceedings of the nato advanced research workshop on synthesis, properties and applications of ultrananocrystalline diamond St. Petersburg, Russia, June 7–10, 2004, NATO Science Series II: Mathematics, Physics and Chemistry, vol. 192. Springer, Netherlands, pp 311–320
38. Eryomenko NK, Obraztsova II, Efimov OA, Korobov Yu A, Safonov Yu N, Sidorin Yu Yu (1997) A technology for nanodiamond separation. Patent RU N2081821 from 20.06.97
39. Chiganov AS, Chiganova GA, Tushko Yu M, Staver AM (1993) Purification of detonation diamond. Patent RU N2004491 from 15.12.93
40. Isakova VG, Isakov VP (2004) Phys Solid State 46:622–624
41. Pavlov EV, Skrjabin JA (1994) Method for removal of impurities of non-diamond carbon and device for its realization. Patent RU N2019502 from 15.09.94
42. Xu K, Xue Q (2004) Phys Solid State 46:649–650
43. Osswald S, Yushin G, Mochalin V, Kucheyev SO, Gogotsi Yu (2006) J Am Chem Soc 128:11635–11642
44. Aleksenski AE, Osipov VYu, Didekin AT, Vul' A Ya, Andriaenssens G, Afanas'ev VV (2000) Tech Phys Lett 26:819–821
45. Vul' A Ya, Golubev VG, Grudinkin SA, Krüger A, Naramoto H (2002) Tech Phys Lett 28:787–789
46. Baidakova MV, Siklitsky VI, Vul A Ya (1999) Chaos 10:2153–2163
47. Krüger A, Kataoka F, Ozawa M, Fujino T, Suzuki Y, Aleksenskii AE, Vul' A Ya, Ōsawa E (2005) Carbon 43:1722–1730
48. Vul' A Ya, Dideykin AT, Tsareva ZG, Korytov MN, Brunkov PN, Zhukov BG, Rozov SI (2006) Tech Phys Lett 32:561–563
49. Osawa E (2008) Pure Appl Chem 80:1365–1379
50. Eidelman ED, Siklitsky VI, Sharonova LV, Yagovkina MA, Vul' A Ya, Takahashi M, Inakuma M, Ozawa M, Ōsawa E (2005) Diam Rel Mater 14:1765–1769
51. Xu K, Xue Q (2007) Diam Rel Mater 16:277–282
52. Shenderova O, Petrov I, Walsh J, Grichko V, Grishko V, Tyler T, Cunningham G (2006) Diam Rel Mater 15:1799–1803
53. Kuznetsov VL, Aleksandrov MN, Zagoruiko IV, Chuvilin AL, Moroz EM, Kolomiichuk VN, Likholobov VA, Brylyakov PM, Sakovitch GV (1991) Carbon 29:665–668
54. Kuznetsov VL, Chuvilin AL, Moroz EM, Kolomiichuk VN, Shaikhutdinov ShK, Butenko YuV (1994) Carbon 32:873–882
55. Aleksenski AE, Badakova MV, Vul' A Ya, Siklitskii VI (1999) Phys Solid State 41:668–671
56. Kuznetsov VL, Chuvilin AL, Butenko YuV, Gutakovskii AK, Stankus SV, Khairulin RA (1998) Chem Phys Lett 289:353–360
57. Prasad BLV, Sato H, Enoki T, Hishiyama Y, Kaburagi Y, Rao AM, Eklund PC, Oshida K, Endo M (2000) Phys Rev B 62:11209–11218
58. Raty J-Y, Galli G, Bostedt C, van Buuren TW, Terminello LJ (2003) Phys Rev Lett 90:037401
59. Baidakova M, Vul' A (2007) J Phys D: Appl Phys 40:1–12
60. Panich AM, Shames AI, Vieth H-M, Osawa E, Takahashi M, Vul' A Ya (2006) Eur Phys J B 52:397–402
61. Jiang T, Xu K (1995) Carbon 33:1663–1671
62. Barnard A, Sternberg M (2007) J Mater Chem 17:4811–4819

63. Barnard AS (2008) J Mater Chem 18:4038–4041
64. Shenderova OA, Zhirnov VV, Brenner DW (2002) Crit Rev Solid State Mater Sci 27:227
65. Kuznetsov VL, Butenko Yu V (2006) In: Shenderova O, Gruen D (eds) Ultra-nanocrystalline diamond: syntheses, properties and applications. William Andrew Publisher, Norwich, NY, USA, pp 405–475
66. Palosz B, Pantea C, Grzanka E, Stelmakh S, Proffen Th, Zerda TW, Palosz W (2006) Diam Rel Mater 15:1813–1818
67. Ōsawa E (2007) Diam Rel Mater 16:2018–2022
68. Krueger A (2008) Chem Eur J 14:1382–1390
69. Shenderova OA, Zhirnov VV, Brenner DW (2002) Crit Rev Solid State Mater Sci 27:227–356
70. Ōsawa, E., Ho, D., Huang, H., Korobov, M. V., Rozhkova, N. N. (2009) Diam. Rel. Mater. 18:904–909.
71. Vul' A Ya, Eydelman ED, Inakuma M, Ōsawa E (2007) Diam Rel Mater 16:2035–2038
72. Pichot V, Comet M, Fousson E, Siegert B, Spitzer D (2008) Proceedings of the third International Conference on "Detonation nanodiamonds: technology, properties and applications", July 1–4 2008. Ioffe Institute, St.-Petersburg, Russia, pp 79–82
73. Alexenskii AE, Baidakova MV, Kempinski W, Osipov VYu, Ōsawa E, Ozawa M, Siklitski VI, Panich AM, Shames AI, Vul' A Ya (2002) J Phys Chem Solids 63:1993–2001
74. Alexenskii AE, Baidakova MV, Yagovkina MA, Siklitski VI, Vul' A Ya, Naramoto H, Lavrentiev VI (2004) Diam Rel Mater 13:2076–2080
75. Alexensky AE, Yagovkina MA, Vul' A Ya (2006) Method of nanodiamond purification. Patent RU N2322389 from 13.10.06
76. Osipov V Yu, Enoki T, Takai K, Takahara K, Endo M, Hayashi T, Hishiyama Y, Kaburagi Y, Vul' A Ya (2006) Carbon 44:1225–1234
77. Osipov VYu, Shames AI, Enoki T, Takai K, Baidakova MV, Vul'A Ya (2007) Diam Rel Mater 16:2035–2038
78. Ando T, Ishii M, Kamo M, Sato Y (1993) Chem Soc Faradey Trans 89:1783–1789
79. Baidakova MV, Osipov V Yu, Katsuyama C, Takai K, Enoki T, Yonemoto A, Touhara H, Vul' A Ya (2007) In: Saito G, Wudl F, Haddon RC, Tanigaki K, Enoki T, Katz HE, Maesato M (eds) Multifunctional conducting molecular materials. The Royal Society of Chemistry, UK, pp 224–231
80. Byrappa K, Yoshimura M, Haber M (2001) Handbook of hydrothermal technology. William Andrew Publisher, Norwich, NY, USA, 893 pp
81. Ji S, Jiang T, Xu K, Li S (1998) Appl Surf Sci 133:231–238
82. Chung P-H, Perevedentseva E, Tu J-S, Chang CC, Cheng C-L (2006) Diam Rel Mater 15:622–625
83. Nakanishi K (1962) In: Infrared absorption spectroscopy: practical. Holden-Day, USA, 233 pp
84. Zhu YW, Shen XQ, Wang BC, Xu XY, Feng ZJ (2004) Phys Solid State 46:681–684
85. Ray MA, Shenderova O, Hook W, Martin A, Grishko V, Tyler T, Cunningham GB, McGuire G (2006) Diam Rel Mater 15:1809–1812
86. Larionova I, Kuznetsov V, Frolov A, Shenderova O, Moseekov S, Mazov I (2006) Diam Rel Mater 15:1804–1808
87. Korolkov VV, Kulakova II, Tarasevich BN, Lisihkin GV (2007) Diam Rel Mater 16:2129–2132
88. Krueger A, Ozawa M, Jarre G, Liang Y, Stegk J, Lu L (2007) Phys Stat Solids A 204:2881–2887
89. Huang H, Pierstorff E, Osawa E, Ho D (2007) NanoLetters 7:3305–3314
90. Grichko V, Shenderova O (2006) In: Shenderova O, Gruen D (eds) Ultra-nanocrystalline diamond: syntheses, properties and applications. William Andrew Publisher, Norwich, NY, USA, pp 529–557
91. Spitsyn BV, Gradoboev MN, Galushko TB, Karpukhina TA, Serebryakova NV, Kulakova II, Melnik NN (2005) In: Shenderova O, Gruen D, Vul' A Ya (eds) Synthesis, properties and applications of ultrananocrystalline diamond, vol. 192. Springer, Dordrecht, pp 241–252
92. Alexensky AE, Vul AYa, Yagovkina MA (2006) 17 European conference on diamond-like materials, carbon nanotubes and nitrides. September 3–8, 2006, Estoril, Portugal. Abstract Book. Abstract 15.2.08

Chapter 4
Detonation Nanodiamond Particles Processing, Modification and Bioapplications

Olga A. Shenderova and Suzanne A. Ciftan Hens

Abstract This chapter will detail the requirements of modern detonation nanodiamonds (DNDs) intended for biomedical applications, beginning with DND material preparations and followed by bio-related applications developed at International Technology Center. DNDs are one of the most commercially promising nanodiamonds with a primary particle size of 4–5 nm, produced by detonation of carbon-containing explosives. The structural diversity of DNDs will be described, which depend upon synthesis conditions, postsynthesis processes, and modifications. Bioapplications reviewed include ballistic delivery of bio-functionalized DND into cells, photoluminescent biolabeling, biotarget capturing and collection by electrophoretic manipulation of DNDs, and health care applications. DNDs are advantageous when compared with the other types of nanoparticles due to DND large scale synthesis, small primary particle size, facile surface functionalization, stable photoluminescence as well as biocompatibility. Currently, biotechnology applications have shown that NDs can be used for bioanalytical purposes such as protein purification or fluorescent biolabeling, while research is in the developing stages for DNDs applied as diagnostic probes, delivery vehicles, enterosorbents and advanced medical device applications.

4.1 Introduction: Types of Nanodiamonds

In the introduction, we give a brief outline on the methods of production and their related characteristic sizes of numerous types of nanodiamond particles currently available in the market. A detailed discussion on the methods of synthesis of diamond structures at the nanoscale is provided in a book [1].

Two approaches of ND *dynamic* synthesis were invented in the beginning of the 1960's by DuPont de Nemours, USA and at VNIITF, Snezinsk, Russia. DuPont

O.A. Shenderova (✉) and S.A. Ciftan Hens
International Technology Center, 8100-120 Brownleigh D, Raleigh, NC, 27617-7300, USA
e-mail: oshenderova@itc-inc.org

D. Ho (ed.), *Nanodiamonds: Applications in Biology and Nanoscale Medicine*,
DOI 10.1007/978-1-4419-0531-4_4, © Springer Science+Business Media, LLC 2010

produced polycrystalline diamond particles using shock wave compression of carbon materials (graphite, carbon black) mixed with catalyst. An approach initiated in Russia was based on the conversion of carbon-containing explosive compounds into diamond during the detonation of explosives in hermetic tanks. The fascinating history of the discovery of the detonation ND particulate was discussed by Danilenko [2]. Other groups of nanocrystalline diamond particles are obtained by the processing of micron-sized monocrystalline diamond particles, which are, in turn, a byproduct of the natural diamond or diamond obtained by static high-pressure high-temperature (HPHT) synthesis. The processing of micron-sized diamond particles into smaller fractions includes grinding, purification and grading of the powder.

Existing commercial diamond nanoparticles can be tentatively categorized into three groups of products according to the primary (smallest monocrystalline) particle sizes: *nanocrystalline* particles, *ultrananocrystalline* particles and *diamondoids*. Characteristic sizes of nanocrystalline particles encompass the size range of tens of nanometers, while sizes of primary particles of ultrananocrystalline diamond are within several nanometers. Diamondoids are well defined hydrogen-terminated molecular forms consisting of several tens of carbon atoms with characteristic sizes ~1–2 nm.

Nanodiamond particles processed from HPHT synthetic diamond as well as from natural diamond powders are available with smallest average particle size around 25 nm (produced, for example, by *Microdiamant AG*). Monocrystalline grinded diamond particles have rather sharp edges compared to other forms of nanodiamond. Synthetic HPHT type Ib monocrystalline diamond powders (containing typically 100 ppm nitrogen atoms) with particle sizes of 100 nm, 35 nm and 25 nm have been used for producing of fluorescent ND by irradiation with protons followed by annealing for bio-labeling applications [3–5]. Polycrystalline nanodiamond powder is processed from micron sized polycrystalline diamond particles obtained by shock synthesis [6] (DuPont de Nemours's method). Polycrystalline particles consist of nanometer-sized diamond grains (~20–25 nm). The finest diamond fraction produced by micronizing followed by grading has an average particle size of ~25 nm (available at *Microdiamant AG*). This type of ND has a high content of impurities and so far has not been used for bioapplications. The shape of primary particles is more platelet-like rather than spherical [7].

Out of several types of ultrananocrystalline diamond particles, only detonation ND has been commercialized. The average size of primary detonation ND particles produced by most vendors lies in the range of 3.5–6 nm. DND particles demonstrate spherical or polyhedral shapes [1]. This type of ND is most frequently used in studies for envisioned bioapplications such as broad drug delivery platforms for nanoscale medicine [8], protein adsorption [9–11], carriers of genetic material [12] in gene gun ballistic delivery [13], as enterosorbents [14, 15], as well as in the other applications described in this book. Indeed, specific surface areas of ND particles with 4 nm and 30 nm in diameter are 428 m^2/g and 57 m^2/g, correspondingly, making noticeable differences in adsorbing and loading capacity of the nanoparticles. The cost factor is also an important issue for the applications of nanodiamond particles. For example, HPHT nanodiamond with an average particle size 25 nm costs $75/gram. However, well purified polydispersed DND powder costs only about $1–2/gram (with 200 nm average aggregate size) and suspensions of completely disaggregated 5 nm DND are currently priced around \$40/gram (Chap. 1).

Higher diamondoids are highly rigid, well-defined hydrogen-terminated diamond species [16]. With more than 3 crystal diamond cages, higher diamondoids are intermediate in size to the adamantane molecule, the smallest species of H-terminated cubic diamond containing only 10 carbon atoms, and ultrananocrystalline diamond particles with sizes more than 3 nm as described above. Higher diamondoids are extracted from petroleum as diamond molecules in the form of nanometer-sized rods, helices, discs, pyramids, etc [17, 18]. So far it has not been possible to synthesize higher diamondoids except antitetramantane, a tetramantane isomer [16]. Certain higher diamondoids can be now available in multigram quantities through *Molecular Diamond Technologies, Inc*. By comparison, lower diamondoids (adamantane, diadamantane and triadamantane), extracted from crude oil much earlier than larger members of the diamondoid series, are currently available in kilograms quantities [16] and can be synthesized. Lower diamondoids such as adamantane derivatives have been used in pharmacology, clinical medicine and biosensing [19].

Since detonation ND is the most popular starting material within the nanodiamond particle family for biomedical applications, its processing and modification are elaborated upon in more detail in Sect. 4.2.

4.2 Detonation Nanodiamond Synthesis, Processing, and Modification

The three major steps in the conversion of carbon-containing explosives to modern DND products include *synthesis, postsynthesis, processing,* and *modification* (Fig. 4.1), which are discussed in detail in the following sections. *Processing* includes purification of detonation soot from the metallic impurities and nondiamond carbon and is typically performed in conjunction with detonation soot synthesis by the same vendor. The result of the processing is DND of a certain purity (presently at a level of incombustible impurity content 0.5–5 wt.% depending on the vendor) that is available on a large scale (thousands of kilograms). DND treatments, tentatively called *modification*, can include additional deep purification, surface functionalizations toward specific applications, and de-agglomeration or size fractionation. In the last 3–4 years, the development of approaches for DND modifications has been the major focus of research activity. Although, initially, DND modifications were performed on a small scale, currently, these steps are being implemented on a large scale on site at DND production centers [20]. Of primary importance is the method of disintegration of nanodiamond aggregates by stirred-media milling with micron-sized ceramic beads suggested by Osawa and coworkers [21], resulting in diamond slurries containing primary 4–5 nm DND particles. The single-digit DND immediately attracted the attention of relatively large number of researchers working in the area of nanoscale materials, which was demonstrated by recent publications in high profile journals [8, 22, 23].

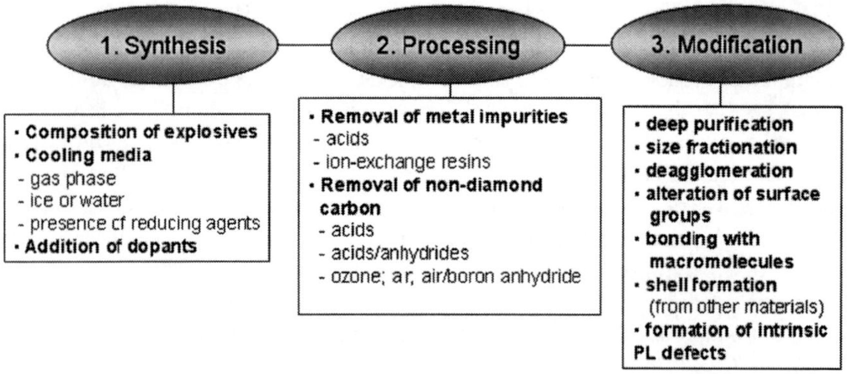

Fig. 4.1 Three major steps of production of detonation nanodiamond

4.2.1 Synthesis

Details on the synthesis of DND are discussed in books [24–27] and reviews [28–32]. In the current section, we emphasize the recent new advances in the methods of synthesis of DND and their envisioned applications approaches. In brief, during the detonation process, DND particles are formed from carbon atoms contained within the explosive molecules, thus only the explosive is used as a precursor material. However, a wide variety of explosive materials (so called CHNO high explosives) can be used, influencing the diamond phase yield in the detonation product as well as, to some extent, primary particles size [26, 30]. The explosion takes place in a nonoxidizing cooling medium of either gas (N_2, CO_2, Ar or other medium that can be under pressure) or water (ice), so called ‘dry’ or ‘wet’ synthesis, respectively. During detonation, the free carbon coagulates into small clusters, which might grow larger by diffusion [33]. The product of detonation synthesis, called detonation soot or diamond blend contains 40–80 wt.% of the diamond phase depending on the detonation conditions [26, 30]. There are two major technical requirements for the DND synthesis using explosives; the composition of the explosives must provide the thermodynamic conditions for diamond formation and the composition of gas atmosphere must provide the necessary quenching rate (by appropriate thermal capacity) to prevent diamond transformation to graphite [30].

For numerous areas of the DND applications including bioapplications, there are several important questions related to revisiting DND synthesis methods and to the development of new generations of DND particles. The major focuses for novel approaches for DND synthesis are as following: (1) possibility of further reduction of the primary particle size, (2) control of level of aggregation of DND at a stage of synthesis, (3) possibility of an increase of sp^3 carbon content in DND (in terms of reduction of noncarbon elements and sp^2 carbon on DND surface) and (4) possibility of doping of DND during synthesis, including control of the amount of substitutional nitrogen in the diamond core.

Reducing DND primary particle size down to the range of 2–3 nm may further increase their capability for penetrating cells and organelle membranes including nuclear pores. ND particles with sizes ~2 nm and smaller show quantum confinement effects [34]. Thus, synthesis of small particles would open perspectives for the development of diamond quantum dots. In principle, the XRD data on the size distribution of primary DND particles within detonation soot, which were reported in a limited number of publications [26, 35], include 2 nm peaks. It was reported by Dolmatov [35] that his novel method of using the reducing agents (for example urea, ammonia) in water cooling media allows the preservation of the DND fraction with a primary particle size less than 3 nm along with a 'typical' fraction of 5 nm primary particles. Also, a method reported by Dolmatov [35] for using reducing agents provides the benefits of increased DND yield from the soot as well as an increased carbon content (up to 96 mass%) within DND particle composition, where the total content of C, H, N and O corresponds to 100 mass%. At the same time, increase of the DND primary particles size to a value about 10 nm would facilitate the development of photoluminescent DND since the formation of NV optical centers is claimed to depend on the ND particle size [36, 37]. In general, production of several classes of DND with average primary particle sizes from the 2 nm to 10 nm size range (for example, classes with 2–3 nm, 4–5 nm and 10 nm average sizes) would be beneficial for various applications.

Another question is related to the possibility of optimizing the detonation process to produce mostly isolated primary particles and only small size aggregates in the detonation soot. This would significantly reduce the cost of the final DND product. In fact, it has been recognized that the dry DND synthesis results in smaller primary DND particle sizes and smaller average aggregate sizes as compared to wet synthesis [7, 38]. This strategy of optimization using dry synthesis for producing small average aggregate sizes has been discussed by Gubarevich [38]. Another factor that influences the aggregation of DNDs during synthesis is the mass of the charge and a ratio between masses of the charge and wet cooling media used [7].

Tailoring of DND electronic and optical properties for specific applications can be possibly done at a stage of synthesis by development of different combinations of explosive materials with possible solid dopants, as well as using the nontraditional cooling media (both gaseous and liquid) with possible additives to alter the DND composition (both bulk defects/doping content and surface groups).

4.2.2 DND Postsynthesis Processing

Biomedical applications set high standards on nanomaterial purity, so the development of DND products of ultrahigh purity remains an important goal. In addition to the diamond phase, the detonation soot contains both graphite-like structures (25–55 wt.%) and incombustible impurities (metals and their oxides – 1–8 wt.%) [30]. The metal impurities originate from a detonator and from the walls of the detonation chamber. The impurity content of nanodiamonds produced by detonation synthesis is higher when compared with the other artificial diamonds (for instance, HPHT diamonds contain no less than 96% carbon). After typical purification steps, powders of DND can be considered a composite consisting of different forms of carbon (~80–89%), nitrogen (~2–3%), hydrogen (~0.5–1.5%), oxygen (up to ~10%) and an incombustible residue (~0.5–8%) [30]. The carbon phase consists of a mixture of diamond (90–99%) and nondiamond carbon (1–10%).

Both nondiamond carbon and metallic, incombustible impurities can be located externally or internally relatively to the tight DND aggregates. In order to remove internal metal impurities and internal nondiamond carbon, tight DND agglomerates should be disintegrated. Currently, detonation soot can be industrially purified to a level of remaining incombustible impurities about 1 wt% (dry synthesis) and 0.8–0.9 wt.% (wet synthesis) [39], demonstrating that not more than 1% of metal impurities is confined in tight aggregates (both ND and graphite phases). After deep purification using acid treatment, the incombustible impurity content in *polydispersed* diamond, for example from *New Technologies* (Chelyabinsk, Russia) can be ~0.2 wt.% as defined using thermal gravimetric analysis (TGA).

In general, methods of DND purification as well as DND purity vary from vendor to vendor [30, 40–46]. 'Classical' purification methods, based upon the use of liquid oxidizers for the removal of metallic impurities, include sulfuric acid, mixture of sulfuric and nitric acids, hydrochloric acid, potassium dichromate in sulfuric acid as well as other schemes [30, 41]. A brief review on the numerous methods of detonation soot purification developed in the former USSR is provided in Petrov '07 [41]. For the oxidation of sp^2 carbons, the purification schemes include KOH/KNO_3, Na_2O_2, CrO_3/H_2SO_4, HNO_3/H_2O_2 under pressure, mixtures of concentrated sulfuric and perchloric acids and other approaches [30, 41, 43–45]. To remove the noncarbon impurities, the chemically purified product is subjected, in some cases, to an additional purification process using ion-exchange and membrane technologies. Currently, the majority of DND vendors use strong liquid oxidizers at elevated temperatures and pressures. However, liquid-phase purification is both hazardous and costly, contributing up to 40% of the product cost. In addition, the expense of waste pretreatment and disposal, which is already high, is expected to increase as governmental policy on environmental protection becomes tighter. Alternatively, DND can be very effectively purified from nondiamond carbon in an environmentally friendly manner by a gas phase treatment using ozone at elevated temperatures [42, 46] to eliminate the need for the use of corrosive liquid oxidizers. In the reactor, the nondiamond by-products of detonation synthesis, mostly graphite, react with ozone at the elevated temperature and are converted to CO_2 or CO. Simultaneously, the surface of detonation nanodiamonds is depleted of nondiamond carbonaceous by-products and enriched with oxygen-containing chemical groups. Ozone is generated using UV light at the input of the reactor and is destroyed at the output of the gas flow. Ozone oxidation is more efficient for sp^2 carbon oxidation than oxygen. The method was first introduced in 1991 [42, 46], and the productivity of the ozone-in-air flow fluidized bed reactor was 6–10 kg/month. Now ozone oxidation is the only gas-phase method for soot purification that has been realized at an industrial scale (by *New Technologies*, Chelyabinsk, Russia). The ozone-modified nanodiamonds (NDO) were found to have a number of distinctive characteristics. The size of polydispersed NDO in water suspensions is about 160–180 nm, as measured by the dynamic light scattering (Malvern Zetasizer). This number is the smallest average size reported for commercially available polydispersed unfractionated DNDs. In addition, the content of the primary, polyhedrally shaped, faceted particles with a size of 3–5 nm in the polydispersed NDO is substantially higher than that of the acid-purified DNDs [42, 47]. The NDO hydrosols were found to possess a very low pH (1.6–2 for 10% hydrosol), low negative zeta potential (−50 mV) for polydispersed sample and down to −100 mV for 20–30 nm fraction [48] and feature a constant zeta potential (~−40 to −50 mV) over a wide pH range of 2–12, apparently due to the enrichment of the surface with oxygen-containing strong-acid groups. The NDO forms stable hydrosols and can be easily fractionated down to a 20-nm average size fraction [48].

There were several efforts underway to purify DND by oxidation of detonation soot with air at elevated temperatures. The method allowed for a significant decrease in the nondiamond carbon content. Larionova et al. purified DND using soot treatment with air at 380–440°C for several hours [52]. Soot purification through a combination of liquid oxidizers and air treatment at temperatures up to 600°C was reported by Mitev [49]. Osswald et al. [40] demonstrated that for the soot (sample UD50 in Ref. [40]), the optimal temperature for the heat treatment in air within several hours is 400–430°C. This process allows for the oxidation of sp^2-bonded carbon, but not that of the sp^3 carbon. The purified DND contained 95% sp^3 carbon, as was measured with XANES, and was substantially lighter in color than the original material [40]. Chiganov purified DNDs from soot through thermal oxidation in air, using boric anhydride in order to selectively oxidize the nondiamond carbon [43].

4.2.3 Detonation Nanodiamond Modification

DND obtained from commercial vendors often requires additional processing and modification, since the content of incombustible impurities and nondiamond carbon can be too high, with an average aggregate size that is too large and with surface chemistry that is not suitable for a specific application. The poor colloidal stability of commercial DND powders after liquid dispersion is a common problem. Besides, there is no universal material called 'detonation nanodiamond', since materials are specific to the synthesis and postsynthesis purification methods adapted by the vendor. Accordingly, the modification strategy would be different for DND obtained from different vendors since compositions of the surface groups are different. For example, DND of wet synthesis purified from soot at the vendor site (VNIITF, Snezinsk) using the mixture of CrO_3/H_2SO_4 was additionally purified with heat treatment in air [50]. This treatment followed by dispersion in water using a high powered sonicator and multi-step ultracentrifugation resulted in stable hydrosols of DND fractions [50]. DND from "dry" synthesis purified from soot at the vendor site (FSPC "Altay") using a mixture of H_2SO_4/HNO_3 was also subjected to additional heat treatment in air [40], resulting in a high purity product. However, the water dispersion of the treated DND [40] was unstable and sedimented easily thereby requiring an additional treatment in HCl to improve it's colloidal property [51]. While both samples contained abundant carboxylic groups after heat treatment in air [40, 50], dissimilarities in other types of surface groups due to different preceding treatments might be a reason for the differences in colloidal stability of these two DND samples. As a means to improve the quality of the DND product, deep purification, fractionation, deaggregation and surface functionalization are discussed below.

4.2.3.1 Deep Purification

Back in the 1990s tons of DND was produced at the Federal Science Production Centre "Altay," Biisk, Siberia, Russia [25]; hundreds of kilograms of DND were produced at VNIITF center, Snezinsk, Ural region, Russia,[53] and at other centers in Russia, Belorussia and the Ukraine; these products are still available in the DND market. At that time, achievement of low incombustible impurities content in DND was not a high priority. The incombustible impurities content in DND produced at Altay and VNIITF was 2.4 wt.% [54] and 1.4 wt.% [50], correspondingly. The purity of commercial DND from a large manufacturer in Lanzhou, China, Gansu Lingyun Nano-Material Co., Ltd, was reported as 1.19 wt% [54]. Additional purification of these DNDs is required for modern applications, especially for biomedical use. It was illustrated that additional treatment with liquid oxidizers reduced the ash content down to 0.5 wt% in the sample (Ch-St DND) that was produced at VNIITF [55]. The sample was purified with HCl, HCl/HNO_3, HF/HCl and H_2O_2–NaOH [55]. The most efficient purification in this series was achieved using treatment with HF followed by HCl acids. Examples of the lowest reported incombustible impurities content obtained on a small scale using hot nitric acid for DND purification include samples containing only 0.08 wt.% [54] and 0.07% [35] of ash.

Heat treatment in air can be considered an efficient means for deep purification of DND from nondiamond carbon, as it was demonstrated in a thorough study by Osswald et al. [40] using the XANES technique for identifying sp^2 and sp^3 carbon content. Gordeev et al. [56] heated DND in air at 440–600°C and obtained DND with a significantly improved stability for their dispersions in water. The effects on stability for DND with an average primary particle size of 8–10 nm, heated in air at a temperature exceeding 557°C, were studied by Xu et al. [57]. Shenderova et al. [50] carried out oxidation of commercial DND products (Ch-St DND) with air at 400–450°C and achieved high colloidal stability and demonstrated an efficient fractionation for DND hydrosols. In an alternative approach of DND purification with gases, a significant decrease in sp^2 carbon content and a significant reduction in Al, Cr, Si and Fe content was achieved by applying Cl_2 treatment at 850°C [58]. Removal of sp^2 carbon in the DND samples was achieved by Yeganeh [59] using atomic H treatment.

4.2.3.2 Fractionation

While the primary DND particle size is 4–5 nm (Fig. 4.2a, b), the primary particles form tightly and loosely bound aggregates. The typical commercial polydispersed

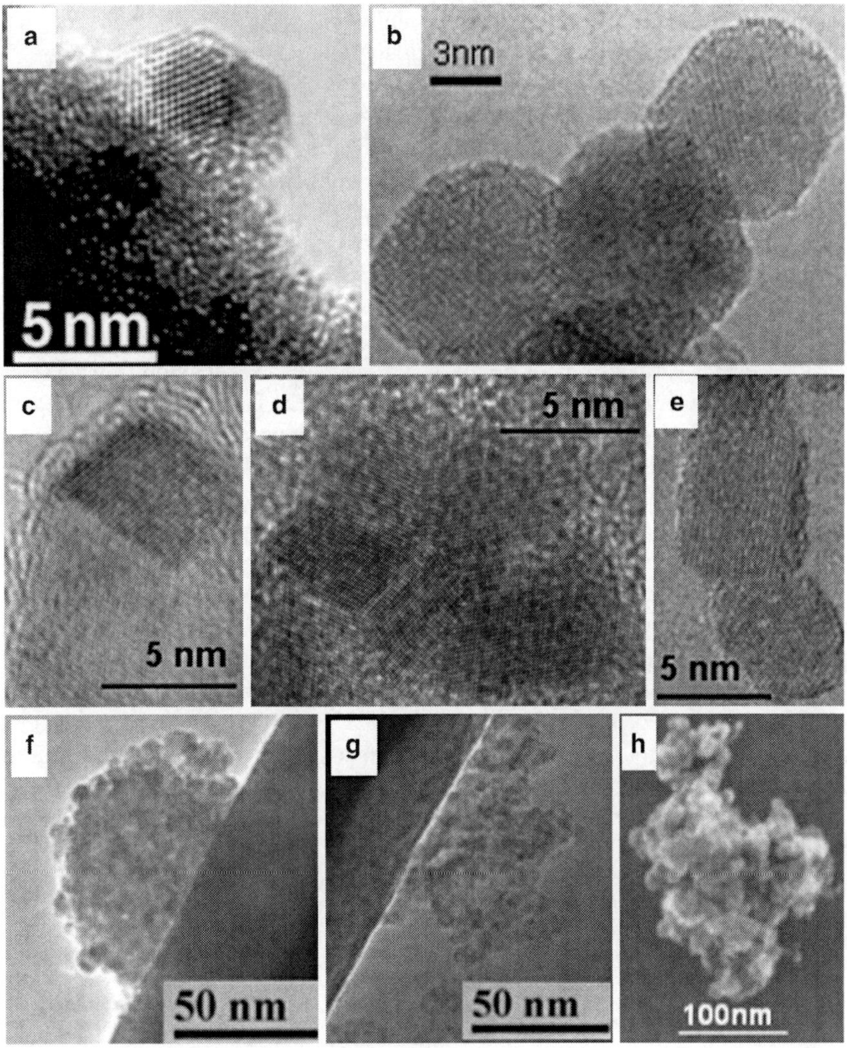

Fig. 4.2 HRTEM images of primary DND particles (**a**,**b**), primary DND particles forming small agglomerates with atomically sharp grain boundaries (**c**–**e**); medium-size agglomerates with high (**f**) and low (**g**) DND packing density and a large aggregate with highly irregular shape (**h**). Images (**a**) and (**f**–**h**) are courtesy of Talmage Tyler, NCSU, ITC. Image (**b**) is courtesy of Bogdan Palosz, IHPP, Warsaw, Poland. Image (**c**) is courtesy of Vladimir Kuznetsov, BIC, Novosibirsk. Image (**d**) is courtesy of O. Lebedev and S. Turner University of Antwerp. Image (**e**) had been adapted from Ref. Iakoubovskii K. et al. 2008 [63] with permission

ND suspensions that are subjected to powerful ultrasound treatments routinely exhibit 200-400 nm average aggregate sizes, which are unbreakable by the ultrasonic treatment (Fig. 4.2f–h). An approach to effectively separate the particles and narrow the size distribution is a centrifugal fractionation [20, 50, 60, 61]. Importantly, DND suspensions must posses high colloidal stability for centrifugal fractionation. It is difficult, if not impossible, to fractionate an unstable suspension.

DND fractionation has several attractive aspects. First, it is a contamination-free approach as compared, for example, with bead milling, which introduces impurities from ceramic beads which require further purification [71]. It is also convenient to be able to fractionate DND into different, narrow distributions of sizes for different niche applications (Fig. 4.3). For example, only DND with aggregate sizes of more than 100 nm can form photonic structures that diffract light in the visible region [62]. Finally, after deep purification or treatment with ozone/air, the content of DND with small-sized aggregates can be significantly increased and the production of suspensions with small fractions of pure DND without added contaminants becomes economically feasible. In fact, suspensions of 5 wt% of 25 nm size DND fraction in water have been produced using the fractionation method [20]. The purity of small aggregates can be very high. First, as it was discussed in the previous subsection, the content of incombustible impurities even in the polycrystalline sample after deep purification can be very low. Second, the small aggregates consisting of several primary DND often have both elongated shapes (Fig. 4.4) and atomically sharp grain boundaries (Fig. 4.2c–e), preventing confinement of amorphous carbon in the inter-grain regions. Thus, it is expected that the content of nondiamond carbon in small fractions after deep purification would be also relatively low. Future characterization of small fractions of DND is required to test these parameters.

The content of individual primary DND particles within the fraction that has small average aggregate sizes is quite substantial (Fig. 4.4a). It is expected that the content of surface groups on these individual primary DND particles is more uniform than for those obtained by bead milling when 'fresh' diamond surfaces are being formed during disintegration of the agglomerates, which can interact with surfactant

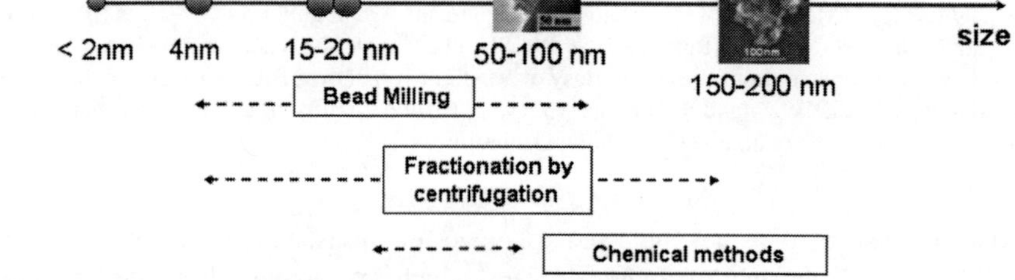

Fig. 4.3 Schematics of the size ranges of DND primary particles and aggregates in relation to different possible areas of biomedical and healthcare applications as well as current approaches for obtaining DND product within these size ranges

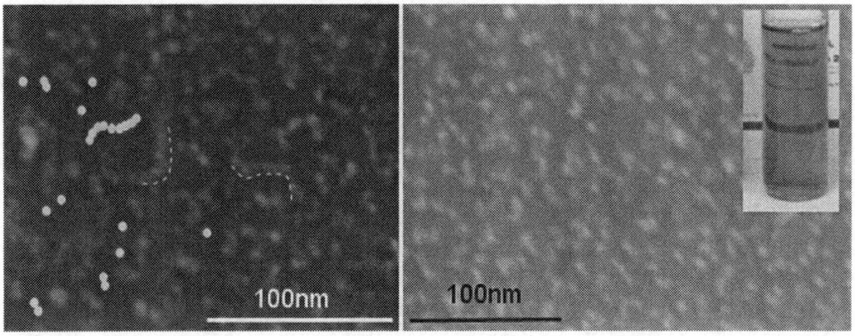

Fig. 4.4 SEM images of the fraction of DND particles with average aggregate size 20 nm (**a**) and 10 nm (**b**) obtained by ultracentrifugation and dispersed over Si substrates. Highlighted in yellow are selected primary particles and their aggregates consisting of 1 or 2 particles as well as elongated worm-like DND aggregates. Inset on (**b**) demonstrates a stable colloidal suspension of DND with 10 nm average aggregate size. Note the high transparency of the suspended sample and amber-like color originating from Raleigh scattering, which is stronger for shorter wavelengths of visible light

and the other molecules from the liquid media [64, 65]. The development of methods of extraction of the primary DND particles from polydispersed DND suspensions would advance DND applications.

The centrifugal fractionation approach has drawbacks that mostly originate from the highly irregular shapes (spherical vs. elongated chains) of the DND aggregates (Figs. 4.4 and 4.2(g,h)). While the centrifugal force depends upon the particle shape, the resulting separation is done based upon the combined size/shape factor rather than solely on the size. Importantly, it is possible to extract only primary DND particles by centrifugal fractionation, including ultracentrifugation [66].

4.2.3.3 Deagglomeration

Deagglomeation of nanodiamond aggregates into individual primary particles is an important goal for biomedical applications. Osawa and coworkers developed methods of mechanical deagglomeration of DND dispersion in suspensions by stirred media milling [54] or bead assisted sonic disintegration [67]. Suspensions of individual 4–5 nm DND particles containing a very small fraction of ~30 nm particles (less than 1 vol.%) have been produced [67]. Since the small fraction of particles with coherent scattering region of more than 5 nm in size was observed in the X-ray diffraction studies [68], the fraction of 'unmilled' particles can originate from these large monocrystalline particles. Micrographs in Fig. 4.2c–e also illustrates primary particles with atomically sharp grain boundaries and high cohesive energies, that can contribute to the 'unbreakable' aggregates. The aggregates of DND particles with ordered grain boundaries can form at the last stage of DND synthesis assuming carbon (diamond) clustering mechanism of detonation synthesis [33, 69, 70].

Undesirable side effects upon bead milling are contamination with bead material and generation of graphitic layers on the particle surface [71]. Attempts to purify bead-milled

DND with liquid oxidizers lead to aggregation of the primary particles [71]. Fortunately, by optimizing the bead milling process, it is possible to minimize the contamination by bead milling media below 0.2% (Chap. 1). Purification from the bead-milled material using NaOH solution in water was considered [72] as well as molten NaOH. In addition, amorphous carbon and metal contaminants confined within DND aggregates and released during bead milling also need to be removed from the resulting suspension of primary particles. Nevertherless, the production in kilograms quantity of the so called "Nanoamando" nanodiamond had been started. Nanoamando has become very popular nanodiamond material for biomedical research and has permitted the successful development of applications thereof. ([71] and related Chapters in this book)

Several other methods have been proposed for DND deagglomeration. Xu et al. developed a two-step deagglomeration procedure that included graphitization of by-products of detonation in N_2 atmosphere at 1,000°C for 1 h followed by their oxidation with air at 450°C for several hours [73]. The final powdered product contained at least 50% DND with the particle size of less than 50 nm. Krueger reported using DND reduced in borane (accompanied by ultrasonic treatment) resulting in significantly smaller sized aggregates. Treatment of DND powder in atmospheric pressure plasma also reduced the average DND aggregate size by ~20% [74].

A concern that requires thorough study is the possible reagglomeration of single digit DND or small sized fractions subjected to further surface functionalization or drying for storage. Typically, during drying of ND from water and other solvents, ND aggregation is further increased due to capillary forces pulling together individual particles. Attractive Van der Waals forces also play a major role in particle agglomeration. Agglomeration during drying makes ND functionalization more difficult, since it often requires dry starting ND material. Recently, a technique of ND fractionation has been developed that allows for drying of the ND to a powder form without concomitant agglomeration [75]. In this case, the size of the ND powder aggregate (20–30 nm) is preserved after dissolving the fractionated powder in a variety of solvents, facilitating the ND functionalization and further processing. Puzyr et al. has developed a modification based upon sonication-assisted treatment of NDs in a NaCl solution [76, 77], which results in the purification of the NDs and possibly the incorporation of Na^+ ions into the ND surface. The attractive feature of the NaCl-treatment method of ND modification [76, 77] is the possibility of drying NDs from a hydrosol to a powder form with subsequent resuspension without agglomeration. DND powder with average aggregate size ~40 nm after dispersion in water was obtained through this method.

4.2.3.4 DND Surface Functionalization

Nanodiamonds are unique among the class of carbon nanoparticles because of their intrinsic hydrophilic surface, which is one of the many reasons that these nanocarbon particles are envisioned for biomolecular applications. The surface of nanodiamond particles contains a complex array of surface groups, including carboxylic acids, esters, ethers, lactones, amines, etc (Fig. 4.5). The amount of functional chemical

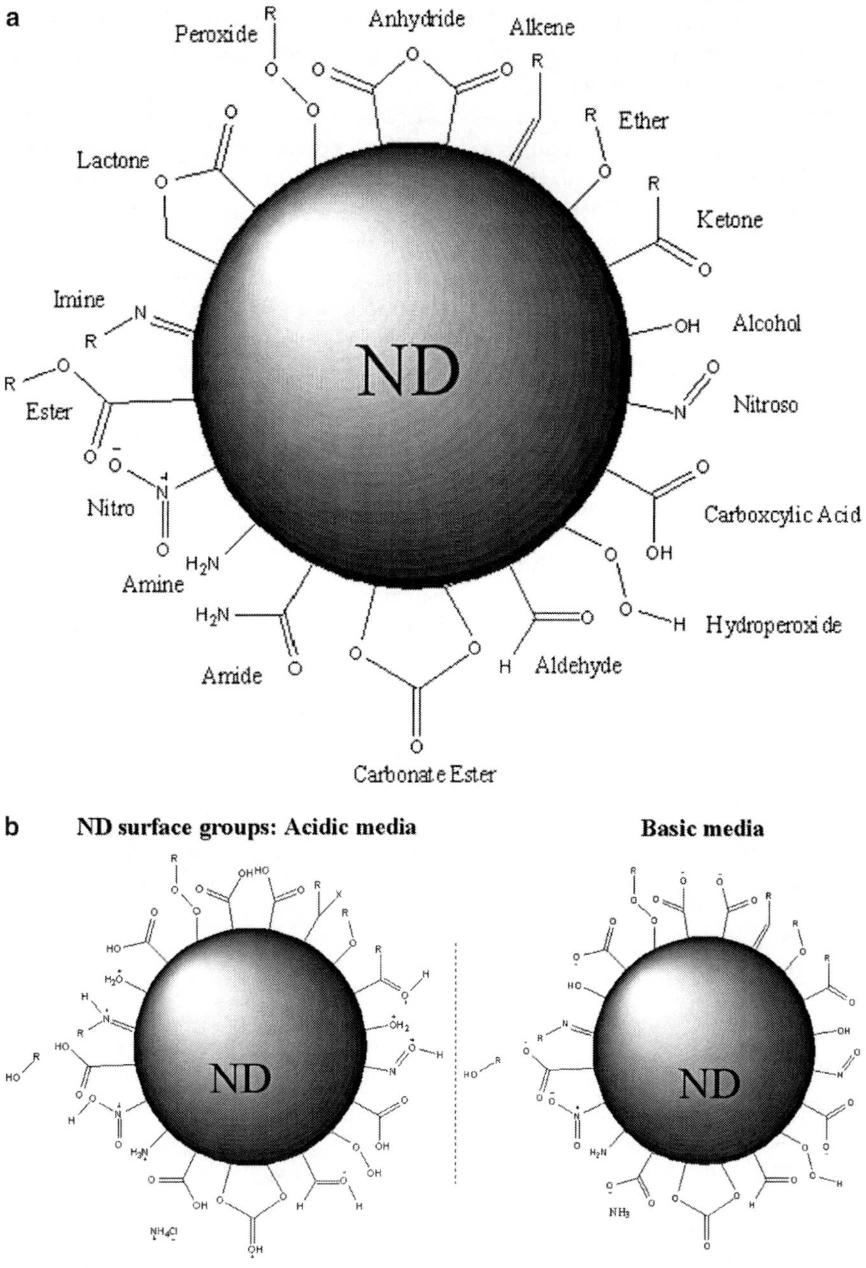

Fig. 4.5 Schematics of surface groups reported for DND after different types of purification/modification (**a**) and possible dissociation of these groups in acidic (HCl) and basic media (**b**). Pictures drawn by Garry Cunningham, ITC

groups per unit mass of DND exceeds by several times the functionalization capacity of nanoscale diamond powders (for example, HPHT NDs with 25–30 nm particle size) due to large surface area.

Nanodiamond surface groups have been altered using various reaction chemistries reviewed in [28, 32, 78] including wet chemistry approaches as well as gas phase methods and atmospheric plasma treatment. Because of the myriad of chemical surface groups found on DNDs, it is necessary to reduce the array of groups to a single functionality in order to decrease the number of functional products resulting from wet synthetic chemical methods, thus increasing the yield of the desired product. In order to achieve this goal, Krueger et al. reduced DND with borane or lithium aluminum hydride to eliminate the surface oxygen groups, such as carboxylic acids, producing a product with increased OH surface groups [23, 79]. Using this product, it was then possible to attach alkyl silanes to the DND surface. Starting with this alkyl silane, a small peptide, as well as biotin were then synthesized onto the DND surface [23, 79]. Lithium aluminum hydride has also been used by Hens et al. to reduce DND surface groups prior to the formation of aminated DND [80]. This aminated DND was attached to both a fluorescent dye and a biotin moiety using their succinimydl ester derivatives [80].

It was also demonstrated that the atmospheric pressure plasma system developed at ITC is a powerful tool for ND functionalization [81], essentially saving time when compared to the approach utilizing a reactive gas flow reactor. Fluorination of ND using atmospheric pressure plasma treatment can result in grams quantity per hour of the product and this can be further scaled up. Based on FTIR spectral analysis of several types of the initial ND, produced by different vendors, it was demonstrated that plasma treatment of ND results in the removal of particular surface groups (such as OH– and C=O, depending on the type of initial ND), as well as in the formation of a variety of carbon–fluorine types of bonding (such as CF, $CF_3(CF_2)$, $C=CF_2$ dependant on the surface chemistry of the initial ND) [81].

4.2.3.5 DND Zeta Potential

The type of charge DND acquires in colloids becomes important in its sorption and electrophoretic applications. Remarkably, there are classes of commercial and modified DNDs with highly positive and highly negative zeta potential values (Fig. 4.6) [55]. The different groups, both acidic and basic found on the ND surface are formed during the chemical treatment of the nanomaterial (Fig. 4.5). DNDs processed with soot oxidation using either singlet oxygen (RUDDM) in NaOH, air treated in the presence of catalyst (Kr series), ozone purified, or oxidized with mixture of H_2SO_4/HNO_3 have negative zeta potentials (Fig. 4.6). Additional oxidation of Ch-st sample (with zeta potential +17 mV) in air resulted in a material with high negative zeta potential, −45 mV too [50]. DND oxidized from soot with HNO_3, CrO_3/H_2SO_4, $NaOH/H_2O_2$ have positive zeta potentials (Fig. 4.6). DND samples after deep purification with HCl, HCl/HNO_3, HF/HCl also exhibit high positive zeta potentials (~ +40 mV) (not shown in Fig. 4.6).

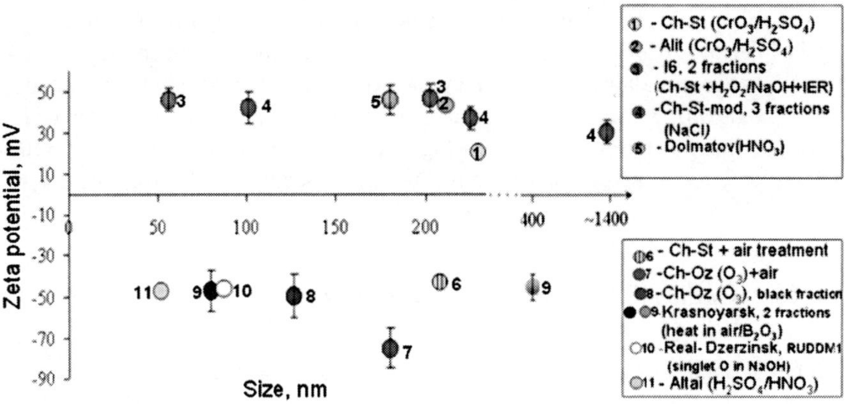

Fig. 4.6 Zeta potential and average aggregate size for different types of commercial DND

Thus, presently it might be suggested that some processes of oxidation of soot or further oxidation of ND containing nondiamond carbon such as the use of singlet oxygen in liquid media, H_2SO_4/HNO_3 mixture or oxygen/ozone in a gas phase result in rather deep oxidation with predominant carboxylic acids groups on the ND surface. Presence of carboxylic groups was supported by taking spectra from ND samples after thorough desorption of water using FTIR vacuum cuvet [20]. When ND is dispersed in DI water, dissociated acidic groups cause a negative charge on ND surface [82]. The amount of carboxylic acid species on the surface of dehydrated ND with negative and positive zeta potential is the major difference observed in FTIR spectra for these groups of nanodiamond [74].

Revealing the origin of positive zeta potential of ND is more complicated. Previously a positive charge on the ND surface was attributed to the protonation of amino groups in acidic media [83]. The spectra were taken in air, so the presence of adsorbed water might interfere and substitute for the signatures from amines [83]. However, FTIR spectra taken in a vacuum cuvet revealed a negligible amount of amino groups on the surface of Ch-series ND with a positive zeta potential [74]. For comparison, the FTIR spectra of aminated DND taken in vacuum demonstrated a very pronounced peak at 3,420 cm^{-1} related to amines [80]. At the same time, the amount of alcoholic groups that might also be responsible for a positive zeta potential is also small [74]. From numerous studies of the nature of oxygen-containing groups on the surface of carbons [84, 85], two families of surface groups have been identified relative to their acidic or basic character in aqueous solutions. Carboxyl groups, lactones, phenol and lactol groups contribute to the acidic character of carbon materials. Several models of basic oxygen-containing functionalities still being debated include chromene structures, diketone or quinone groups, pyrone-like groups and electrostatic interactions of protons with the π-electron system of the graphene structures [84–86]. Quantum chemical calculations on a large series of polycyclic pyrone-like model compounds demonstrated high relevance of the model to carbon basicity [85, 87]. Pyrone-like structures are the combinations of non-neighboring carbonyl and ether oxygen atoms at the edges of a graphene layer. In principle, the model which attributes the basicity of the carbon surface to pyrone-like structures, can be adapted to nanodiamonds to explain the positive zeta potential [74]. It is known, that the sp^2-like carbon shell might present at the surface of ND providing sites for pyrone formation (Fig. 4.2c). Additional studies using titration [84] are required to reveal the nature of the positive zeta potential for DND.

4.2.4 "Modern" Detonation Nanodiamond

DND specification has been significantly tightened lately due to the growth of interest in their use for new applications, specifically in the biomedical field. Major structural characteristics that need to be carefully controlled are the size of DND particles (both primary particles and their aggregates), their 'external' composition (surface groups) and 'internal' lattice defects responsible for the important physical properties such as photoluminescence.

To date, suspensions of DND primary particles 4–5 nm in size obtained by bead milling as well as fractions of ~20 nm DND agglomerates obtained by centrifugal fractionation have been developed. Both approaches have positive and negative facets in terms of purity of bead milled DND and consistency of DND aggregate shapes within fractions as was discussed in Sect. 4.2.3.2. Fractions of ND aggregates with small size can have high purity since highly purified diamond can be used for fractionation, ultracentrifugation does not introduce contamination and small fractions do not contain the confined metal or amorphous carbon inclusions. It was also demonstrated that full deagglomeration of DND should not be an ultimate goal, since a wide variety of applications require particle sizes much larger than the 4–5 nm size of the primary particles (Fig. 4.3), such as the use of DND for efficient UV protection [88] or formation of photonic structures [62]. Logically, two methodologies should be advanced in parallel: development of slurries of DND primary particles free from contamination and attempts to obtain DND fractions with a more narrow size distribution for niche applications. Importantly, the strategy of optimization of DND deagglomeration shifted from the stage of modification to steps of DND synthesis and purification from the soot.

Other key requirements for the chemical composition of the modern DND are purity (absence of incombustible impurities and nondiamond carbon) and uniformity of surface groups. It was demonstrated at the laboratory scale that incombustible impurity content in DND can be extremely low, less than 0.1 wt% (Sect. 4.2.3.1). Previously, using X-ray diffraction and small angle X-ray scattering, it was postulated that a DND cluster in detonation soot has a complex structure consisting of a diamond core and a shell made up of sp^2-coordinated carbon atoms implying that the content of nondiamond carbon in DND is relatively high [89]. At the same time, HRTEM images demonstrate that DND particles can be purified to a level where sp^2 carbon on DND surface is visually absent (Fig. 4.2a, b); this was also demonstrated by Iakubovslii et al. [63]. Very low sp^2 carbon content in DND heat treated in air was also demonstrated by Osswald et al. [40]. Using HRTEM and EELS analysis, Turner et al. [90] demonstrated that depending on the purification technique, sp^2 carbon content at a surface of individual 5 nm particles separated by centrifugation can be noticeably decreased. These data imply that the high purity DND can be available in the nanodiamond market.

Inhomogeneity of pristine DND surface groups hampers control of the surface functionalization including modification with biologically active moieties [79]. However, the question 'can homogeneity of DND surface functional groups be achieved?' might have a negative answer. A nanodiamond particle of 4–5 nm in size with a shape, close to a sphere has multiple low-index facets (mostly (001), (011) and (111) type) exposed at its surface (Fig. 4.7). Surface energies of these facets are very different, involving possible surface reconstructions for (111) and (001) surfaces. Binding energies with different surface groups can be rather different as well [91–93]. For example, as reported by Kern et al. [94], the hydrogen adsorption energy differs by 0.4 eV for C(111) and C(100) diamond surfaces. Petrini et al. [92, 93] performed comprehensive first principles simulations on the stability of hydrogen and oxygen groups (-OH, ether and ketone) on (100) and (111) unreconstructed and reconstructed diamond surfaces. The calculated adsorption energies for H, O, and OH are −4.53, −5.28, and −4.15 eV, respectively, for 100% terminated diamond (111)-1 × 1 surfaces and −3.29, −3.82, and −2.77 eV, respectively, for the diamond (111)-2 × 1 surfaces. Ketone group is most energetically favorable at (111)-1 × 1 surface, while ether is most preferable at (111)-2 × 1 diamond surfaces. The OH groups showed less-favorable adsorption energies in comparison to H and O on (111) surface. Regarding (001) surface, it was demonstrated that a replacement of hydrogen adsorbates with hydroxyl groups is slightly disfavored, whereas a corresponding replacement with oxygen atoms in ketone formations is energetically preferred. The adsorption of oxygen atoms in ether positions become energetically preferred over full H coverage only at a level of coverage more than 50%. The adsorption energy for the functional groups (at 100% surface coverage) is −4.13, −4.30, −5.95, and −6.21 eV for H, OH, ketone, and ether species, correspondingly [92, 93]. Thus, the binding of different surface groups can be energetically preferable for

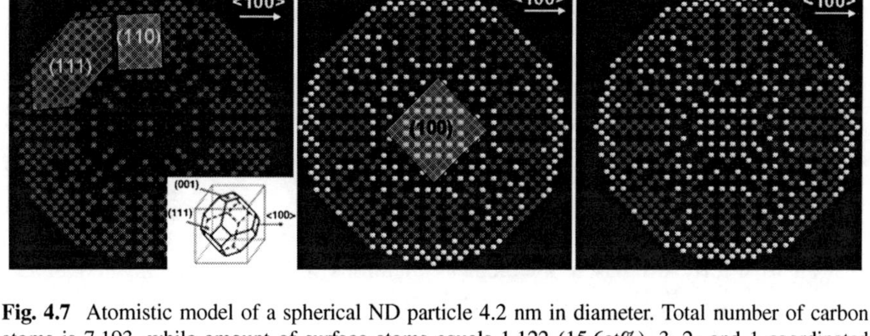

Fig. 4.7 Atomistic model of a spherical ND particle 4.2 nm in diameter. Total number of carbon atoms is 7,193, while amount of surface atoms equals 1,122 (15.6at%). 3-,2- and 1 coordinated atoms are denoted in blue (708 atoms), yellow (378 atoms) and purple (36 atoms), correspondingly

different DND facets. While simulations by Petrini et al. [92, 93] were performed for the bulk diamond surface, to extend these results to a nanodiamond particle, buckyfication of the (111) nanodiamond surface needs to be included to the analysis. It is also important to simulate and compare energetic stability of (111) ND facets with Pandey chain termination [93] and sp^2 C bucky shell structures, since the area of (111) facets on 4–5 nm particles is quite larger than the sizes of 1–2 nm particles used in the first principle simulations revealing buckyfication [34]. Over size, energetic preference of different surface termination, in principle, can be changed. In general, it can be expected that even after an attempt to homogenate the DND surface, it still will be polyfunctional including hydrophilic (oxygen containing) and hydrophobic (hydrocarbons) groups. As it was pointed out by Chiganova [82], the hydrophilic-hydrophobic mosaic structure of the surface of DND particles has a strong influence on the aggregation behavior of the low concentration DND aqueous dispersions.

Zeta potential of DND suspensions is another characteristic depending on the DND surface composition and will be defined by the majority of the surface groups (basic vs. acidic species). Modern commercial DND may have positive or negative charge in solvents (Sect. 4.2.3.5) depending on the majority of the surface groups, which facilitates applications where the type of the charge is important (adsorption of biomolecules, for example). Recently, Barnard et al. [95] using density functional tight binding simulations revealed that morphology of bare (all-carbon) diamond clusters influence the sign of electrostatic potential. Thus, the (100) surfaces consistently exhibited a strong positive electrostatic potential, while extended areas of (111) surface graphitization possessed a strong negative electrostatic potential. A mosaic structure of the particles surface charge would enhance the agglomeration. However, absence of the functional groups on DND surface in reality is unlikely.

Finally, the internal structure of the DND core and the development of methods for its control also need to be addressed. An important aspect of DND bio-related applications is a question of DNDs strong internal photoluminescence, originating, particularly, from lattice structural defects. It is important to have a well pronounced unambiguous zero phonon line (ZPL) signal in the spectra rather then than just a broad PL emission peak. So far, the ZPL in the photoluminescent spectra of DND with 4–5 nm in size has not been observed, to the best of our knowledge, and the development of bright, intrinsically photoluminescent DND remains an important target. At the same time, DND core contains an abundance nitrogen [90], which is responsible for the photoluminescent properties of bulk diamond.

4.3 Selected Biomedical Applications

In this section, we will outline the research topics under investigation in our laboratory as well as briefly discuss biomedical applications including ballistic delivery of bio-functionalized DND into cells, photoluminescent biolabeling, biomolecule target capturing and collection by electrophoretic manipulation of DND, use of DND in sunscreens and other bio-related research.

4.3.1 DND Optical Labels

The photoluminescent (PL) properties of nanodiamonds of static synthesis are outstanding as compared to small molecule dyes [3–5, 36]. Nanodiamonds have high quantum yields, little photobleaching, no photoluminescent blinking characteristics, and long luminescence lifetimes. These superior PL properties along with their diamond-like chemical, physical, and biological stability coupled with their noncytotoxicity have contributed to the intense interest of applying these particles to numerous biotechnological and medical applications.

4.3.1.1 DND Photoluminescence

One of the unique physical properties of nanodiamond that is inherited from bulk diamond is its intrinsic photoluminescence originating from structural defects and impurities (dopants). Of particular importance for the intrinsic PL properties of diamond are N-related defects (complexes of substitutional N atoms and vacancies). Nitrogen is incorporated to HPHT diamond during synthesis at a level of 100 ppm. Nitrogen vacancy centers are typically formed under proton, helium or electron irradiation followed by annealing [3, 4, 96, 97]. Very bright PL with well pronounced ZPL have been observed from N-V centers created by proton irradiation and annealing of HPHT diamond nanoparticles 25 nm in size [3–5]. In this regard, significant efforts were directed toward clarification if N is incorporated inside ND core of detonation nanodiamonds. In general it is expected that N should be incorporated in the DND lattice since explosives plays a role of source material for detonation ND synthesis contain significant amounts of N (TNT – 14 at% of N, RDX – 28 at% and BTF – 33 at%). So far, EPR spectra did not reveal the presence of substitutional N in detonation nanodiamond [30]. First principle simulations suggested that substitutional nitrogen is unstable in several-nm diamond particles [98] and vacancies would diffuse out of the nanodiamond particle core during synthesis or irradiation unless appropriate surface passivation is realized [99]. In experiments by Borjanovich et al. on DND powder irradiated with protons followed by annealing only a broad structureless emission peak was observed [100]. At the same time, Kvit et al. performed series of EELS probing across an individual DND particle and concluded that the content of nitrogen in a DND core is high [101]. While in experiments by Kvit et al. nitrogen content was only detected in the EELS spectra taken from the central part of a DND particle and not at its periphery, a possibility that the detected nitrogen was a part of a surface group that was not excluded. Recently, Turner et al. [90] unambiguously proved that N is located in the core of DND particles using STEM-EELS analysis. First, N from the inter grain region was excluded by particles deagglomeration to primary ND size and all surface, and subsurface N in the sample was removed by the graphitization/liquid oxidation process. In the graphitization/liquid oxidation process applied to the DND sample, all original surface groups on the nanodiamond material were desorbed and the outer diamond layer was converted to an sp^2 shell during vacuum high temperature annealing. The sp^2 carbon shell was subsequently removed using liquid oxidizer not containing nitrogen, so that formation of new nitrogen functional groups on the surface of the diamond cores was excluded. Then, EELS experiments were performed on these particles free from surface nitrogen. A careful STEM-EELS investigation probed *individual* particles of the nanodiamond sample in both central and periphery parts of the particles. In the core-loss spectra, a clear nitrogen signal having a half maximum at 403 eV and peaking at 406 eV was present in all samples. No π^* prepeak was present at 397 eV, a first indication that all nitrogen present in the samples is incorporated into a sp^3 coordinated surrounding and thus in the diamond core. A second indication that the nitrogen is embedded within the diamond cores was given by the quantification of the atomic nitrogen to carbon ratio in the DND samples. Model-based quantification indicated a nitrogen content of around three atomic percent in all samples. Thus the EELS measurements therefore indicate that the nitrogen impurities that are deemed to contribute to the photoluminescent effect in DND samples are embedded within the diamond nanoparticles core in a tetrahedral coordination (sp^3), and are not positioned close to the particle surface or within the graphitic shell. Feasibility of formation of optical centers based on N complexes with vacancies requires further investigation.

Chemical modification of the DND surface can enhance photoluminescence. Several DND surface functionalizations performed in our lab have demonstrated that it is possible enhance to the PL intensity of original DND powder by an order of magnitude (Fig. 4.8) by attaching small molecules to their surface. These small molecules have no intrinsic PL characteristics on their own. In similar work, it was recently reported by Mochalin et al. that the chemical modification of DND surfaces with octadecylamine resulted in a highly fluorescent material [51].

From work with HPHT nanodiamonds, it has been shown that irradiation of high energy beams greatly enhances the PL properties of nanodiamond. In analogous work, we found that proton irradiation of DND embedded into the rubber polydimethylsiloxane (PDMS) resulted in a strong dose-dependent photoluminescence. PDMS without DND and pure ND powder did not show the same effect, even at higher irradiation doses. We suggested that this photoluminescence may arise from defects formed at the interface of DND/PDMS matrix [100].

4.3.1.2 Fluorescently Labeled DND

The development of nanodiamond optical labels has tremendous potential in many areas of biomedicine and biotechnology. Optical labeling nanodiamonds directly with fluorophore tags may improve the nanoparticle tracing through cells or for biotagging. The advantage of these optical labels is that it is possible to fabricate nanodiamond particles with a desired spectral property using a library of photoluminescent dyes. It was demonstrated by Huang and coworkers that Alexa Fluor dye could be conjugated directly to poly-L-lysine, which was physisorbed onto nanodiamonds [10].

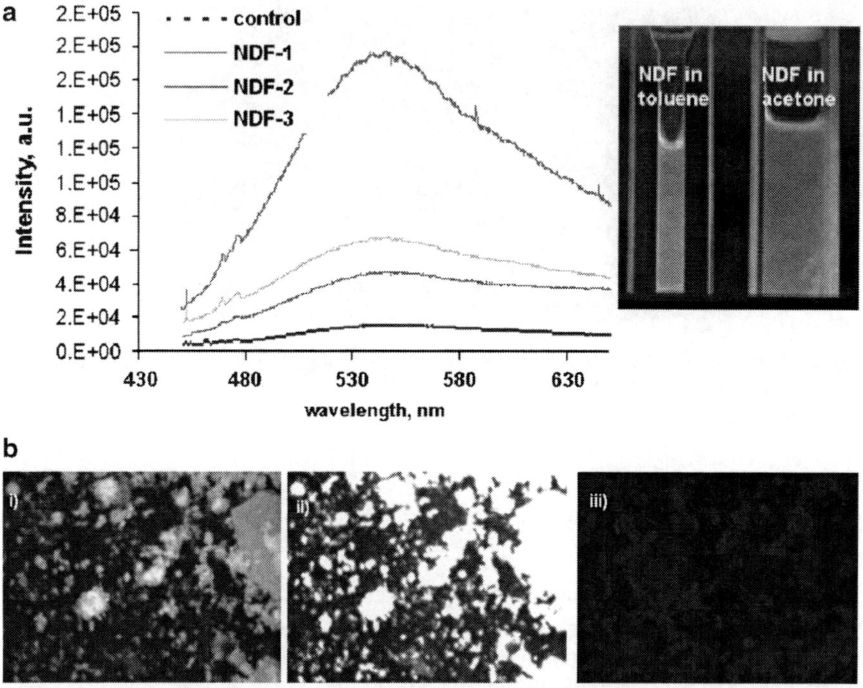

Fig. 4.8 Emission spectra for the pristine ND powder (control) and 3 types of functionalized ND (NDF) (**a**). Excitation wavelength is 442 nm. Photo on the right (**a**) illustrates photoluminescence of the NDF sample under UV lamp. Illustration of intrinsic photoluminescence under illumination with green light for the ND powders inspected under fluorescent microscope (**b**): NDF-2 at 500 ms (1) and 1500 ms (2) exposure time, correspondingly and ND pristine powder at 1,500 ms exposure time (3). Images are taken with 10× objective

In this way, chemical functionalization of nanodiamonds to the abundance of amines on the polylysine chain provides a high loading efficiency of dyes to the particle surface. This experimental method was used for the localization of nanodiamond using biolistic delivery methods [13]. Our group showed that dyes can be directly conjugated to the chemical species on the surface of nanodiamonds. We chemically functionalized the nanodiamond surface groups to achieve a high density of amine groups, which were then linked to the NHS ester of TAMRA forming a stable amide bond, see Fig. 4.9 [80]. This TAMRA-ND conjugate was found useful to trace the capture of biomolecules in vitro and for fluorescent cellular tracing of nanodiamonds in cytotoxicity studies. The cytotoxicity studies showed that the conjugate was chemically stable, being viewed over a 24 h period, and was noncytotoxic while the particle localized in the cytoplasm after being transiently located in the lysosome region – it did not penetrate the cell nucleus [102, 103]. For a larger scale reaction, FITC was conjugated to ND using the acid chloride derivative of ND, see Fig. 4.9 [104].

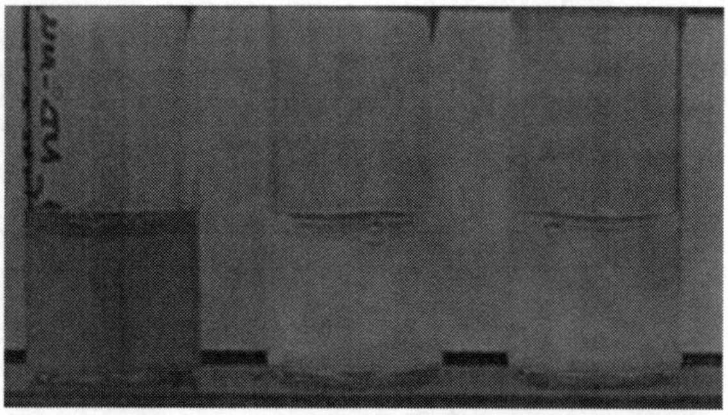

Fig. 4.9 Photograph of vials containing colloidal suspensions of ND and ND-dye conjugates, whereby the FITC and TAMRA are chemically attached to the surface of the nanoparticle. The advantage of ND-dye conjugates is their intense, tunable photoluminescence properties

4.3.2 DND Biofunctional Applications

Recent advances in the chemical manipulation of nanodiamonds have shown the practicality and uniqueness of NDs as compared with the other well studied nanoparticles. For example, nanodiamonds are chemically and physically stable, but their surface can be chemically modified. ND optical properties include transparency in the visible wavelength range and have photoluminescence with nonblinking and no photobleaching properties. NDs can now be fractionated into stable colloidal solutions of narrow sizes, while the smallest primary DND particle of 5 nm has recently been made available [71]. Although the use of NDs in biotechnology is in its infancy, recent nanodiamond studies highlight the utility of these particles for bioanalytical applications.

4.3.2.1 DND Electromanipulation

Two electrically induced phenomena can be used to manipulate the dielectric and/or charged nanoparticles in suspension: dielectrophoresis and electrophoresis [105, 106]. In dielectrophoresis, a dielectric particle is suspended in a spatially *nonuniform* electric field; the applied field induces a dipole in the particle. Due to the presence of a field gradient, electric forces acting on the charges induced on each side of the dipole are not equal, resulting in a net force and particle movement. The dielectrophoretic particle movement depends on a particle volume, relative complex permittivities of the media and a particle, electric field gradient and angular frequency of the applied electric field. The overall particle charge does not affect its dielectrophoretic mobility. On the other hand, in electrophoresis, electrically charged particles in suspension move in an applied electric field. The electrophoretic mobility is directly proportional to the magnitude of the charge on the particle, and is inversely proportional to the size of the particle. In practice, quite often both phenomena, electrophoresis and dielectrophoresis, take place simultaneously. Electromanipulation of nanoparticles plays an important role in the concentration of colloids from solution, nanoparticle separation, transport, formation of coatings on a substrate, including the formation of micropatterns in biosensors [105, 106].

Electromanipulation of DND is useful in a variety of applications. For example, the combined dielectrophoretic/electrophoretic deposition of DND has been used by Alimova et al. to coat microscopic silicon tips with nanodiamond for cold cathode applications [107]. Electrophoretic seeding of substrates with submicrometer diamond particles for the growth of diamond films was also explored [108]. Modern commercial DND are available with both positive and negative zeta potentials, which allows for both cathodic and anodic electrodeposition of DND.

The electrophoretic collection of proteins using DND probes was explored by Hens et al. [80]. In this work, it was found that the electrophoretic collection of DND on a silicon substrate was dependent upon both the field strength and the time of electrophoresis (Fig. 4.10). The substrate saturates rapidly showing a concomitant drop in current from the electrode (Fig. 4.10). This method was used for the collection of target bound NDs, whereby the target streptavidin was collected using a ND-biotin probe, by applying a potential across an electrically conductive field tip array (FTA) substrate [80].

It is anticipated that the electrophoretic and dielectrophoretic manipulation of nanodiamond will afford many applications for trapping, manipulation and separation of biomolecules. In addition, because of the high surface charge of DND and relatively high dielectric constant, DND movement in the electric field can be controlled with high precision by properly designing the electrode configurations [105]. Furthermore, it may be possible to fractionate the DNDs according to their size by electrophoresis methods.

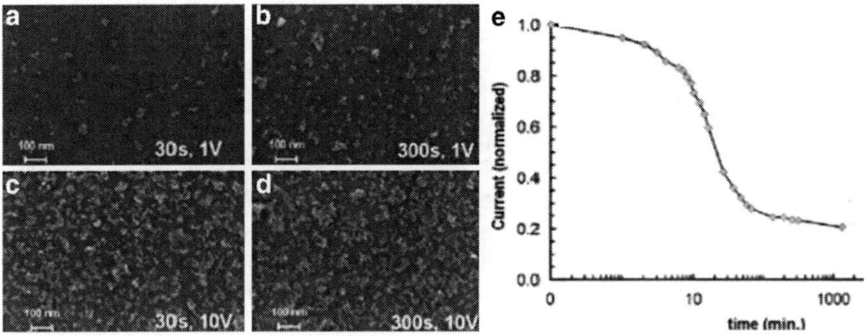

Fig. 4.10 SEM photographs are shown for the time and potential dependence of ND collection on the surface of a conductive Si substrate (**a–d**). After a certain threshold of ND adsorption is reached, the electrode becomes more insulating as can be seen from the dependence of electrode current with time (**e**). Figure is partially adapted from Ref. [80] by Hens et al. with permission

4.3.2.2 DND Solid Phase Supports

Early research showed that DND nanoparticles are capable of adsorbing the recombinant apoobelin and luciferase proteins [9]. Thus, DND may be used as a solid phase capture probe for the physisorption of biological molecules. It has been found that DND has a tenfold greater binding capacity for the protein lysozyme as compared to traditional chromatographic substrate – nanosilica, which is probably due to the greater surface area of porous nanodiamond [109]. More recently, DND has been applied to the binding of small molecules as a replacement of alumina silicate enterosorbants for the removal of aflotoxin during the feeding of corn to livestock [14, 15].

Selective binding of proteins may be accomplished through the conjugation of small molecules to the surface of DND. One way of achieving high binding affinity to the surface of DND is to physisorb positively charged poly-L-lysine, which has a multitude of reactive amine groups, to negatively charged DND. Succinimydl conjugate dyes and proteins may then be chemically linked to these amine groups on the DND surface [10]. Heterofunctional linkers have also been used to bind the protein cytochrome c to poly-L-lysine coated DND, providing up to two molecules of protein for a 5 nm particle.

In the last several years, advances have been made in modifying DND surface groups using standard wet synthetic methods. For example, the functional group OH has been enriched on DND followed by functionalizing the surface with a peptide [79]. The advantage of direct chemical attachment to the DND surfaces affords greater robustness of the biomolecular tag since it does not rely on electrostatic binding. It has recently been shown in our laboratory and others that it is possible to directly functionalize DND surface groups to produce biotinylated DND for the capture of streptavidin protein, see Fig. 4.11 [22, 23, 80].

It is also possible to functionalize DND with a DNA oligomer to capture the complementary DNA target strand in solution. To do this, we functionalized ozonated DND with an amine reactive DNA oligomer using the coupling reagent EDC to form an amide bond, see Fig. 4.12. We then used the complementary DNA strand, which has a biotinylation label on the end. After the DND-ssDNA probe hybridized to its complementary DNA strand, we detected this hybridization by adding fluorescent streptavidin [104]. In this way, DND can serve as a solid support matrix for the DNA detection methods.

In addition to DND serving as a solid support matrix, we found that the collection of the support and target complex was possible by electrophoresis, see Sect. 4.3.2.1. This result is important for the quantitative analysis of target biomolecules in a sample, for example in diagnostics. The ability to collect the captured biomolecular target in samples allows one to not only quantify the sample, but also to remove the impurities which would interfere with sample quantification. This method of target capture and collection was demonstrated using a biotinylated nanodiamond conjugate probe (ND-biotin) that captured streptavidin and was collected onto a substrate by electrophoresis [80]. A schematic diagram of this method is shown in Fig. 4.13. One of the advantages of using this method is that it can be performed at long distance. For example, we have collected DND by electrophoresis over many centimeters in

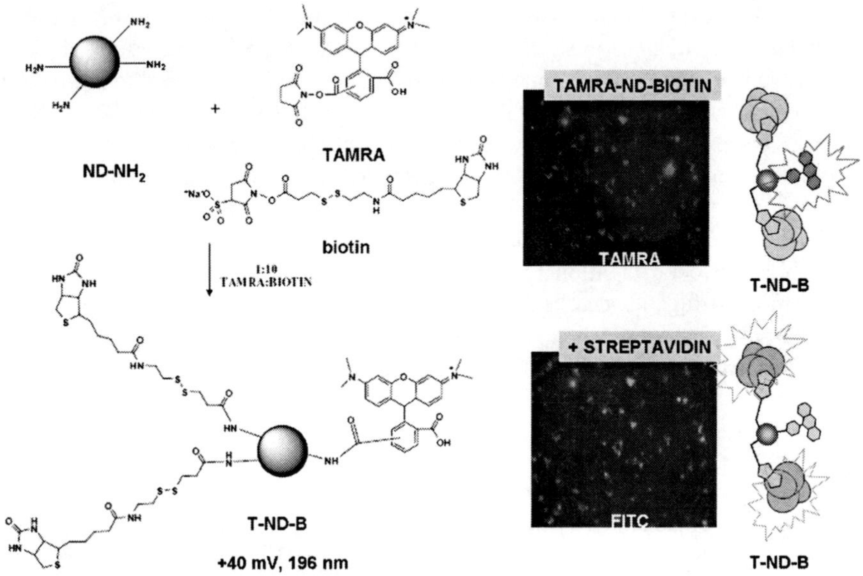

Fig. 4.11 A schematic representation of aminated ND surface coupled with the conjugates of TAMRA-NHS dye and biotin-NHS, making a visible nanodiamond protein-probe complex. By fluorescence microscopy (1,000× magnification), the red fluorescence emission of TAMRA localizes ND particles (*top photo*), while the green fluorescence emission of FITC identifies ND-bound streptavidin (*bottom photo*). Figure is partially adapted from Ref. [80] by Hens et al. with permission

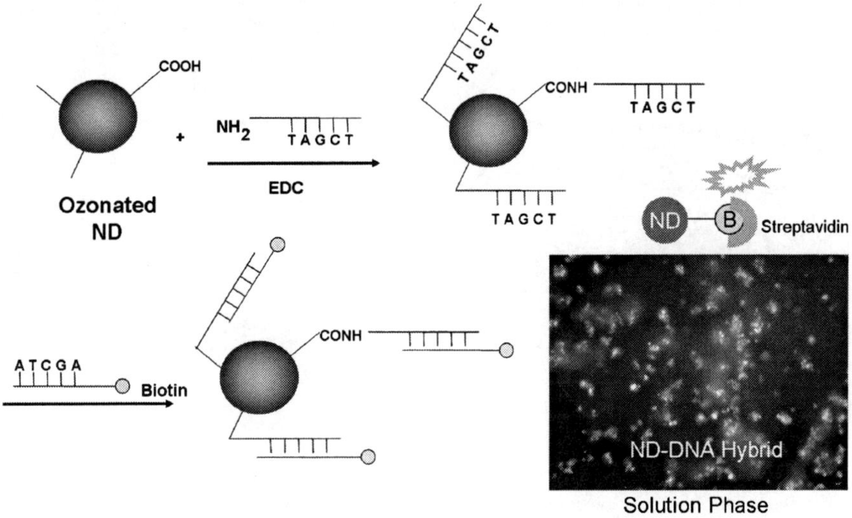

Fig. 4.12 Schematic for the reactivity of carboxylated nanodiamond with an aminated DNA oligomer strand using the coupling reagent EDC. Once the conjugate DND-ssDNA probe is constructed, hybridization to its complementary DNA strand may be detected through the binding of streptavidin to the biotinylated DNA target moiety (photo taken with fluorescence microscope at 1,000× magnification)

distance. However, this method does require that the DND probe forms a stable colloidal solution, since these electrophoresis methods are completed in solution. As was shown previously, this method of collection may even be used to fabricate a bioactive substrate for sensor applications or diagnostics such that the DND probe complex is substrate bound prior to target collection. In this way, it would be possible to develop diagnostic tools for protein or gene detection. For example, this method may one day be used to fabricate the DNA microarrays, see Fig. 4.14.

Scientists at ITC observed that DND particles are capable of forming ordered assemblies by centrifugation [62]. While using a narrow size fraction of DND ~150 nm, centrifugation causes the particles to form ordered structures with different inter-particle distances, which causes a rainbow of colors along the length of an eppendorf tube; we call these "DND Rainbows." Analogous to "ND Rainbows,"

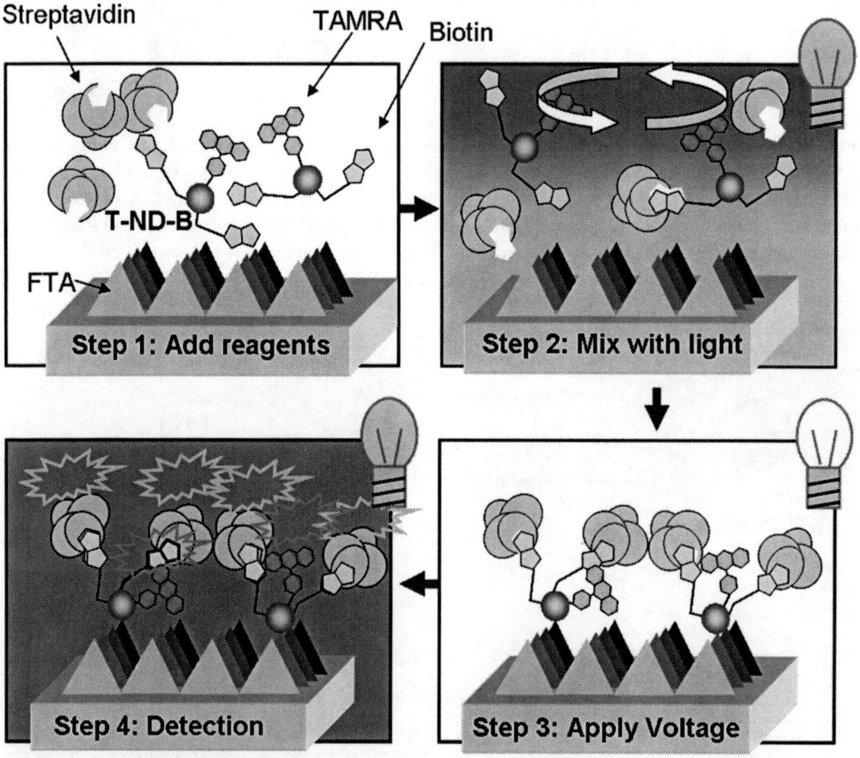

Fig. 4.13 A schematic representation for the ND capture probe process. After the addition of reagents (*step 1*) the solution is mixed (*step 2*) to capture targets onto the nanodiamond-probe complex. The solution may be mixed in a remote fashion by using a visible light source for a closed system such as a microfluidic platform. Next, the targets are collected by electrophoresis (*step 3*) on a conductive surface from which the capture efficiency of targets can be identified by fluorescence emission (*step 4*), for example. Figure is partially adapted from Ref. [80] by Hens et al. with permission

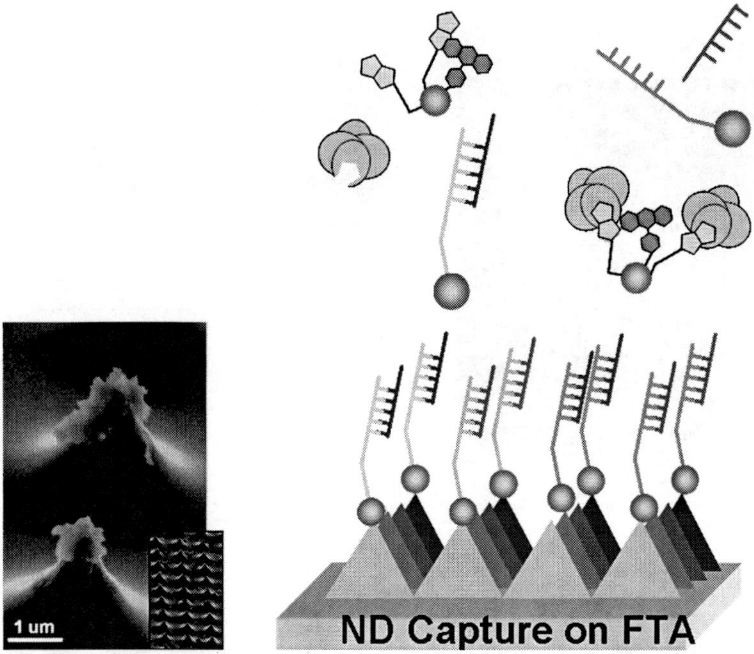

Fig. 4.14 Schematic of a field tip array (FTA) that can be used to collect solution targets using a ND-probe conjugate by applying a bias on the electrode (*right*). A photograph of nanodiamonds collected on FTAs by applying a modest potential (*left*)

size selection of DND particles also produced ordered photonic crystals by centrifugation that were millimeter in size [62]; we named these "ND Christmas Lights." These ordered DND particle assemblies may also have applications in biodiagnostics, see Fig. 4.15 [62]. One can imagine that centrifugation on a chip may use the DND particles that distribute themselves in an ordered arrangement, providing a colorimetric signal for the presence or absence of target biomolecules. In addition, the robust chemical and physical properties of DND may be exploited to advance the biosensor field [110, 111].

4.3.2.3 DMSO Solvent for DND

Along with others, scientists at ITC have found that formulation of DND in DMSO prove to be useful in a number of applications, most notably with DND seeding for CVD diamond growth. Although DMSO is known as one of the best solvents, we found a number of unanticipated benefits in using this solvent for DND processing and for biotechnological applications.

In general, it is expected that DMSO formulations of DND will allow for greater reactivity in biological media, since DMSO is a superior solvent as compared to water for biologicals such as proteins, enzymes, etc. DMSO solvents are miscible in water,

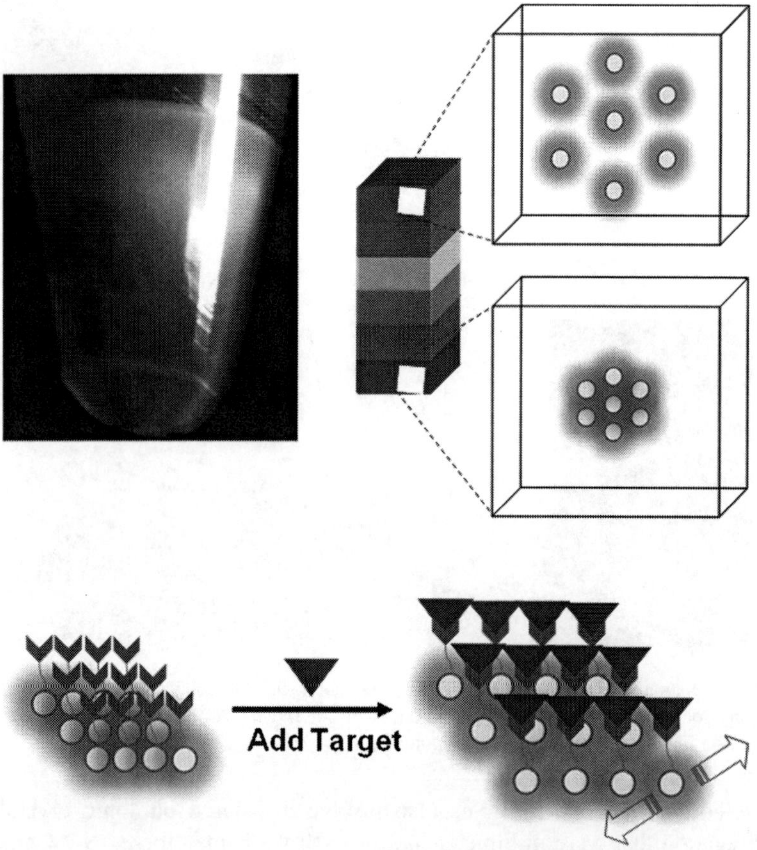

Fig. 4.15 Nanodiamonds form a rainbow of colors in an eppendorf tube depending upon the spacing between nanodiamond particles (**a**). Future applications of DNDs may employ nanodiamonds that change color when the spacing is increased in the presence of a target analyte (**b**). DND photo was taken from Ref. by Grichko et.al. [62] with permission

thus DNDs suspended in DMSO can readily be transferred to aqueous solution systems. DMSO may also act as a penetration enhancer, namely to introduce antivirals, steroids, and antibiotics through the skin [112]. Thus, DND in DMSO may be used to penetrate a biological matrix, to introduce ND into cell membranes and epidermal layers and may be used for biological tagging and drug delivery applications.

For micromixing applications, we found that DND-DMSO formulations allow for light-activated mixing by thermophoresis in microfluidic cells [113]. This property may be used to mix solutions together in a remote fashion and may be used to disrupt laminar flow in fluidic devices. Nanoparticles have further been applied as nanoheat-sinks in microchannel laminar flow devices [114].

In drug delivery applications, methods of using supercritical CO_2 in DMSO solvents are expected to allow for the precipitation of proteins on DND using the solvent/antisolvent method. Thus, because DND is readily soluble in DMSO, this and other supercritical CO_2 applications are feasible. This is important because proteins precipitated under these conditions retain their activity [115]. These and the other methods may lend themselves for the use of DND in pharmaceutical applications since industrial pharmaceuticals have widely accepted supercritical CO_2 processing methods. Pharmaceuticals could potentially use DND as a delivery vehicle or as a drug filler, which maintains the bioactivity of mixed components. In addition the large surface area to volume ratio is expected to allow for a slow release formulation.

Scientists at ITC have found that stable colloidal solutions of DMSO can be made at very high concentrations of DND. Thus, a ND-DMSO formulation may be used as a highly concentrated organic adsorbent. Applications include the collection of toxins (enterosorbents), proteins, peptides, polysaccharides, antibodies, nucleic acids, viruses, bacteria, etc. are used as a highly concentrated biological receptor made of specifically targeted ND conjugates. Other applications are expected to benefit from DND suspensions in DMSO. For example, it is possible that it may be useful for the preparation of chromatography columns because DND suspensions in DMSO are highly concentrated and would allow for efficient packing of the column. Also, our proposed use of DND for sunscreen applications would benefit from a DMSO formulation that is compatible with sunscreen solvents and forms a highly stable colloidal solution of DND, which could allow for greater loading of DND as a physical UV absorber.

4.3.2.4 Thermophoresis of DND

The illumination of DND suspensions with a light beam leads to an active directional movement of the DND. When realized in microliter volumes, rigorous turbulent flows are formed resulting in efficient mixing of DND with the other components dispersed in the suspension. A possible explanation for this observation may be a process called thermophoresis, or thermodiffusion, the effect of a temperature gradient on mixtures of particles [113]. During the illumination of nanoparticles, which can interact with EM radiation and acquire local heating, thermal gradients arise at the boundary of the illuminated and nonilluminated nanoparticles causing particle movement. This phenomenon may be used to efficiently mix DND target probes, which is of special importance in microfluidic platforms since laminar flow prevents mixing. To do this, the DND solution located within the microfluidic cavities would be exposed to light, providing a simplified means of remotely controlling the mixing location and duration without modifying the microchannel architecture. Researchers from ITC have used this method of mixing DND in DMSO solutions located within glass capillary columns.

The application of DND mixing would be particularly useful with DND probes, allowing for efficient collection of targets within the solution. However, DND mixing application may be extended in more general terms as a mean of physisorbing biomolecules and small molecules within closed cavities. The utility of DNDs used purely as a mixing reagent would require that these particles be coated in an inert film to prevent adsorption of solution organics or biomolecules.

4.3.3 DND Medical Applications

It has been proposed that DND may be applied for medicinal purposes, either alone or biofunctionalized, in diverse medical fields, including oncology, cardiology, gastroenterology, and dermatology [116]. DND conjugates are chemically stable and do not alter the activity of bound biomolecules [8, 117, 118]. For example, DNDs with attached enzymes or drugs have effectively killed bacteria [118] and cancer cells [8]. There may be several means for which DND may be used in medicine, as either the particle itself or in a film.

4.3.3.1 DND Enterosorbents

Carbon and clay-containing adsorbents are commonly used in medical and pharmacological industries to bind ingested toxins, which can be potentially fatal to both animals and humans [119]. As a remedy after the accidental ingestion of mycotoxins, which are low molecular weight by-products of mold growth, enterosorbents are designed to specifically bind the toxins while remaining biocompatible, well dispersed in aqueous solution, and stable for transportation and administration. Currently, enterosorbents that are safe and practical are not readily available [119].

DNDs are considered potential enterosorbents for binding a group of mycotoxins called aflotoxins, due to their high biocompatibility, small size, large surface area, rich surface chemistry, inexpensive production, and variety of available samples [15]. A positive zeta potential is the main property that contributes to the greatest aflotoxin adsorption onto DNDs [14]. Zeta potential of DND after different purifications may vary between +20 and +50 mV [74]. It was also found [74] that both the direction of the titration (from pH 1 to 12 or pH 12 to 1) and the concentration of the sample had an effect on the zeta potential. At 0.1 wt% concentrations of DND in water slurries, the two samples with adsorptive capacities of over 100 mg of aflotoxin per kg of ND were purified detonation NDs 185 nm and 300 nm in size with surface charges of +47 and +18, respectively. Further controlling the ND surface chemistry, determining optimal concentrations and reducing the size of the ND aggregates for applications in low pH environments such as the stomach would be desirable.

4.3.3.2 DND in Ballistic Delivery

The usefulness of DND as a general, robust solid phase support was demonstrated in experiments using the Bio-Rad PDS-1000/He instrument used for the delivery of bio-active molecules to yeast, fall armyworms, cacti, and bananas [117]. In comparison to other cell delivery methods, the combination of ND bioconjugates and ballistic bombardment instrumentation, which uses He pressure and vacuum circuits to generate a shock wave, can precisely target cells with bio-active molecules over a wide range of velocities and distances. One example included the use of ND-assisted ballistic delivery of diphenylcyclopropenone, an ethylene antagonist, which was used to prevent the ripening of bananas. Compared to the most potent ethylene antagonists that are currently used to control fruit ripening, flower senescence and biotic/abiotic stress, the ND bioconjugates are less toxic, nonexplosive, nonvolatile, water soluble, cost efficient, and potent.

The immobilization of DNA onto 1–2 μm diamond particles via covalent attachment [120] and more recently onto NDs led to its use in plasmid delivery via ballistic bombardment [117]. After ~50% of plasmid DNA containing ampicillin-resistant genes was adsorbed onto the surface of 5 nm NDs or their 100–200 nm aggregates, *E. coli* were bombarded, transformed, and bioluminescent or blue colonies were selected due to their conferred resistance to the antibiotic. The particular advantages of using ND as a carrier for DNA include its ability to stabilize DNA without nicking under longer storage times compared to traditionally used heavy metal nanoparticles. Further optimization of the ND surface could not only increase the efficiency of DNA binding, but introduce multifunctionality for extending its applications, possibly for DNA delivery for gene therapy applications. This work on biolistic delivery of nanodiamonds has been published by a single team of scientists, however there is great potential for extending these applications to vaccine delivery and gene therapeutics.

4.3.3.3 DND for Seeding of CVD Diamond Films

Several other demonstrated or emerging applications for DNDs include their use in the CVD growth of nanocrystalline diamond films on medical implants or biosensor electrodes. Different types of carbon coatings (among them diamond-like carbon (DLC) and diamond) deposited on the surface of implants increase the implant biocompatibility, haemocompatibility and serve the role of a barrier against corrosion [121]. Nanocrystalline (NC) and ultrananocrystalline diamond (UNCD) coatings on suitable substrates are promising materials for medical orthopedic implants, for example, hip and knee joint implants [122] and for coating of certain components of artificial heart valves [123] due to their extremely high chemical inertness, surface smoothness and good adhesion of the coatings to the substrate. These films are able to coat complex geometrical shapes with good conformal accuracy and with smooth surfaces to produce hermetic bioinert protective coatings, or to provide surfaces for cell grafting through appropriate functionalization. However, recent studies have shown that diamond powders are not bioinactive at the molecular level; they affect the cellular gene expression and inhibit stress (oxidative, cellular, genotoxic) [124]. UNCD is a more suitable material than silicon for fabrication of bioMEMS. For example, in the case of a Si microchip implantable in the eye as the main component of an artificial retina to restore sight to people blinded by retina degeneration, Si cannot survive long term implantation, while UNCD bioinert encapsulation can successfully replace the Si microchip implantation in the eye [125].

As a seed material for CVD diamond growth, DND particles play an important role for the development of medical implants [126, 127]. Seeding with DND allows obtaining coatings with small grain size and, therefore, smooth surfaces that are important for medical implant applications. Until recently, the preparation of the seeding slurry was performed in-house in the laboratories. After advances in DND deagglomeration and fractionation, seeding slurries with consistent properties have become available in the market. Williams et al. [126, 127] is developing methods for improving the nucleation density of nanocrystalline diamond film growth using bead-milled DND that is processed to form a stable aqueous colloidal solution of primary particles. This colloid was applied to various substrates to yield a high density of individually spaced diamond nanoparticles (greater than 10^{11} cm^{-2}) [126, 127]. Seeding slurries of DND with average aggregate size 10–15 nm in DMSO have been developed at ITC (www.itc-inc.org). Prior to seeding, DMSO-based DND slurry is mixed with methanol or ethanol [128] and the substrate intended for seeding is ultrasonically treated in the slurry, resulting in a monolayer of primary DND particles and small DND aggregates.

4.3.3.4 DND in Health Care Products

Nanodiamonds are also considered for cosmetics and health care applications due to their ability to bond with biological materials, improve durability and robustness of a composition, provide protection from harmful UV light, possibly scavenge free radicals and protect against viruses and bacteria [88, 129–131]. Sung Chien-Min et al. [129] hypothesize that nanodiamonds dispersed in a biologically acceptable carrier and bonded with biological materials may improve skin cleansing and exfoliation, add mechanical strength and provide treatment of adverse conditions. The authors consider additions of DND to deodorant, toothpaste, shampoo, antibiotics, dermal strips, DNA test strips, skin cleanser, dental filler, nail polish, eyeliner, lip gloss, and exfoliants. The authors report that nanodiamonds can provide increased resistance to chipping and wear of a nail polish so that it can last from about three to ten times longer than the typical nail lacquer formulations. No other specific results, demonstrating improvement in the performance/appearance after addition of ND to healthcare/cosmetic formulations, however, have been demonstrated by Sung Chien-Min et al. [129]. Lunkin et al. [130] studied the influence of DND formulations on viscoelastic properties of tissue by measuring the velocity of surface acoustic waves in skin. Dynamic shear modulus of human skin was increased by 15% after treatment with a skin cream containing 10^{-9} g/g of DND. In experiments with white mouse with skin burn wounds, addition 10^{-10} g/g of DND to the treatment gel accelerated the time of healing by twofold [130]. The Environmental Working Group [132] lists eight cosmetic products currently on the market containing diamond powder including nail polishes, nail treatments, anti-aging formulation and facial cleanser. The diamond powder is listed in the formulations as a material with a low hazard score (1 out of 10).

Another potential application of nanodiamond in the healthcare products is protection from ultraviolet radiation (UVR), as discussed by Shenderova et al. [88]. Ultraviolet radiation causes severe damage in humans [133] and natural and synthetic materials. Chemical sunscreens reduce the damaging effects of UVR via absorption, whereas physical sunscreens cause reflection and scattering of UVR [134]. In the case of inorganic sunscreens such as barium sulfate, strontium carbonate, titanium dioxide, and zinc oxide nanosuspensions have found applications in sunscreens made for human use [135]. DND has high refractive index (~2.4) and it efficiently scatters light. The attenuation of UVR by DND can be enhanced due to the absorption by the sp^2 carbons on the DND surface. In addition, many lattice defects and impurities found in natural and synthetic diamonds absorb UVR [136], contributing to the radiation attenuation. Photoluminescent defects that can be excited by UV radiation, emit light in the 'safe' visible spectral range. Shenderova et al. demonstrated that the detonation nanodiamonds with sizes ranging from ~50 to 100 nm effectively attenuate UVR without compromising the transparency of a sample in the visible spectral region and, thus, can be used in sunscreens in a variety of applications [88]. The ability of DND to absorb and scatter UVR depends on the concentration and size of nanodiamond particles and their surface composition, which may be related to the presence of sp^2 carbon, lattice defects and impurities. DND is a very attractive candidate as a physical sunscreen because they are biocompatible, nontoxic, and have excellent physical performance. Human trials are needed, though, to make a final conclusion of applicability of ND in healthcare and cosmetic products.

4.4 Future of DND Biomedical Applications

There are a multitude of nanoparticles currently being examined for biomedical applications (i.e. imaging, diagnostics, drug or gene delivery), however many have limitations to overcome such as controlling surface properties, increasing dispersion in physiological solutions, expanding on biocompatibility criteria, and effectively targeting the intracellular locations. Thus, the interactions of nanoparticles with biological systems will depend highly on the form (i.e. airborne particulate, aqueous suspension, protective coating, solid substrate), presence of impurities (i.e. metal, amorphous carbon), cell type or animal model, and route of administration and exit. Fortunately, nanodiamonds appear to be ideal candidates for biomedical use due to their small primary particle size (~4–5 nm), purity, facile functionalization, and retention of high cellular viability. However, the qualities of NDs that must be controlled include their aggregation state, surface chemistry, as well as their localization and accumulation behavior within the body. Therefore, advances in their functionalization and comparisons between in vitro and in vivo testing for biocompatibility [137] will be required before the full biomedical potential of any nanoparticle, including ND, can be realized. The anticipated major areas for ND biomedical use are in cancer therapeutics (i.e. drug delivery), analytical diagnostics, and imaging. Other clinical uses of NDs for instruments in cell surgery, prosthetic devices for retinal implants, and platforms for nerve stimulation will also undoubtedly impact the field of nanomedicine [138].

The highly intense research into ND chemistry and biomedicine in the last several years has lead to a new quality of nanodiamond that is far superior to its predecessors and may be almost unrecognizable. This focused work has lead to ND products that rival other nanoparticles in purity, size selectivity, prevention of aggregation, colloidal stability, surface functionality, and photoluminescence. Thus, the authors of this work expect that in the near future, NDs will become more widely recognized for their unique surface and diamond core properties and will attract further attention, bring about new applications, and possibly create new scientific fields of research and applications. We believe that forming close working relationships with investigators outside the nanodiamond community is paramount to the successful development of nanodiamond products and is critical in promoting and exploiting their potential in medicine and biotechnology.

Acknowledgments The authors would like to acknowledge the help of V. Kuznetsov and B. Palosz for providing illustrations for this chapter, as well as helpful discussions with Amanda Schrand and G. McGuire.

References

1. Shenderova O, Gruen D (2006) Ultrananocrystalline diamond. William-Andrew, New York
2. Danilenko VV (2004) Phys Solid State 46:595–599
3. Yu SJ, Kang MW, Chang HC, Chen KM, Yu YC (2005) J Am Chem Soc 127:17604–17605
4. Fu CC, Lee HY, Chen K, Lim TS, Wu HY, Lin PK, Wei PK, Tsao PH, Chang HC, Fann W (2007) Proc Natl Acad Sci U S A 104:727–732
5. Sonnefraud Y, Cuche A, Faklaris O, Boudou JP, Sauvage T, Roch JF, Treussart F, Huant S (2008) Opt Lett 33:611–613
6. Decarli PS, Jamieson JC (1961) Science 133:1821–1822
7. Danilenko VV (in press) Solid State Phys
8. Huang H, Pierstorff E, Osawa E, Ho D (2007) Nano Lett 7:3305–3314
9. Bondar VS, Pozdnyakova IO, Puzyr AP (2004) Phys Solid State 46:758–760
10. Huang LC, Chang HC (2004) Langmuir 20:5879–5884
11. Kong XL, Huang LC, Hsu CM, Chen WH, Han CC, Chang HC (2005) Anal Chem 77:259–265
12. Kong X, Huang LC, Liau SC, Han CC, Chang HC (2005) Anal Chem 77:4273–4277
13. Grichko V, Grishko V, Shenderova O (2007) Nanobiotechnology 2:37–42
14. Gibson N, Shenderova O, Puzyr A, Purtov K, Grichko V, Luo TJM, Fitgerald Z, Bondar V, Brenner D (2007) For detoxification. In: Technical proceedings of the 2007 NSTI NanoTechnology Conference and Trade Show
15. Puzyr AP, Purtov KV, Shenderova OA, Luo M, Brenner DW, Bondar VS (2007) Dokl BiochemBiophys 417:299–301
16. Schwertfeger H, Fokin AA, Schreiner PR (2008) Angew Chem Int Ed Engl 47:1022–1036
17. Dahl JE, Liu SG, Carlson RMK (2003) Science 299:96–99
18. Carlson RMK, Dahl JEP, Liu SG (2005) Diamond molecules found in petroleum. In: Gruen DM, Vul A, Shenderova OA (eds) Synthesis, properties, and applications of ultrananocrystalline diamond. Dordrecht, The Netherlands, Springer
19. Freitas RA (2003)Nanomedicine. vol IIA. Landes Bioscience: Texas, pp 348
20. Larionova I, Kuznetsov V, Frolov A, Shenderova O, Moseenkov S, Mazov I (2006) Diam Relat Mater 15:1804–1808
21. Krueger A, Boedeker T (2008) Diam Relat Mater 17:1367–1370

22. Neugart F, Zappe A, Jelezko F, Tietz C, Boudou JP, Krueger A, Wrachtrup J (2007) Nano Letters 7:2588–3591
23. Krueger A, Stegk J, Liang YJ, Lu L, Jarre G (2008) Langmuir 24:4200–4204
24. Dolmatov V (2003) Ultradispersed Diamonds of Detonation Synthesis. SPbGTU, Sank-Petersburg
25. Vereschagin AL (2001) Barnaul, Russian Federation, Altai State Technical University; Vereschagin AL (2005) Altay Region, Barnaul State Technical University
26. Danilenko, V.V. (2003), ed. Energoatomizdat.
27. Gruen DM, Shenderova OA, Vul AY (2005) Synthesis, properties, and applications of ultrananocrystalline diamond. Springer, Dordrecht, Netherlands
28. Schrand AM, Hens SC, Shenderova OA (2009) Crit Rev Solid State Mater Sci vol 34, 18–74
29. Shenderova OA, Zhirnov VV, Brenner DW (2002) Crit Rev Solid State Mater Sci 27:227–356
30. Dolmatov VY (2001) Russ Chem Rev 70:607–626
31. Holt KB (2007) Philos Transact A Math Phys Eng Sci 365:2845–2861
32. Krueger A (2008) Adv Mater 20:2445
33. Viecelli JA, Ree FH (2000) J Appl Phys 88:683–690
34. Raty JY, Galli G, Bostedt C, Van Buuren TW, Terminello LJ (2003) Phys Rev Lett 90:037401
35. Dolmatov V (2008) In: 3rd international symposium detonation nanodiamonds: technology, properties and applications, St. Petersburg, Russia
36. Rabeau JR, Stacey A, Rabeau A, Prawer S, Jelezko F, Mirza I, Wrachtrup J (2007) Nano Letters 7:3433–3437
37. Smith BR, Inglis DW, Sandnes B, Rabeau JR, Zvyagin AV, Gruber D, Noble CJ, Vogel R, Osawa E, Plakhotnik T vol 5, 1649–1653
38. Gubarevich (2008) 3rd international symposium detonation nanodiamonds: technology, properties and applications, St. Petersburg, Russia
39. Padalko V (private communication)
40. Osswald S, Yushin G, Mochalin V, Kucheyev SO, Gogotsi Y (2006) J Am Chem Soc 128:11635–11642
41. Petrov I, Shenderova O (2006) Chapter 16: history of Russian patents on detonation nanodiamonds. In: Shenderova O, Gruen D (eds) Ultrananocrystalline diamond. Norwich, UK, William-Andrew
42. Petrov I, Shenderova O, Grishko V, Grichko V, Tyler T, Cunningham G, Mcguire G (2007) Diam Relat Mater 16:2098–2103
43. Chiganov AS (2004) Phys Solid State 46:620–621
44. Dolmatov VY, Veretennikova MV, Marchukov VA, Sushchev VG (2004) Phys Solid State 46:611–615
45. Gubarevich T, Larionova IS, Kostukova IN, Ryzko LS, Tyricuna VF, (1992) RU 1770272
46. Pavlov EV, Skrjabin JA (1994) Method for removal of impurities of non-diamond carbon and device for its realization. 1994: Russia
47. Cunningham G, Panich AM, Shames AI, Petrov I, Shenderova O (2008) Diam Relat Mater 17:650–654
48. Shenderova O (unpublished)
49. Mitev D, Dimitrova R, Spassova M, Minchev C, Stavrev S (2007) Diam Relat Mater 16:776–780
50. Shenderova O, Petrov I, Walsh J, Grichko V, Grishko V, Tyler T, Cunningham G (2006) Diam Relat Mater 15:1799–1803
51. Mochalin V, Behler K, Stravato A, Giammarco J, Gogotsi Y, Picardi C, Kalter M (2008) 3rd international symposium detonation nanodiamonds: technology, properties and applications, St. Petersburg, Russia
52. Larionova IS, Molostov IN, Kulagina LS, Komarov VF, RU 2168462.
53. Timofeev VT, Detkov PY (2005) Atom 4:1
54. Krueger A, Kataoka F, Ozawa M, Fujino T, Suzuki Y, Aleksenskii AE, Vul AY, Osawa E (2005) Carbon 43:1722–1730

55. Shenderova O, Larinova I, Petrov I, Hens S et al (in preparation)
56. Gordeev SK, Kruglikova S, Gordeev SK, Kruglikova S (2004) Superhard Mater 6:34
57. Xu XY, Yu ZM, Zhu YW, Wang BC (2005) J Solid State Chem 178:688–693
58. Spitsyn B, Davidson J, Gradoboev M, Galushko T, Serebryakova N, Karpukhina T, Kulakova I, Melnik M (2006) Diam Relat Mater 15:296
59. Yeganeh M, Coxon PR, Brieva AC, Dhanak VR, Siller L, Butenko YV (2007) Phys Rev B 75:155404
60. Chukhaeva SI, Detkov P, Tkachenko A, Toropov A (1998) Sverkhtv Mater 4:29
61. Chukhaeva SI (2004) Phys Solid State 46:625–628
62. Grichko V, Tyler T, Grishko VI, Shenderova O (2008) Nanotechnology 19:225201
63. Iakoubovskii K, Mitsuishi K, Furuya K (2008) Nanotechnology 19:155705
64. Krueger A, Ozawa M, Jarre G, Liang Y, Stegk J, Lu L (2007) Phys Status Solidi A 204:2881–2887
65. Zhu YW, Xu F, Shen JL, Wang BC, Xu XY, Shao JB (2007) J Mater Sci Tech 23:599–603
66. Morita Y, Takimoto T, Yamanaka H, Kumekawa K, Morino S, Aonuma S, Kimura T, Komatsu N (2008) Small 12:2154–2157
67. Ozawa M, Inaguma M, Takahashi M, Kataoka F, Kruger A, Osawa E (2007) Adv Mater 19:1201
68. Ozerin A, Kurkin TS, Ozerina LA, Dolmatov VY (2008) Crystallogr Rep 53:60
69. Titov VM, Tolochko BP, Ten KA, Lukyanchikov LA, Pruuel ER (2007) Diam Relat Mater 16:2009–2013
70. Danilenko VV (2006) Superhard Mater N5:9
71. Osawa E (2007) Diam Relat Mater 16:2018–2022
72. Huang HJ, Dai LM, Wang DH, Tan LS, Osawa E (2008) J Mater Chem 18:1347–1352
73. Xu K, Xue QJ (2007) Diam Relat Mater 16:277–282
74. Gibson N, Shenderova O, Luo TJM, Moseenkov S, Bondar V, Puzyr A, Purtov K, Fitzgerald Z, Brenner D (2008) Diam Relat Mater 2009 vol 18, 620–626
75. Hens, S., Wallen, S., and Shenderova, O., (2007) *U.S. Patent Application: Nanodiamond fractionation and products thereof.*
76. Bondar VS, Puzyr AP (2004) Phys Solid State 46:716–719
77. Puzyr AP, Bondar VS (2003) RU patent 2252192
78. Krueger A (2008) Chem Eur J 14:1382–1390
79. Krueger A, Liang YJ, Jarre G, Stegk J (2006) J Mater Chem 16:2322–2328
80. Hens SC, Cunningham G, Tyler T, Moseenkov S, Kuznetsov V, Shenderova O (2008) Diam Relat Mater 17:1858–1866
81. Ray MA, Tyler T, Hook B, Martin A, Cunningham G, Shenderova O, Davidson JL, Howell M, Kang WP, Mcguire G (2007) Diam Relat Mater 16:2087–2089
82. Chiganova GA (2000) Colloid Journal 62:238–243
83. Xu X, Yu Z, Zhu Y, Wang B (2005) Diam Relat Mater 14:206–212
84. Boehm HP (2002) Carbon 40:145–149
85. Fuente E, Menendez JA, Suarez D, Montes-Moran MA (2003) Langmuir 19:3505–3511
86. Donnet JB, Boehm HP, Stoeckli F (2002) Carbon 40:145–149
87. Montes-Moran MA, Suarez D, Menendez JA, Fuente E (2004) Carbon 42:1219–1225
88. Shenderova O, Grichko V, Hens S, Walsh J (2007) Diam Relat Mater 16:2003–2008
89. Aleksenskii AE, Baidakova MV, Vul AY, Siklitskii VI (1999) Phys Solid State 41:668–671
90. Turner S, Lebedev OI, Shenderova O, Vasov II, Verbeeck J, Tendeloo GV (2009) Adv Funct Mater, 19:2116–2124
91. Sque S, Jones R, Briddon P (2006) Phys Rev B 73:85313
92. Petrini D, Larsson K (2007) J Phys Chem C 111:796–801
93. Petrini D, Larsson K (2008) J Phys Chem C 112:4811–4819
94. Kern G, Hafner J (1997) Phys Rev B 56:4203
95. Barnard AS, Stenberg M (2007) J Mater Chem 17:4811–4819
96. Gruber A, Drabenstedt A, Tietz C, Fleury L, Wrachtrup J, Vonborczyskowski C (1997) Science 276:2012–2014

97. Chang YR, Lee HY, Chen K, Chang CC, Tsai DS, Fu CC, Lim TS, Tzeng YK, Fang CY, Han CC, Chang HC, Fann W (2008) Nat Nanotechnol 3:284–288
98. Barnard AS, Sternberg M (2007) Diam Relat Mater 16:2078–2082
99. Barnard AS, Sternberg M (2008) J Comput Theor Nanoscience 5:1–7
100. Borjanovic V, Lawrence WG, Hens S, Jaksic M, Zamboni I, Edson C, Vlasov V, Vlasov V, Shenderova O, Mcguire G (2008) Nanotechnology 19(45):455701
101. Kvit AV, Zhirnov VV, Tyler T, Hren JJ (2004) Compos B Eng 35:163–166
102. Schrand AM (2007) Characterization and in vitro biocompatibility of engineered nanomaterials in The School of Engineering. 2007. University of Dayton, Dayton, p 276
103. Schrand A, Braydich-Stolle Laura K, Schlager John J, Hussain Saber M, Liming Dai (2008)
104. Hens SC, Cunningham G, Grichko V, Tyler T, Moseenkov S, Kuznetsov V, Shenderova GMO (2008) 3rd international symposium detonation nanodiamonds: technology, properties and applications, St. Petersburg, Russia
105. Hughes MP (2000) Nanotechnology 11:124–132
106. Zhitomirsky I (2002) Adv Colloid Interface Sci 97:279–317
107. Alimova AN, Chubun NN, Belobrov PI, Detkov PY, Zhirnov VV (1999) J Vac Sci Tech B 17:715–718
108. Zhitomirsky I (1998) Mater Lett 37:72–78
109. Wu VWK (2006) Chem Lett 35:1380–1381
110. Yang WS, Auciello O, Butler JE, Cai W, Carlisle JA, Gerbi J, Gruen DM, Knickerbocker T, Lasseter TL, Russell JN, Smith LM, Hamers RJ (2002) Nat Mater 1:253–257
111. Nebel CE, Rezek B, Shin D, Uetsuka H, Yang N (2007) J Phys D Appl Phys 40:6443–6466
112. Williams AC, Barry BW (2004) Adv Drug Deliv Rev 56:603–618
113. Koo J, Kleinstreuer C (2005) Int Comm Heat Mass Tran 32:1111–1118
114. Koo J, Kleinstreuer C (2005) Int J Heat Mass Tran 48:2652–2661
115. Winters MA, Knutson BL, Debenedetti PG, Sparks HG, Przybycien TM, Stevenson CL, Prestrelski SJ (1996) J Pharm Sci 85:586–594
116. Freitas Ra J (2003) *Nanomedicine Volume IIA: Biocompatibility* 2003. Landes Bioscience, Georgetown, TX
117. Grichko V, Grishko V, Shenderova O (2006) Nanobiotechnology 2:37–42
118. Perevedentseva E, Cheng CY, Chung PH, Tu JS, Hsieh YH, Cheng CL (2007) Nanotechnology 18:315102
119. Phillips TD (1999) Toxicol Sci 52:118–126
120. Ushizawa K, Sato Y, Mitsumori T, Machinami T, Ueda T, Ando T (2002) Chem Phys Lett 351:105–108
121. Hauert R (2003) Diam Relat Mater 12:583–589
122. Amaral M, Abreu CS (2007) Diam Relat Mater 16:790–795
123. Mitura S, Mitura A, Niedzielski P, Couvrat P (1999) Chaos, Solitons Fractals 10:2165–2176
124. Bakowicz-Mitura K, Bartosz G, Mitura S (2007) Surf Coating Techn 201:6131–6135
125. Xiao XC, Wang J, Liu C, Carlisle JA, Mech B, Greenberg R, Guven D, Freda R, Humayun MS, Weiland J, Auciello O (2006) J Biomed Mater Res B 77B:273–281
126. Daenen M, Williams OA, D'haen J, Haenen K, Nesladek M (2006) Phys Status Solidi A 203:3005–3010
127. Williams OA, Douheret O, Daenen M, Haenen K, Osawa E, Takahashi M (2007) Chem Phys Lett 445:255–258
128. Feygelson TI, Shenderova O, Hens S, Cunningham G, Hobart KD, Butler JE (2008) 3rd international symposium detonation nanodiamonds: technology, properties and applications, St. Petersburg, Russia
129. Chien-Min S, Michael S, Emily S, Patent US 7, 294, 340
130. Lunkin VV Patent RU 2 257 889
131. Dolmatov VY (2006) Application of detonation nanodiamond. In: Shenderova OA, Gruen DM (eds) Ultra nanocrystalline diamond: synthesis, properties, and applications. William Andrew, Norwich, NY, USA, pp 513–527

132. Environmental Working Group, www.cosmeticdatabase.com.
133. Gasparro FP, Mitchnick M, Nash JF (1998) Photochem Photobiol 68:243–256
134. Cockell CS, Knowland J (1999) Biol Rev Cambridge Philosophical Soc 74:311–345
135. Nash JF (2006) Dermatol Clin 24:35
136. Zaitsev AM (2001) vol 348. Springer
137. Sayes CM, Reed KL, Warheit DB (2007) Toxicol Sci 97:163–180
138. Han SW, Nakamura C, Obataya I, Nakamura N, Miyake J (2005) Biochem Biophys Res Commun 332:633–639

Chapter 5
Functionalization of Nanodiamond for Specific Biorecognition

Weng Siang Yeap and Kian Ping Loh

5.1 Introduction

Diamond has grown increasingly important in science and technology due to its extreme hardness, chemical inertness, high thermal conductivities, wide optical transparency and other unique properties [1–4]. In 2005, Hasegawa of AIST reported the low temperature growth (~90°C) of nanocrystalline diamond in the European Diamond Conference in Toulouse [5]. This significant breakthrough affords the promise of low temperature deposition of diamond on plastics and polymer, which can be the basis for many new technological applications [6]. Parallel to this development, there are also exciting developments in the purification and applications of detonation nanodiamond powder. Detonation synthesis has now made nanodiamond powder commercially available in ton quantities, thus enabling many engineering applications [5]. It is now possible to produce bulk quantities of fluorescent nanodiamond via electron beam generation of nitrogen-vacancy defect centers [7]. In order to realize the practical applications of nanodiamond particles, surface functionalization of the nanodiamond is necessary in order to achieve specific functions such as bioaffinity and solution-processability, the latter is especially important in applications ranging from polymer blends to composite films [8–13].

5.2 Production of Nanoscale Diamond

There are several methods to produce nanoscale diamond particles. The simplest method is the milling of larger synthetic or natural microdiamonds and the separation of the smaller fraction by centrifugation. Another method is the circular shock-waves transformation of graphitic material (usually graphite dust) into diamond crystallites. This method applies the ignition of an explosive which can lead to the

W. Siang Yeap and K. Ping Loh (✉)
Department of Chemistry, National University of Singapore, 3 Science Drive 3, Singapore, 117543
e-mail: chmlohkp@nus.edu.sg

D. Ho (ed.), *Nanodiamonds: Applications in Biology and Nanoscale Medicine*,
DOI 10.1007/978-1-4419-0531-4_5,

propagation of a circular shock wave that compresses the driving tube, sp^2 carbon material is transformed into sintered nanodiamond particles [14] in the process. Another technique for bulk-scale production of nanodiamond is called detonation synthesis. A mixture of trinitrotoluene (TNT), hexogen and octogen is used for the detonation process [7]. Nanodiamond (ND) powders prepared by this explosive technique present a novel class of nanomaterials possessing unique surface properties. The lack of oxygen in the combustion of the explosive led to a high percentage of diamond particles in the soot residue. The soot residue also contains a variety of impurities, including metal and concrete debris from the reaction chamber and a significant amount of nondiamond carbon.

5.3 The Structure of Detonation Nanodiamond

Milled synthetic microdiamond particles exhibit pronounced facets while shockwave and detonation diamond usually posses rather rounded shapes without pronounced facets. The detonation ND has the smallest particle sizes among particulate synthetic diamonds, with size ranging from 4 to 5 nm [5, 15, 16]. An individual nanodiamond grain measuring 4 nm in width consists of about 7,200 carbon atoms, out of which, nearly 1,100 atoms are located on the surface [15]. Oxygen-containing groups are usually present on the particles surface [5]. These oxygen-containing groups can be chemical functional groups like hydroxyl, carboxylic, lactones, ketones and ethers (Fig. 5.1). In addition to functional groups, the surface of ND particles usually contains graphitic material. Due to the tendency to aggregate via interparticle bonding, the dispersal of detonation diamond powder can be challenging.

Loose agglomeration, due to electrostatic interactions, can be easily overcome by ultrasonic treatment. However, core agglomerates are not affected by this treatment. Core agglomerates are strongly bound by graphitic soot-like structures [17, 18]. Ozawa et al. [19] reported that the agglomerates can be destroyed mechanically

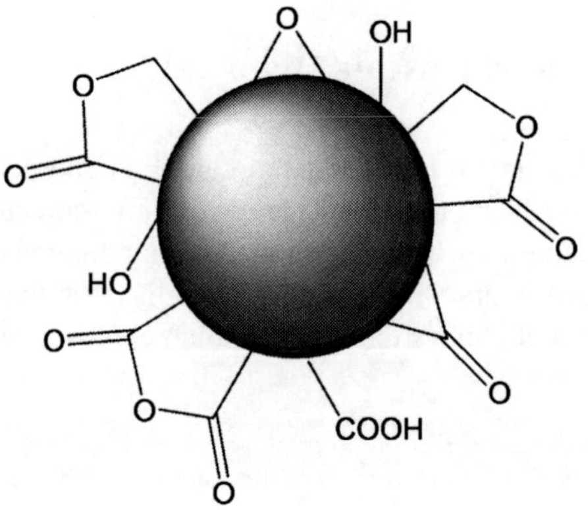

Fig. 5.1 The detonation nanodiamond (ND) surface is covered with a variety of functional groups [5]

using shear force which was inflicted by small zirconia milling beads. The treatment can be accelerated either in a fast stirring mill or in the cavitation of strong ultrasound. The resulting ND particles form stable colloids in a variety of polar solvents, such as water, ethanol, and DMSO. However, rapid precipitation occurs in nonpolar organic solvents. Gogotsi et al. carried out the oxidation of detonation ND in air which successfully removed graphitic and amorphous carbon leading to the breaking of large agglomerates [20].

5.4 Applications of Nanodiamond

There have been several reports on the use of ND as an adsorbent for large biomolecules like proteins. One application is the use of diamond as an adsorption platform for the preconcentration of biomolecules in dilute solution, the diamond platform can then be used in MALDI-TOF mass spectrometry for the direct detection of the biomolecules [21]. The noncovalent adsorption of biomolecules on ND can be very effective due to a combination of hydrogen bonding and electrostatic attraction forces. Highly efficient, nonspecific capture of proteins, such as cytochrome C, myoglobin and albumin can occur. After coating with poly-L-lysine, ND particles can also be applied for the detection of DNA oligonucleotides [22]. ND particles have also been considered as a vehicle for gene and drug delivery. Dean Ho et al. [23] demonstrated the loading of doxorubicin hydrochloride (DOX) on ND, the resultant complex serves as a highly efficient chemotherapeutic drug delivery agent for murine macrophages as well as human colorectal carcinoma cells. Doxorubicin hydrochloride (DOX) is an apoptosis-inducing drug widely used in chemotherapy. The DOX-ND agent can be introduced into cells, and DOX can be reversibly released from ND by regulating Cl^- ion concentration.

Volgin et al. [24] reported that microdispersed sintered ND can serve as a stationary phase in high-performance liquid chromatography. The electrochemical properties of detonation nanodiamond have also been explored [25]. Due to its large specific surface area, the use of ND as modifiers in electrode material is attractive.

5.5 Functionalization of Nanodiamond with Aminophenylboronic Acid

Aminophenylboronic acid is an affinity ligand that is commonly used in boronate affinity chromatography. The retention is based on the interaction between boronic acids and *cis*-diol compounds (Fig. 5.2). In the 1940s, the specific interaction between borate and *cis*-diol had been employed as the biorecognition mechanism in the analysis of carbohydrates. In the 1950s, borate/*cis*-diol interactions were used for seperations in zone electrophoresis. In the 1960s, such separation was applied to ion-exchange chromatography and in the 1970s, researchers developed immobilized boronate columns [26].

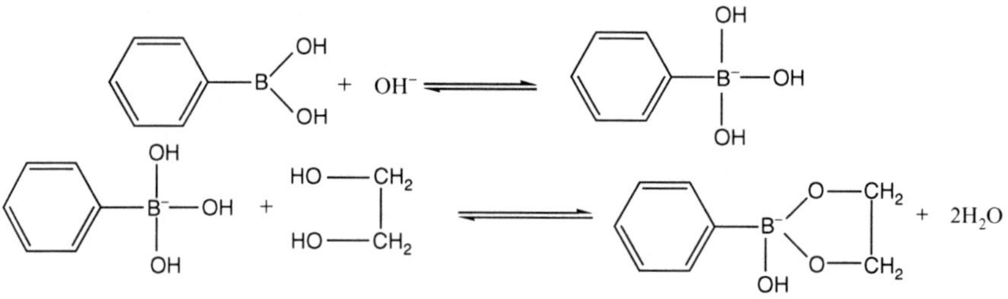

Fig. 5.2 The interaction between a boronic acid and *cis*-diol in aqueous solution

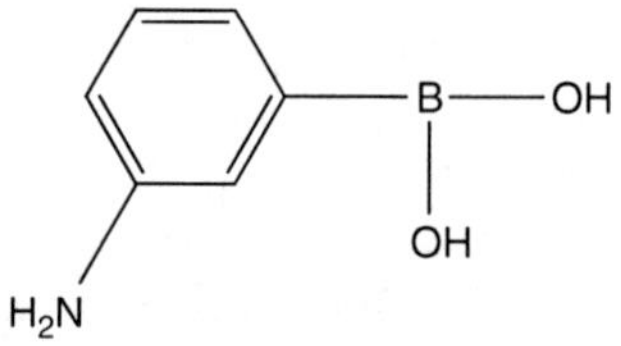

Fig. 5.3 Structure of 3-aminophenylboronic acid (3-APBA)

Since then, boronate affinity columns have been employed for the separation of sugars as well as a wide variety of *cis*-diol compounds, including nucleosides, nucleotides, nucleic acids, carbohydrates, glycoprotein and enzyme. As shown in Fig. 5.2, the key interaction between boronate and analyte is the esterification reaction that occurs between a boronate ligand and a *cis*-diol compound. In aqueous solution, and under basic conditions, the boronate is hydroxylated and goes from a trigonal coplanar form to a tedtrahedral boronate anion, which can form esters with *cis*-diols (Fig. 5.2). 3-Aminophenylboronic acid, also known as 3-APBA (Fig. 5.3), can bind specifically to the glycoprotein or glycated proteins.

Efforts have been made to immobilize aminophenylboronic acid on various supports such as silica [27], polymers [28], agarose matrices [29], and hydrogel beads [30] for improved separations. The supports were selected based on properties such as ligand capacity, mechanical stability, hydrophilicity, porosity, and cost.

In order to achieve specific bonding to glycoproteins, we have selected aminophenyl boronic acid (APBA) as the recognition motif to be constructed on ND. Glycoproteins are produced by the covalent addition of sugar moieties called glycans to proteins in cellular posttranslational modification events. Glycans play important roles in protein folding, cell–cell recognition, cancer metastasis, immune system and therapeutics [31]. It is quite challenging to study glycoprotein due to their extreme diversities, therefore, the development of a sensitive and specific technique for their elucidation is required. A full characterization of glycoprotein component in complex protein mixtures or contaminated solution is a challenging task. Separation of the glycated proteins from nonglycated proteins, through efficient enrichment, can increase the sensitivity of the glycation assay. Micrometer-sized silica beads made for affinity chromatography columns had been used to capture glycoproteins and proteins of interest in unfractionated sample solutions [27].

However, the interference from the beads in ion formation decreased the mass resolution and mass accuracy.

In this regard, ND is particularly suited as a surface-enhanced platform for protein immobilization because of its ability to blend well with matrixes used in Matrix-asssited Laser Desorption Ionization (MALDI). The optical transparency of ND to the ionizing laser [21] means minimum substrate interference. Preconcentration of biomolecules on ND is especially relevant to the analysis of biomolecules in dilute solutions using the technique of MALDI, a laser-based soft ionization method [21] which has proven to be one of the most successful ionization methods for mass spectrometric analysis and investigation of large molecules. The sample analysis in MALDI requires a matrix to absorb the laser light energy and transfer the ionization energy to the target substrate. With this motivation in mind, we consider the chemistry needed to derivatize ND with aminophenyl boronic acid (APBA), in order to achieve specific bonding with glycoprotein [32, 33]. We found that by inserting an alkyl spacer chain between the terminal boronic acid moieties and the ND, very high specific uptake of glycoproteins, i.e., of up to 350 mg/g of ND could be obtained, compared to the maximum reported value of 200 mg/g reported previously for nonspecific adsorption. Successful assay by MALDI could be achieved [7, 9].

The equations for the various steps in the chemical functionalization of the ND are shown in Fig. 5.4. The phenylboronic acid group can be immobilized directly on acid-treated diamond via carboimdide chemistry. However, while this allows the end groups to bond covalently with glycoproteins, the binding has to compete with nonspecific interactions due to the short distance between the charged oxygen functionalities on the ND surfaces and the proteins. ND has been found to exhibit tenacious binding interactions with proteins due to the interplay of hydrophobic, hydrogen-bonding and electrostatic interactions. To suppress these nonspecific interactions, the strategy we employed is to extend an alkyl spacer group of about 20 nm between the ND and the phenylboronic acid functionality, by inserting an alkyl linker chain using APTES and succinic anhydride. Figure 5.4 shows a schematic drawing illustrating the (a) direct immobilization of APBA on nanodiamond for covalent bonding to glycoprotein, and (b) extension of a spacer group between APBA and diamond before bonding with glycoprotein. For the spacer chain extension, amino-alkyl terminated spacer chain was attached to the hydroxylated ND via silanization, subsequently, the amino-alkyl group was extended further using succinic anhydride, such that a carboxylic terminated methylene chain was obtained at the end. Our silanization procedure is distinguished by the application of ultrasonic irradiation (20 kHz) to assist the reaction. In this case, the reaction could be completed in an hour, with very good yield compared to previous methods [34]. The outwardly extending spacer chain forms an exclusion shell around the ND, therefore in principle, electrostatic interactions with nontargeted proteins should decrease, since the distance between the ND and nonspecific proteins is increased. Although nonspecific binding of proteins cannot be totally eliminated, we showed that insertion of an alkyl spacer chain to form a molecular shell around the ND can improve specific adsorption, as opposed to direct functionalization of the carboxylated ND, which resulted in a much higher degree of nonspecific

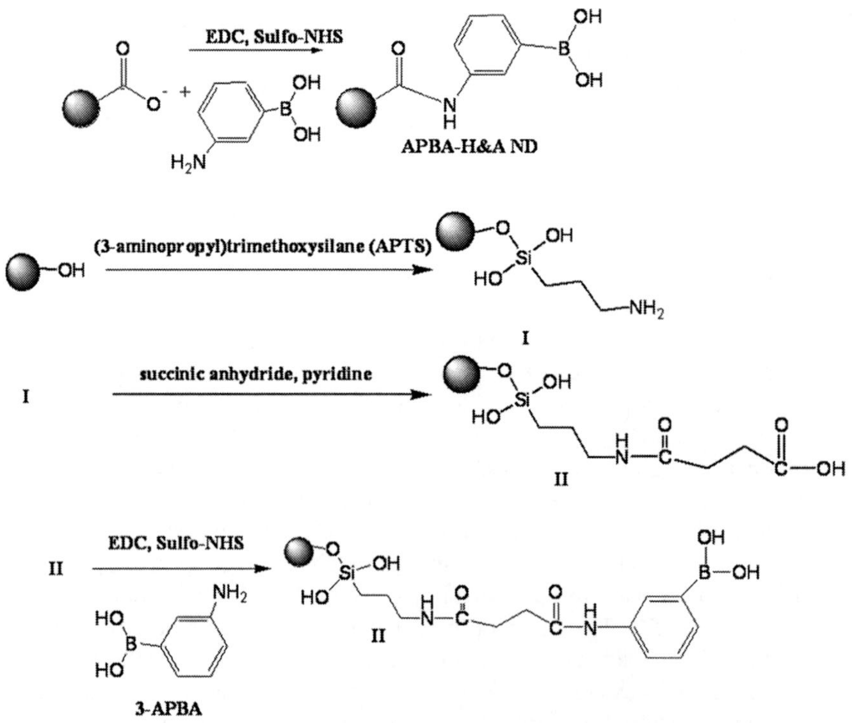

Fig. 5.4 Schematic showing the chemistry employed for the functionalization of nanodiamond (ND) to generate either ND-APBA or ND-spacer-APBA. APBA, aminophenyl boronic acid)

binding. We demonstrated that ND functionalized with the alkyl spacer chain and APBA can be used to preconcentrate glycoproteins from unfractionated mixtures because it showed selective binding affinity for glycoprotein, and it can be blended with the matrix and used directly in MALDI analysis without extra pretreatment steps. Because of its combination of specificity and high extraction efficiency, functionalized ND can be very useful in proteomics research in combination with techniques like MALDI.

5.5.1 Suzuki Coupling Reaction

To study charge transfer interactions between the nanodiamond and organic molecules, it is important to consider the functionalization of nanodiamond with aryl organic groups, since the conjugated chains in these molecules facilitates charge transfer. Suzuki coupling is one of the most widely used generic methods for the C-C coupling of biaryls. Suzuki coupling reaction is a catalyst-catalyzed cross coupling between aryl halides and aryl boronic acids to form biaryls. The most common catalyst used

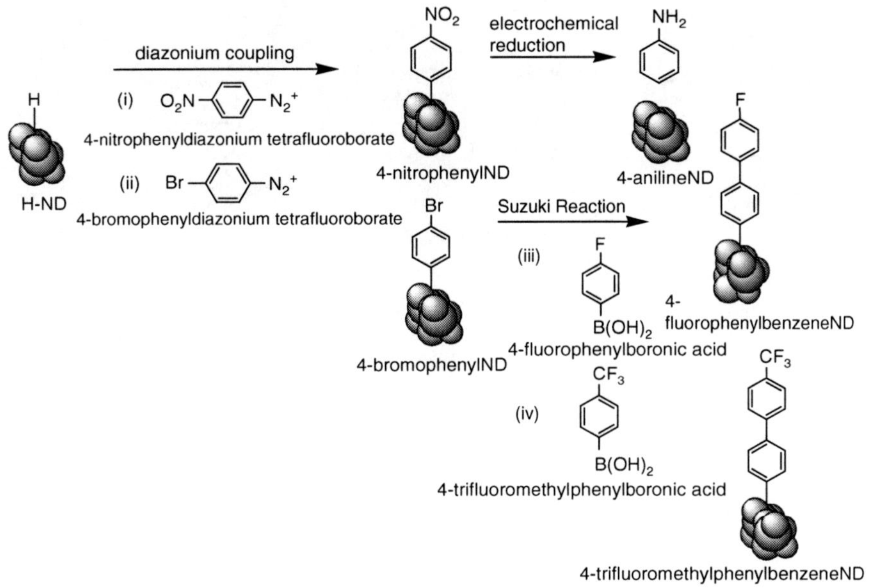

Fig. 5.5 Schematic showing the coupling of diazonium salts on nanodiamond to generate the 4-nitrophenyl or 4-bromophenyl ND, which serves as a synthon for Suzuki coupling of 4 fluro-phenylboronic acid or 4-trifluorophenylboronic acid [37]. ND, nanodiamond

is palladium complex. This reaction has emerged as an extremely powerful tool in organic synthesis [35] for C–C coupling. This cross-coupling reaction can be used to couple a wide range of reagents and hence, has wide applicability, ranging from materials science to pharmaceuticals. For example, with respect to pharmaceuticals, the biaryl group is the key feature in the sartans family of drugs for high blood pressure [36].

In a recent work [37], we demonstrate the functionalization of nanoscale diamond particles with aryl organics using Suzuki coupling reactions. In route one, hydrogenated nanodiamond is derivatized with aryl diazonium to form the bromophenyl-nanodiamond complex, this is subsequently reacted with phenyl boronic acid to generate the biphenyl adduct. In route two, the nanodiamond is first derivatized with boronic acid groups to form the boronic acid–nanodiamond complex, this is followed by Suzuki cross-coupling with arenediazonium tetrafluroborate salts to generate the biphenyl product. Good chemoselectivity can be obtained in both routes. The efficiencies of the Suzuki coupling reaction can be further improved by performing the chemistry in a microreactor, where electro-osmotic flow accelerates the mixing of reactants. Using the Suzuki coupling reactions, we can functionalize nanodiamond with trifluoroaryls and increase the solubilities of nanodiamond in ethanol and hexane. Fluorescent nanodiamond can be generated by the Suzuki-coupling of pyrene to nanodiamond (Figs. 5.5 and 5.6).

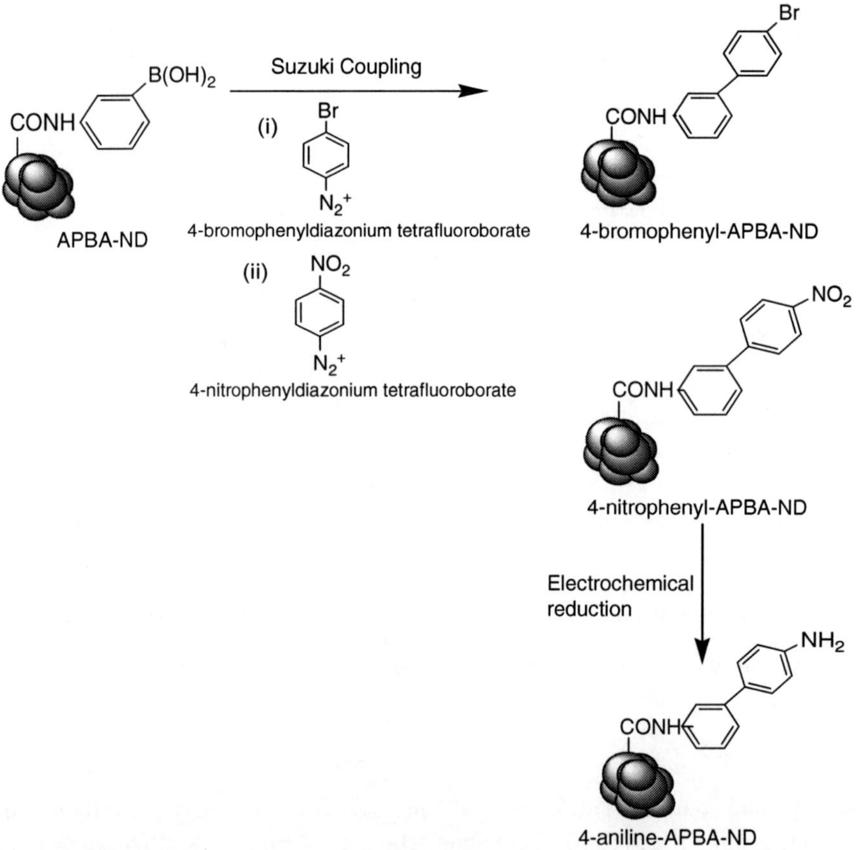

Fig. 5.6 Schematic diagram showing Suzuki coupling of arenediazonium salts on nanodiamond particles that were pretreated with aminophenyl boronic acid diazonium salts (APBA-ND), to generate biphenyl adducts terminating in either the bromo (4-bromophneyl-APBA-ND) or nitro groups (4-nitrophenyl-APBA-ND). 4-nitrophenyl-APBA-ND can be electrochemically reduced further to 4-aniline-APBA-ND [37].

5.6 Conclusion

In summary, we have shown that the functionalization of nanodiamond can be readily carried out using conventional solution chemistry. Functionalized nanodiamond can readily be used as a preconcentration platform for biomolecules due to its high extraction efficiencies, and we demonstrated that in combination with techniques such as MALDI, it can be useful in proteomics research. Nanodiamond can also be used as a versatile organic platform for performing C–C cross-coupling reactions, opening up potential applications as functional components in polymer blends and carbon fibers.

References

1. Tang L, Tsai C, Gerberich WW, Kruckeberg L, Kania DR (1995) Biomaterials 16:483
2. Hauert R (2003) Diamond Relat. Mater 12:583
3. Wei J, Yates JT Jr (1995) Crit Rev Surf Chem 5:1
4. Bigelow LK, D'Evelyn MP (2002) Surf Sci 500:986
5. Dolmatov VYu (2001) Russ Chem Rev 70:607–626
6. Zhang QX, Naito K, Tanaka Y, Kagawa Y (2007) Macromolecules 41:536–538
7. Huang LCL, Chang HC (2004) Langmuir 20:5879–5884
8. Chung PH, Perevedentseva E, Tu JS, Chang CC, Cheng CL (2006) Diam Relat Mater 15:622–625
9. Nguyen TTB, Chang HC, Wu VWK (2007) Diam Relat Mater 18:872–876
10. Kong XL, Huang LCL, Hsu CM, Chen WH, Han CC, Chang HC (2005) Anal Chem 77:259–265
11. Krueger A, Stegk J, Liang YJ, Lu Li, Jarre G (2008) Langmuir 24:4200–4204
12. Ozawa M, Inaguma M, Takahashi M, Kataoka F, Kruger A (2007) Adv Mater 19:1201–1206
13. Liu Y, Gu ZN, Margrave JL, Khabashesku VN (2004) Chem Mater 16:3924–3930
14. DeCarli P, Jamieson J (1961) Science 133:1821
15. Aleksenskii AE, Baidakova ME, Vul' AYa, Siklikskii VI (1999) Phys Solid Stat 41:668–671
16. Greiner NR, Philips DS, Johnson JD, Volk F (1988) Nature 333:440
17. Kruger A, Ozawa M, Kataoka F, Fujino T, Suzuki Y, Aleksenskii AE, Vul' AY, Osawa E (2005) Carbon 43:1722–1730
18. Kruger A (2008) J Mater Chem 18:1485–1492
19. Ozawa M, Inaguma M, Takahashi M, Kataoka F, Kruger A, Osawa E (2007) Adv Mater 19:1201–1206
20. Osswald S, Yushin G, Mochalin V, Kucheyev SO, Gogotsi Y (2006) J Am Chem Soc 128:11635–11642
21. Kong XL, Huang LCL, Hsu CM, Chen WH, Han CC, Chang HC (2005) Anal Chem 77:259–265
22. Kong XL, Huang LCL, Vivian Liau SC, Han CC, Chang HC (2005) Anal Chem 77:4273–4277
23. Huang HJ, Pierstoff E, Osawa E, Ho D (2007) Nano Lett, 7(11):3305–14
24. Nesterenko PN, Fedyanina ON, Volgin YV (2007) Analyst 132:403–405
25. Article can be updated: Huang HJ, Pierstoff E, Osawa E, Ho D (2007) Nano Lett, 7(11) 3305–14
26. Zang JB, Wang YH, Zhao SZ, Bian LY, Lu J (2006) Diam Relat Mater 16:16–20
27. Liu XC, Scouten WH (2006) Taylor & Francis, London, UK, pp. 216–229
28. Li FL, Zhao XJ, Wang WZ, Xu GW (2006) Anal Chim Acta 580:181–187
29. Koyama T, Terauchi KI (1996) J Chromatogr B 679:31–40
30. Bouriotis V, Galpin IA, Dean PDG (1981) J Chromatogr A 210:267–278
31. Camli ST, Senel S, Tuncel A (2002) Colloids Surf A Physicochem Eng Asp 207:127–137
32. Gottschalk A (1972) Glycoproteins. Elsevier, Amsterdam
33. Lee JH, Kim Y, Ha MY, Lee EK, Choo J (2006) J Am Soc Mass Spectrom 16:1456–1460
34. Yeap WS, Tan YY, Loh KP (2008) Anal Chem 80:4659–4665
35. Kruger A, Liang Y, Jarne G, Stegk J (2006) J Mater Chem 16:2322
36. Miyaura N, Suzuki A (1995) Chem Rev 95:2457–2483
37. George BS, George CD, David LH, Anthony OK, Thomas RV (1994) J Org Chem 59:8151–8156
38. Yeap WS, Chen SM, Loh KP (2009) Langmuir 25:185–191

Chapter 6
Development and Use of Fluorescent Nanodiamonds as Cellular Markers*

Huan-Cheng Chang

Abstract Diamond is an allotrope of carbon. A unique property that distinguishes it from the other carbon materials is that diamond is optically transparent and often contains point defects as color centers. Nitrogen vacancy (N-V) defects are the most noteworthy color centers in diamond. These centers can be produced reproducibly by ion beam irradiation, followed by thermal annealing, and can emit strong and stable fluorescence when excited by visible light. This unique optical property combined with the non-cytotoxicity and good surface functionalizability characteristics of the material makes nanoscale diamonds a promising fluorescent probe for bioimaging applications in cellular environments. This article summarizes the results of our efforts in production and characterization of bright, multicolored (red and green) fluorescent nanodiamonds (FNDs) and their use as cellular markers. Notable advancement of technologies along this line includes mass production of FNDs and real time tracking of a single 35-nm red FND particle in three dimensions in live cells. We envision that further development of the material will provide an increased sensitivity and improved capability for fruitful applications of FNDs in biology and medicine.

6.1 Introduction

The application of nanotechnology to life sciences is an exciting, challenging, and rapidly evolving field which holds great promise for developing new types of diagnostic and therapeutic tools for biomedical use [1–3]. The utilization of nanoparticle-based technologies, in particular, is expected to advance our understanding of biological processes at the molecular, cellular, and even whole animal levels. Novel optical properties, for example, can be incorporated into the nanostructures

---

H.-C. Chang
Institute of Atomic and Molecular Sciences, Academia Sinica, Taipei, 106, Taiwan
e-mail: hcchang@po.iams.sinica.edu.tw

* This article is dedicated to the memory of Wunshain Fann (1961–2008).

D. Ho (ed.), *Nanodiamonds: Applications in Biology and Nanoscale Medicine*,
DOI 10.1007/978-1-4419-0531-4_6,

to probe and manipulate bioactive molecules at the nanometer scale. One of the researches in this direction is focused on the development of high sensitivity and high specificity nanoprobes that can circumvent the intrinsic limitations of organic dyes and fluorescent proteins in bioimaging applications [4]. Much effort has been dedicated to the study of fluorescent dye doped nanoparticles [5], quantum dots [6] and metallic nanoclusters [7]. After covalent conjugation with biomolecules of interest such as peptides, antibodies, nucleic acids, or small-molecule ligands, these nanoparticles are prospective to serve as fluorescent cellular markers. Of these, quantum dots (QDs) stand out as the most promising candidate because, in comparison with organic fluorophores, they have larger absorption coefficients and much higher levels of brightness and photostability [8]. However, the inherent toxicity of the metal ions that QDs are composed of and the interference of QDs with cellular processes retard their in vivo applications [9]. Although a variety of synthesis, core masking, solubilization, and surface functionalization protocols have been established [8], QDs still display adverse effects at high concentrations and long exposure times [9]. Another negative aspect of QDs is the fluorescence intermittency [10], or photoblinking, which renders continuous tracking of individual nanoparticles in three dimensions difficult [11]. It is therefore highly desirable to develop new types of nanomaterials that can overcome these limitations.

Being an allotrope of carbon, diamond offers great potential for biomedical applications. While unique properties such as high thermal conductivity, wide optical transparency range, and excellent chemical stability have extended its applicability to several areas of science and technology [12], the use of diamond in life science research is just beginning [13–15]. Due to its high rigidity, low chemical reactivity, and good biocompatibility, diamond has been considered the biomaterial of the twenty-first century [16]. Recently, there is an increasing interest towards fabricating surface functionalized diamond films for biosensor applications [13, 17, 18] and utilizing nanoscale diamond particles for chemotherapeutic delivery [19] and cellular imaging [20–29]. A large number of experiments demonstrate that diamond surfaces can be readily derivatized with various functional groups for DNA as well as protein immobilization [30–39]. Cytotoxicity tests with cancer cell lines show that diamond nanoparticles have the highest biocompatibility of all carbon-based nanomaterials including carbon blacks, multi-walled nanotubes, single-walled nanotubes, and fullerenes [40–43]. Moreover, particulates of diamond (refractive index of 2.42) are applicable as scattering optical labels in cells [24–26] and radiation damaged diamond nanocrystallites can emit multicolor fluorescence from nitrogen vacancy (N-V) defect centers by laser excitation [20–23, 27–29]. The latter, known as fluorescent nanodiamond (FND), represents an attractive alternative to QDs for in vivo applications.

The study of the fluorescence properties of diamond began in the 1950s, when man-made diamond was first produced by high pressure and high temperature (HPHT) methods [44]. Despite its high chemical purity (>99.9% C), synthetic HPHT diamond always contains nitrogen atoms as point defects with a concentration up to 300 ppm by mass, classified as type Ib diamond [45]. These atomically dispersed nitrogen impurities absorb light at wavelength below 500 nm, giving the material a yellowish color. Radiation damage of the HPHT diamond with high energy particles (e.g. 2 MeV electrons), followed by thermal annealing at high temperatures (e.g. 800ºC), produces an easily observable sharp line at 1.945 eV (or 637 nm) in the low-temperature absorption spectrum [46]. Intensive red fluorescence emerges when the irradiated/annealed diamond specimen is exposed to green yellow light. The fluorophore responsible for the observation is later identified as the negatively charged nitrogen vacancy center, $(N\text{-}V)^-$, which exhibits a zero-phonon line (ZPL) at 637 nm for the electronic excitation, $^3A \rightarrow {}^3E$ [47]. The ZPL is accompanied with a broad phonon sideband peaking at ~560 nm with an absorption cross section of $\sim 4 \times 10^{-17}\,cm^2$/molecule [48]. Its fluorescence occurs at ~700 nm with a quantum yield close to 1 [49, 50]. The center is the dominant end product of the irradiation/annealing treatment; therefore, it is thermally stable up to 1,000°C and resistant to photobleaching even under continuous laser excitation with intensity in the range of 5 MW/cm^2 [51]. This remarkable feature has enabled the observation of a single $(N\text{-}V)^-$ center in diamond at room temperature with confocal fluorescence microscopy [51] and utilizing it as a stable single photon source [52] for quantum cryptography [53] and quantum computation [54].

Apart from the red fluorescence, diamond can also emit green photoluminescence as well as other color variations [55]. As illustrated in Table 6.1, the green fluorescence arises from photoexcitation of the nitrogen vacancy complex, N-V-N, which originates from the A aggregate in type Ia diamond typically containing up to 1,000 ppm of nitrogen as impurities [56]. This special type of color center, designated as H3, can be generated as well by radiation damage of type IaA diamond with 2 MeV electrons and subsequently by heat treatment at temperatures above 800 °C in vacuum. The center absorbs light strongly at ~470 nm (absorption cross section $\sim 2 \times 10^{-17}\,cm^2$/molecule [23, 57, 58] for the phonon sideband), with a ZPL centered at 503 nm. It emits light at 530 nm with a quantum yield close to 1 [59]. Similar to $(N\text{-}V)^-$, H3 shows no sign of photobleaching and, moreover, the radiative lifetime of the excited state does not degrade even at 500°C [58]. These unique photophysical properties have led to the development of H3-containing diamond single crystal as a lasing medium for room temperature color center lasers [57, 58].

Inspired by the remarkable properties of the N-V centers, our group set goals to develop bright and multicolored FNDs (particle size < 100 nm) for in vivo imaging applications [20–24]. A simple calculation for type Ib diamond indicates that if all nitrogen atoms (typically 100 ppm) embedded in the lattice of diamond are converted to color centers by the aforementioned irradiation and annealing treatments, the number of $(N\text{-}V)^-$ centers created is about 10,000 for a single 100-nm FND particle [20]. Such a large number of defect centers, each of which has fluorescence brightness only about 5-fold lower than that of rhodamine dyes, should allow easy visualization of individual FND particles by epifluorescence. In the case of smaller FNDs, which have sizes in the range of 10 nm, single particle detection is still achievable with confocal fluorescence microscopy. At the concentration of 100 ppm and, assuming that all the nitrogen atoms are atomically dispersed in the particle, the distance between the nearest neighboring $(N\text{-}V)^-$ centers is ~5 nm.

Table 6.1 Zero-phonon lines (ZPL), emission maxima (λ_{em}), emission lifetimes (τ), and quantum efficiencies (φ) of some vacancy-related defect centers in diamond[a]

Diamond	Defect center	ZPL (nm)	λ_{em} (nm)	τ (ns)	φ
Type IIa	V^0 (GR1)	741.2	898	2.55	0.014
Type Ib	$(N\text{-}V)^-$	637.6	685	11.6	0.99
Type Ib	$(N\text{-}V)^0$	575.4	600	–	–
Type IaA	N-V-N (H3)	503.5	531	16	0.95
Type IaB	N_3+V (N3)	415.4	445	41	0.29

[a]Reference [45] and references therein.

This feature chapter provides an overview of the efforts that we have been making in the past few years to develop bright and multicolored FNDs as cellular markers. Several challenges have been met and overcome. First, we found that FNDs are non-toxic [20], and their surfaces can be readily functionalized with carboxyl groups or other functional groups for specific or nonspecific binding with nucleic acids and proteins [32]. Second, we developed a method for mass production of FNDs using a medium-energy ion beam, and the irradiation can be operated safely and routinely in general laboratories [22, 23, 48]. Third, we demonstrated that both red and green FNDs (designated as rFNDs and gFNDs, respectively) can be produced in large quantities, and they are suitable for multicolor imaging applications [23]. Fourth, we showed that single particle detection in cells can be readily achieved for 35-nm rFNDs, which emit fluorescence at a spectral range well separated from that of cell autofluorescence [21]. Finally, we successfully performed a three-dimensional, long-term imaging and tracking of a single 35-nm rFND in a live cell [22], thanks to the excellent photostability (no photobleaching and photoblinking) of the material. In this article, some salient features of these results are presented, and promises and challenges of this novel material for biomedical applications are discussed.

6.2 Materials and Methods

6.2.1 Nanodiamond Samples

6.2.1.1 Synthetic Nanodiamonds

Synthetic type Ib diamond powders with sizes in the range of 140 nm and 35 nm were obtained from Element Six (USA) and Microdiamant (Switzerland), respectively. The former contains a nitrogen concentration of ~100 ppm, as revealed by the infrared spectroscopy measurements [48] for the C–N stretches of diamond single crystals fabricated by the same manufacture. The corresponding concentration of nitrogen in the latter sample, however, was not determined.

6.2.1.2 Natural Nanodiamonds

Natural type Ia diamond powders with sizes in the range of 140 nm and 70 nm were obtained from Microdiamant (Switzerland). The concentration of nitrogen atoms in these samples was undetermined.

6.2.2 *Production of FNDs*

6.2.2.1 High Energy Ion Irradiation

Diamond powders were radiation damaged with 3-MeV H^+ generated from a high energy particle accelerator. Prior to the ion irradiation treatment, the sample was purified in concentrated H_2SO_4-HNO_3 solution (3:1, v/v) at 90°C for 30 min (or by microwave cleaning in strong oxidiative acids (*vide infra*)). After washing extensively in deionized water, an aliquot of the diamond suspension containing ~5 mg of the powders was deposited on a silicon wafer and dried in air to form a thin film (area ~ 0.5 cm^2). The dried diamond film had a thickness of ~50 μm, matching closely with the penetration depth of 3-MeV H^+ into diamond [23, 60]. It was then exposed to a 3-MeV proton beam from an NEC tandem accelerator (9SDH-2, National Electrostatics Corporation) at a dose of $\sim 1 \times 10^{16}$ ions/cm^2. As the typical current of the ions delivered by the tandem particle accelerator was in the range of 0.1 μA ($\sim 6 \times 10^{11}$ ions/s), the time required to complete the irradiation was ~4 h.

6.2.2.2 Medium Energy Ion Irradiation

Helium ions with an energy of 40 keV were generated from an ion beam apparatus built in-house [22, 23]. Prior to the ion irradiation treatment, diamond powders were purified in concentrated H_2SO_4–HNO_3 solution (3:1, v/v) at 100°C in a microwave reactor (Discover BenchMate, CEM) for 3 h. After separation by centrifugation and rinsing extensively with deionized water, the purified powders were suspended in 95% ethanol. A thin diamond film (~0.2 μm thick) was prepared by spreading the diamond suspension over a long copper tape (20-m long and 35-mm wide) and dried in air. The film was then exposed to the 40-keV He^+ ions by inserting the nanodiamond loaded copper tape in the path of the He^+ beam and rolling the tape inside the vacuum chamber using an external stepper motor. Figure 6.1 shows a schematic diagram of the experimental setup, in which the He^+ ions were generated by discharge of pure He in a radio frequency (RF) positive ion source (National Electrostatics Corporation) and accelerated to 40 keV by a high voltage acceleration tube (National Electrostatics Corporation). The typical current of the unfocused He^+ beam without mass discrimination was ~16 μA and the area-averaged flux at the sample target was $\sim 7 \times 10^{12}$ He^+/cm^{2}s. Since the 40-keV He^+ ion penetrates through diamond for only ~0.2 μm and each ion can create up to 40 vacancies in the crystal

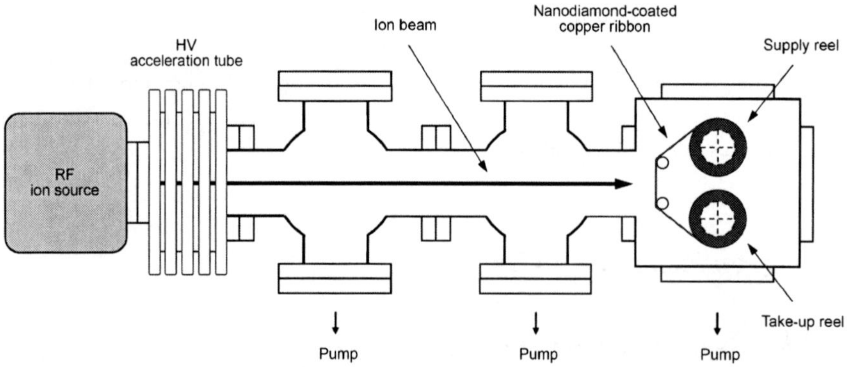

Fig. 6.1 Schematic of the experimental setup for mass production of FNDs with a medium energy ion beam. Both H^+ and He^+ are typical ions produced by the RF ion source. Adapted from [23]

lattice [60], the optimum dosage for the irradiation was estimated to be in the range of ~1×10^{13} He^+/cm^2 and ~ 1×10^{14} He^+/cm^2 for type Ib and type Ia diamonds, respectively [23]. To produce FNDs in a larger quantity, a total of 0.7 g of the nanodiamond powders were deposited to cover the entire copper tape. With the tape moving at a speed of ~0.5 cm/s, the total time required to scan the full length tape was ~1 h. In comparison, it would take more than 100 h to prepare the same amount of FNDs if a 3-MeV proton beam facility had been used for irradiation.

In both protocols as described above, after the ion irradiation treatment, formation of $(N\text{-}V)^-$ (or H3) defect centers in type Ib (or type IaA) diamond was facilitated by annealing the proton beam treated substrates in vacuum at 800°C for 2 h. As graphitic structures inevitably formed on the diamond surfaces during the irradiation/annealing treatment, the freshly prepared FNDs were additionally oxidized in air at 490°C for 2 h [61], followed by microwave cleaning in concentrated H_2SO_4–HNO_3 (3:1, v/v) at 100°C for 3 h and, finally, extensive rinsing with deionized water. The FNDs (either green or red), after going through all these cleaning procedures, are surface functionalized with a variety of oxygen-containing groups including the carboxyl moiety [32, 36].

6.2.3 Spectroscopic Characterization of FNDs

6.2.3.1 Ensemble Measurements

The fluorescence properties of the freshly prepared FNDs were first characterized by an ensemble-averaged measurement in a cuvette. Figure 6.2 shows an optical layout of the experimental setup used to acquire the fluorescence spectra of FNDs suspended in water. These optical components are arranged in such a way that the adverse effect arising from the strong light scattering at diamond-water interfaces

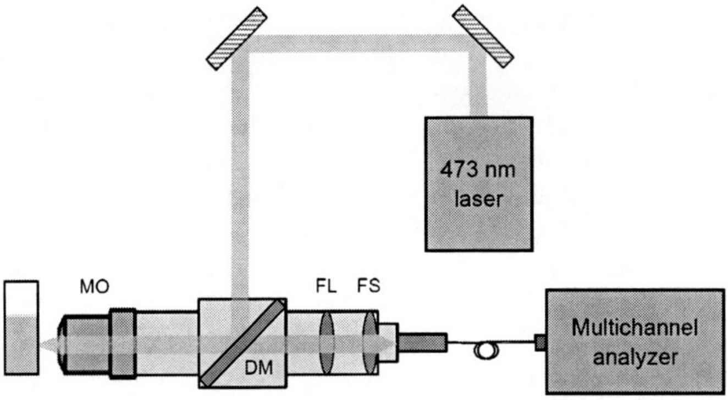

Fig. 6.2 An optical layout of the experimental setup for fluorescence measurements of FNDs suspended in water. In this particular experiment, a cw 473-nm solid-state laser was used as the excitation source to obtain the spectra of gFNDs. *MO* microscope objective, *DM* dichroic mirror, *FL* focusing lens, *FS* filter set. Adapted from [23]

is avoided [22, 24]. For gFNDs, the system consisted of a continuous-wave (cw) 473-nm laser (LDC-1500, Photop Technologies), a dichroic beam splitter (505DCLP, Chroma Technology), a microscope objective lens (10×, NA 0.3, Nikon), a long wave pass edge filter (LP02-488RU, Semrock), and a multichannel spectrometer (C7473, Hamamatsu). The same system was used to collect the fluorescence of rFNDs, after replacement of the excitation light source with a 532 nm laser and the mirrors and filters by appropriate optics.

6.2.3.2 Single Particle Measurements

Fluorescence images and spectra of single rFNDs were acquired for particles dispersed on a glass plate or internalized by cells. In the former experiment, a coverglass slide spin-coated with rFNDs was mounted on a modified confocal optical microscope (E600; Nikon) for inspection. Excitation of the particles was made through a ×100 objective (Plan Fluor, NA 1.3 oil; Nikon) with a solid state laser (JL-LD532-GTE; Jetlaser) operating at 532 nm. Epifluorescence was collected through a 565-nm long pass filter (E565lp; Chroma Tech) and detected by an avalanche photodiode (SPCM-AQR-15; PerkinElmer). Fluorescence images of the specimen were first obtained by raster scanning with a piezo-driven nanopositioning and scanning system (E-710.4CL and P-734.2CL; Physik Instrument). After acquisition of the images, each rFND particle was moved consecutively to the focal point of the objective to record the time evolution of the fluorescence intensity. The corresponding spectra were acquired by using a monochromator (SP300i; Acton Research) equipped with a liquid-nitrogen-cooled CCD camera (LN/CCD-1100-PB; Princeton Instruments). For fluorescence lifetime measurements, a frequency-doubled picosecond Nd:YAG laser (IC-532-30; High Q Laser) served as the excitation source.

The corresponding fluorescence decays were measured with a time-correlated single photon counting module (SPC-600; Becker and Hickl). The same microscope system was used to obtain the fluorescence images of rFNDs in cells.

An apparatus used for three-dimensional tracking of single rFND particles in cells is illustrated schematically in Fig. 6.3. It consisted of an inverted microscope (IX71; Olympus), an oil immersion objective (Plan-Apo 100×, NA 1.4; Olympus), and a diode-pumped solid state laser (Verdi DPSS; Coherent) operating at 532 nm as the excitation source. The resulting fluorescence, after passing through a dichroic mirror (565dclp; Chroma Technology) and an emission filter (HQ560LP; Chroma Technology), was detected by an electron multiplier CCD camera (EMCCD, DV887DCS-BV; Andor Technology). A user developed computer program analyzed the fluorescence image data and maintained the maximum intensity and contrast of the image of interest by adjusting the focal plane of the microscope

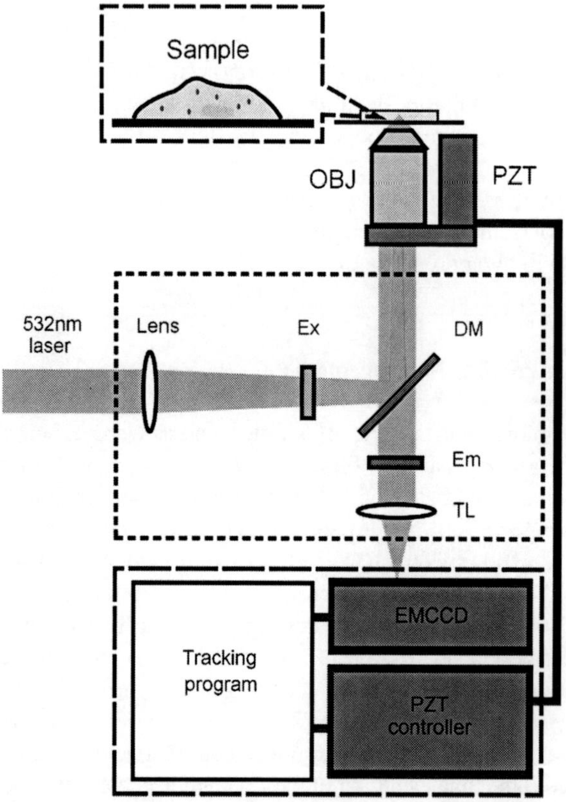

Fig. 6.3 Schematic of the experimental setup for three-dimensional tracking of single FNDs in live cells by wide-field fluorescence microscopy. Abbreviations: *OBJ* objective, *Ex* excitation filter, *DM* dichroic mirror, *Em* emission filter, *TL* tube lens, *PZT* piezoelectric translational stage. Adapted from [22]

objective mounted vertically on a *z*-motion piezoelectric translational stage (MCLS F100; Mad City Labs). With this feedback servo control system, it has been possible to monitor continuously the targeted fluorophores within the depth of field of the microscope objective for hours.

6.2.4 Biological Use of FNDs

6.2.4.1 Cell Culture

HeLa cells (~10^5 cells/mL) were cultured in Dulbecco's modified Eagle's medium (DMEM) supplemented with 10% fetal bovine serum. Suspensions of FNDs were first mixed with serum-free DMEM and diluted to desired concentrations. After sonication for 30 min to ensure the complete dispersion of FNDs in the medium, an aliquot (typically 0.2 mL) of the suspension was added to the wells of a chamber glass slide (Lab-Tek II, Nunc) containing the HeLa cells pre-cultured for 24 h. The chamber slide was then incubated at 37°C in a 5% CO_2 incubator for 5 h to facilitate cellular uptake. The FNDs not uptaken by the cells were removed by washing the sample thoroughly with warm phosphate buffer saline prior to flow cytometry analysis and fluorescence microscopy inspection.

6.2.4.2 Flow Cytometry

The cellular uptake of gFNDs by HeLa cells was characterized by flow cytometry to quantify the level of the nanoparticles uptaken by the cells as a function of particle concentration and incubation time. In these measurements, growing HeLa cells were first plated in a Petri dish for 16 h and then incubated with gFNDs at a concentration of 1–100 μg/mL in serum-free DMEM at 37°C. After incubation for a time period of 1–27 h, the cells were detached from the dish surfaces by treatment with trypsin at 37°C for 1 min, followed by washing three times with warm phosphate buffer saline to remove the exterior gFNDs. A fluorescence activated cell sorter (FACS Calibur, Becton Dickinson), operated at the laser excitation wavelength of 488 nm and the fluorescence detection window of 515–545 nm (i.e. the FL1 channel), analyzed the cells individually. A minimum of 10,000 events per sample were analyzed.

6.2.4.3 Fluorescence Microscopy

Fluorescence images of FNDs internalized by HeLa cells were acquired using a fluorescence microscope (TE2000, Nikon) equipped with an electron multiplying CCD camera (iXon, Andor). With the electron multiplying gain set at 200, the typical CCD exposure time per pixel was 0.3 s when the particles were imaged under mercury lamp excitation. To enhance the fluorescence image quality of the cells, the nuclei were stained with Hoechst Dye 33258 (861405, Sigma) and excited with the UV lines (λ_{ex} = 330 – 380 nm) from the same mercury lamp. A Nikon oil immersion DIC objective lens (100×, NA 1.4, PlanApo) was further employed to obtain the differential interference contrast (DIC) images of the cells. Aside from the high performance CCD camera, the fluorescence microscope was also equipped with a laser scanning confocal system (C1, Nikon), a 488-nm Ar ion laser (or a 543-nm HeNe laser), and a filter set for collection of emission at the wavelength of 505–545 nm (or >590 nm) for fluorescence imaging of gFNDs (or rFNDs). Both the laser excitation and the emission collection were typically made through a 100× oil immersion objective with a numerical aperture of 1.4. To confirm that the FNDs were really internalized by the cells, not just attached to the cell surfaces, optical cross sectional images of the cells were taken. The typical spacing of the optical sections was ~0.25 μm.

6.3 Results and Discussion

6.3.1 Mass-produced FNDs

A key step to promote wide applications of FND in biological research is to produce this nanomaterial in an economic and scalable manner. Conventionally, the vacancies in diamond are created by bombarding the material at room temperature with a high energy electron beam (typically 2 MeV) from a van der Graaff accelerator [46, 56] or with a high energy proton beam (typically 3 MeV) from a tandem particle accelerator [20, 29]. These vacancies become mobile at high temperatures (ca. above 600°C) and are subsequently trapped by substitutional nitrogen atoms to form N-V centers during annealing. The need for such highly sophisticated and costly equipments to create vacancies, however, severely hampers the easy availability of FNDs. To overcome this hurdle, our group has been exploring the possibility of using a medium energy, low-mass ion beam as the damage agent [22, 23]. It involves the use of a high fluence 40-keV He^+ beam generated by an RF ion source to irradiate the nanodiamond particles. As described in detail in Sect. 6.2.2.2, the current delivered by this ion source is typically in the range of 10 μA, which are about two orders of magnitude higher than that of the 3-MeV H^+ beam. Mass production of FNDs is thus possible and practical with the medium energy ion beam irradiation.

To ensure that the mass-produced FNDs are of high quality in terms of brightness, we compare the fluorescence spectra of rFNDs prepared by 40-keV He^+ and 3-MeV H^+ irradiations. Figure 6.4 shows an ensemble emission spectrum (λ_{max} = 680 nm) of 35-nm rFNDs prepared with the 40-keV He^+ irradiation and excited with a cw 532 nm laser for particles in a suspension (inset of Fig. 6.4). The spectrum reveals that there are two types of N-V centers in this type Ib diamond material: $(N\text{-}V)^0$ with a zero-phonon line at 575 nm and $(N\text{-}V)^-$ with a

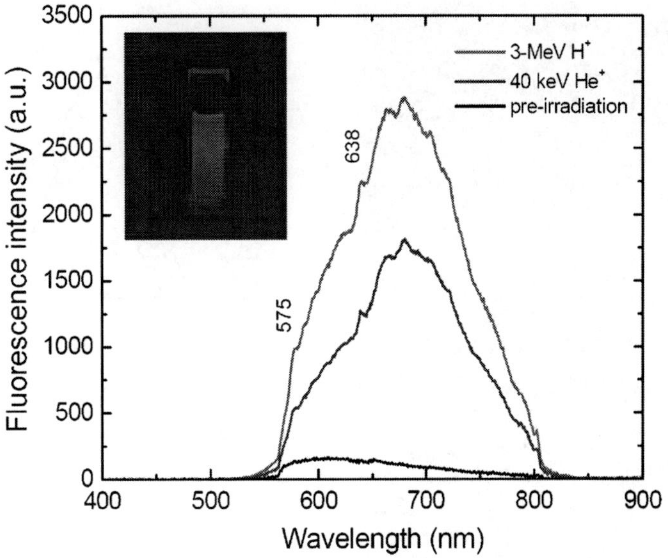

Fig. 6.4 Fluorescence spectra of 35-nm rFNDs, prepared with either 40-keV He^+ or 3-MeV H^+ irradiation. *Inset*: a photo showing the fluorescence image of a cuvette containing rFND suspension (1 mg/mL) excited by a 532-nm laser. Adapted from [22]

zero-phonon line at 638 nm (cf. Table 6.1). Comparing it with the spectrum of another sample from the same batch of crude diamond particles but prepared with the 3-MeV H^+ irradiation indicates that the fluorescence spectra of the rFNDs prepared by these two drastically different conditions are essentially the same, except that their intensities differ by ~30%. This difference is acceptable for the majority of our applications. Similar results were obtained for the gFNDs, which yielded green fluorescence at 530 nm from the H3 centers upon laser excitation at 473 nm [23].

6.3.2 *Multicolored FNDs*

6.3.2.1 Red FNDs

Figure 6.5a shows a bright field image of neat 140-nm rFNDs deposited on a coverglass slide [20]. Exposure of the particles and their aggregates to the yellow lines ($\lambda_{ex} = 510-560$ nm) from a 100 W mercury vapor lamp yields intense red emission (Fig. 6.5b). Unlike that of 35-nm rFNDs, which show spectral features characteristic of both $(N\text{-}V)^0$ and $(N\text{-}V)^-$ (Fig. 6.4), the spectrum of the 140-nm particles is predominated by the $(N\text{-}V)^-$ center (Fig. 6.5c). No photobleaching occurred after 8 h of continuous excitation with the mercury lamp (Fig. 6.5d). By contrast, the 0.1-μm red fluorescent polystyrene nanospheres (F8801, Molecular Probes), with excitation/emission maxima of 580/605 nm and containing ~10^4 dye equivalents,

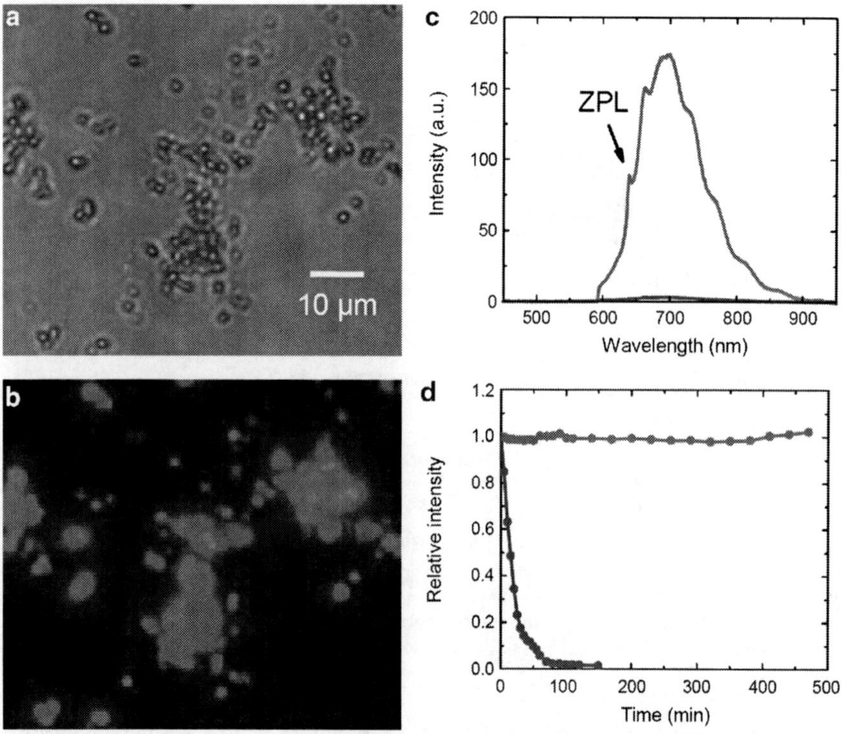

Fig. 6.5 Microscopic inspection and photostability test of 140-nm rFNDs and their aggregates. (**a**) Bright-field and (**b**) epifluorescence images of rFNDs. Both images were obtained with a 40× objective. (**c**) Fluorescence spectra of annealed nanodiamonds with (*red*) or without (*blue*) proton beam irradiation. The excitation was made with mercury lamp light at λ_{ex} = 510–560 nm and the emission was collected at λ_{em} > 590 nm. (**d**) Photostability tests of rFND (*red*) and 0.1-µm fluorescent polystyrene nanospheres (*blue*) excited under the same conditions. The fluorescence intensity was obtained by integrating over the wavelength range of 590–900 nm for each sample. Adapted from [20]

photobleached within 0.5 h under the same excitation conditions. The comparison clearly demonstrates the superb photostability of rFNDs, which are ideal for long-term imaging and tracking applications.

6.3.2.2 Green FNDs

In Fig. 6.6, we also compare the fluorescence spectra of 140-nm and 70-nm gFNDs prepared in the same way as rFNDs by the 40-keV He^+ beam irradiation [23]. The spectra were obtained for particles suspended in water (inset of Fig. 6.6). Both the gFND particles exhibit a spectrum contributed predominantly by the H3 centers, showing a prominent peak at 530 nm. However, an additional broader and weaker

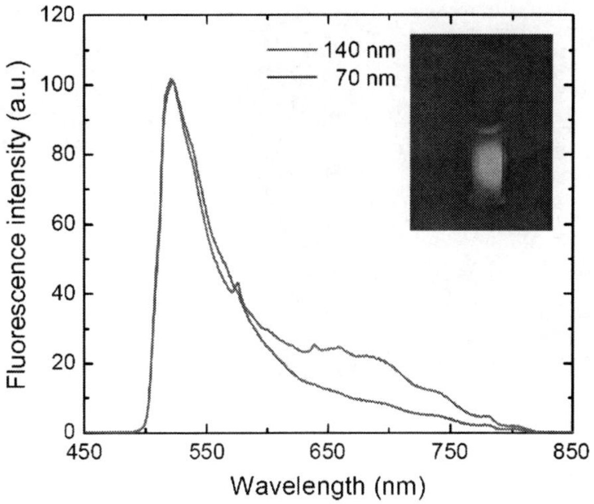

Fig. 6.6 Fluorescence images and spectra of gFNDs with sizes in the range of 70 nm and 140 nm. Both the samples are prepared by 40-keV He^+ ion beam irradiation and 800°C annealing. *Inset*: Photo showing the fluorescence image of a cuvette containing gFND suspension excited by a 473-nm laser. Adapted from [23]

band spanning from 600 nm to 800 nm, contributed by the $(N\text{-}V)^-$ centers, can also be found for the 140-nm particles. The appearance of the latter is not too surprising since natural diamonds are often mixtures of type Ia and type Ib materials [62].

A bright-field image of neat 70-nm gFNDs deposited on a coverglass slide is shown in Fig. 6.7a. Intense green emission emerges when the particles are excited with the blue lines (λ_{ex} = 450–490 nm) from a mercury lamp (Fig. 6.7b). Similar to that of rFNDs, no sign of photobleaching was found for the gFNDs even after 5 h of continuous excitation (Fig. 6.7c). In comparison, the 0.1-μm green fluorescent polystyrene beads (FluoSpheres yellow-green, Molecular Probes), with excitation/emission maxima of 505/515 nm, photobleached within 10 min under the same excitation conditions. It indicates that the excellent photostability is preserved by the gFNDs, which are also well suited for long-term imaging and tracking applications.

6.3.3 *Single FNDs*

6.3.3.1 Fluorescence Spectra

As illustrated in the introduction section, an aim of this work is to develop high brightness FNDs as cellular markers and so their imaging can be performed promptly at the single particle level in cells. Fig. 6.8a displays a confocal scanning image of 35-nm rFNDs dispersed on a bare glass substrate with a density of ~5 particles per 10 × 10 μm². The full width at half maximum (FWHM) of each peak

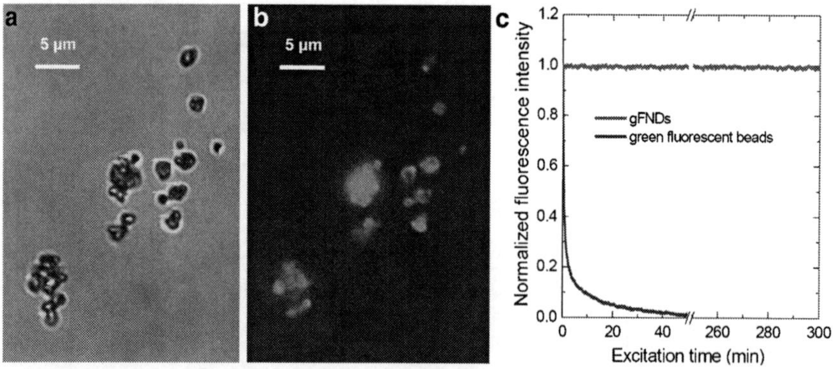

Fig. 6.7 Microscopic inspection and photostability test of 70-nm gFNDs and their aggregates. (**a**) Bright-field image, (**b**) epifluorescence image, and (**c**) time traces of gFNDs and 0.1-μm green fluorescent polystyrene beads excited under the same conditions. The excitation was made with mercury lamp light (λ_{ex} = 450–490 nm) and the resulting fluorescence was collected over the wavelength range of λ_{em} = 505–545 nm. Images in (**a**) and (**b**) were both obtained with a 100× microscope objective. Adapted from [23]

in the image is 2–3 pixels, which correspond to a physical distance of 400–600 nm, in coincidence with the diffraction limit of the optical microscope. The observation of these diffraction limited spots strongly suggests that they are derived from single isolated FND particles. The suggestion is confirmed independently by scanning electron microscopy, as reported in [21].

The implication of single particle detection is also supported by dispersed fluorescence studies. As shown in Fig. 6.8b, each 35-nm rFND exhibits a distinct spectrum in the wavelength region of 550–800 nm, a characteristic consistent with single particle detection. While the observed fluorescence spectra are markedly different in profile, two distinct ZPLs from the $(N\text{-}V)^0$ and $(N\text{-}V)^-$ centers are always identifiable for all particles interrogated. It is noted that the fluorescence of the rFNDs occurs in the red region, where both scattering and autofluorescence are much reduced and photons are allowed for deeper penetration into cells and tissues [63]. The good signal-to-noise ratio of the spectra thus warrants imaging of single rFND particles in these complex biological environments.

6.3.3.2 Photostability and Fluorescence Brightness

An outstanding feature that distinguishes FND from other fluorophores is the extreme photostability. To demonstrate the photostability on a single-particle basis, we monitored the fluorescence intensities of the individual 35-nm rFNDs over an extended period of time (Fig. 6.9a). In accord with the earlier finding for an ensemble of 140-nm rFNDs (Fig. 6.5d), excellent photostability was observed for the individual 35-nm particles. Under the excitation with 532-nm light at a power density of 8×10^3 W/cm^2, the fluorescence intensity of the rFND investigated stays essentially

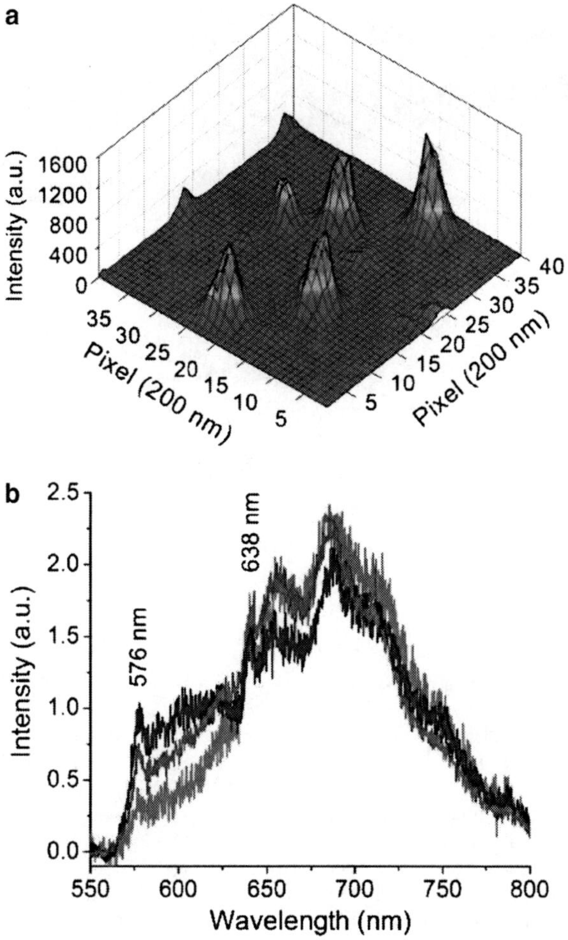

Fig. 6.8 Fluorescence images and spectra of single 35-nm rFNDs. (**a**) Confocal scanning image of 35-nm rFNDs dispersed on a coverglass slide. The FWHM of each peak is 2–3 pixels, corresponding to a physical distance of 400–600 nm. (**b**) Fluorescence spectra of three different 35-nm rFND particles. Adapted from [21]

the same over a time period of 300 s. No sign of photoblinking was detected within the time resolution of 1 ms for both 35-nm and 140-nm rFNDs (Fig. 6.9b). In contrast, single dye molecules such as Alexa Flour 546 covalently linked to dsDNA blinked randomly and photobleached rapidly within 12 s (Fig. 6.9a). The exceptional photostability of the rFNDs is also reflected in the power dependence measurement for the fluorescence intensity. As shown in Fig. 6.9c, the observed intensity scales nearly linearly with the laser power density from 1×10^2 to 1×10^5 W/cm^2 and gradually reaches its saturation at 1×10^6 W/cm^2. Notably, both the rFNDs retain their fluorescence characteristics even under such a high-power laser excitation for several minutes [21].

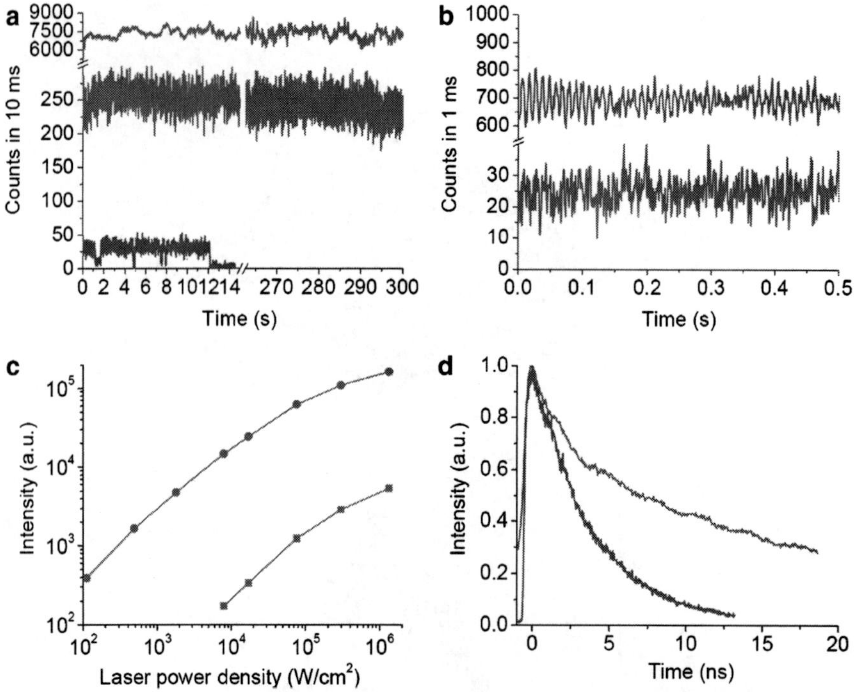

Fig. 6.9 Spectroscopic characterization of 35-nm and 140-nm rFNDs. (**a**) Typical time traces of the fluorescence from a single 140-nm rFND (*green*), a single 35-nm rFND (*red*), and a single Alexa Fluor 546 dye molecule attached to a single dsDNA molecule (*blue*). (**b**) Time traces of the fluorescence from a single 140-nm rFND (*green*) and a single 35-nm rFND (*red*) acquired with a time resolution of 1 ms, showing no photoblinking behavior. (**c**) Plot of the fluorescence intensity as a function of the laser power density over the range of 1×10^2–1×10^6 W/cm^2 for 35-nm (*red*) and 140-nm (*green*) rFNDs. (**d**) Fluorescence lifetime measurements of 140-nm rFNDs (*green*) and Alexa Fluor 546 dye molecules (*blue*). Fitting the time traces with two exponential decays for the rFND reveals a fast component of 1.7 ns (4%) and a slow component of 17 ns (96%). The latter is ~4 times longer than that (~4 ns) of Alexa Fluor 546. Adapted from [21]

It is instructive to compare the fluorescence brightness of FNDs, dye molecules, and quantum dots under the same experimental conditions. From a measurement of 30 molecules or particles for each sample, we estimated that the average fluorescence intensity of the 35-nm rFNDs is higher than that of the dye molecules such as Alexa Flour 546 by roughly one order of magnitude, but is about a factor of 2 lower than that of quantum dots such as CdSe/ZnS emitting light in the same wavelength region. We have recently performed photon correlation spectroscopy measurements to quantify precisely the number of $(N\text{-}V)^-$ centers in the individual rFND particles [64]. We determined that a single rFND particle with an average diameter of 28 nm can contain up to 8 $(N\text{-}V)^-$ centers. This number of color centers translates to a concentration of 4 ppm, which differs from the typical concentration (~100 ppm) of nitrogen in type Ib diamond by about 25-folds. Such a discrepancy suggests that there is still plenty of room for further improvement of the brightness of the rFNDs by properly choosing the crude diamond material and optimizing the sample preparation conditions. The bioimaging application would benefit tremendously from this 25-fold enhancement in fluorescence intensity.

6.3.3.3 Fluorescence Lifetimes

Another parameter that can be manipulated to enhance the image contrast of single FNDs in biological environments is the fluorescence lifetime. Figure 6.9d displays the result of fluorescence lifetime measurements for 140-nm rFNDs on a glass plate. The displayed fluorescence decay was obtained by co-adding the time traces of 30 individual particles. The major component of the decay has a lifetime of 17 ns, which is close to that (11.6 ns) measured for bulk diamonds [49] but is substantially longer than that of dye molecules (~4 ns for Alexa Flour 546 in Fig. 6.9d). This distinct difference in fluorescence lifetime allows for easy isolation of the rFND's emission from other backgrounds, such as cellular endogenous fluorescence, using various time-gating techniques including the fluorescence lifetime imaging microscopy (FLIM) [65].

6.3.4 Applications of FNDs as Cellular Markers

6.3.4.1 Cellular Uptake

To demonstrate that FND is a promising biomarker candidate for in vivo imaging and diagnosis, 35-nm rFNDs were incubated with HeLa cells cultured in DMEM at 37°C on a chamber slide. For the HeLa cell, intense cell autofluorescence was observed at 510–560 nm when it was exposed to blue light. Switching the excitation wavelength to 532 nm and collecting the emission at 650–720 nm, where the rFND fluorescence resides, greatly reduce the background signals [21]. Displayed in Fig. 6.10 are the fluorescence images of a HeLa cell after incubation with 35-nm rFNDs. Many particles are seen to be internalized by the cell, confirmed by axial cross-sectional imaging of the specimen, and they appear as bright red spots in the fluorescence image. It is noticed that most of the internalized rFNDs form aggregates in the cytoplasm and do not enter the cell nucleus as a result that the 35-nm particles are too large to cross nuclear pores [66].

In addition to rFNDs, gFNDs are also useful as cellular markers for bioimaging applications. Figure 6.11a and b display the DIC and fluorescence images of a HeLa cell after incubation with 70-nm gFNDs. With the cell nucleus stained in blue, the internalized gFNDs can be seen quite clearly in the cytoplasm. To attain a more quantitative understanding for the cellar uptake process, we analyzed the level of gFNDs internalized by the individual live cells with flow cytometry [67]. For the gFND, it has a fluorescence spectrum matching closely into the green detection channel (e.g. 530 ± 15 nm) of flow cytometers available commercially and,

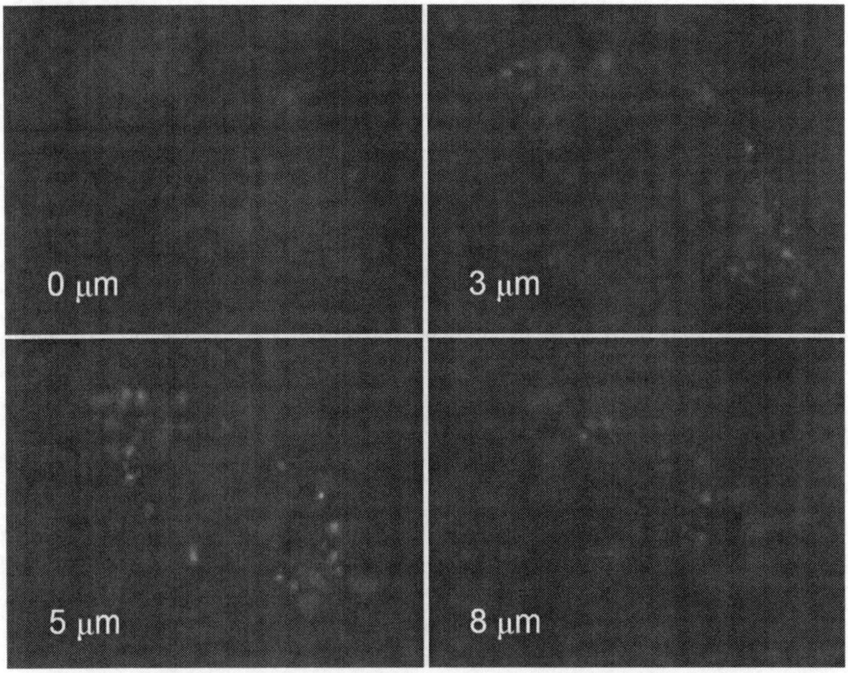

Fig. 6.10 Vertical cross section scans (0–8 μm) of the wide-field epifluorescence image of a single HeLa cell after incubation with 35-nm rFNDs. No rFNDs could be found in the central region where the nucleus resided. The particles reappeared when the vertical scan passed across this region (3–5 μm). Adapted from [21]

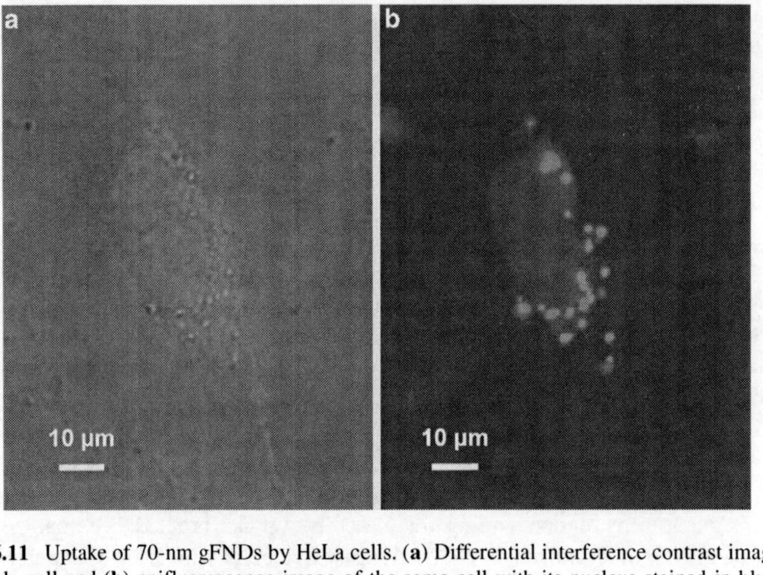

Fig. 6.11 Uptake of 70-nm gFNDs by HeLa cells. (**a**) Differential interference contrast image of a single cell and (**b**) epifluorescence image of the same cell with its nucleus stained in blue by Hoechst Dye 33258. In this particular experiment, the cells were incubated with 70-nm gFNDs for 5 h at a particle concentration of 10 μg/mL. Adapted from [23]

therefore, is amenable to this type of analysis. Figure 6.12a shows a typical result of the time-dependent flow cytometry analysis for HeLa cells incubated with gFNDs at a concentration of 50 μg/mL in serum-free DMEM. The mean of the fluorescence intensity measurements increases steadily with the incubation time and gradually levels off at the fifth hour, with a particle uptake half-life of ~3 h (Fig. 6.12b). The observation is consistent with the mechanism of endocytosis, which has been identified for 74-nm gold nanoparticles [68, 69] and 150-nm mesoporous silica nanoparticles [70] uptaken by the same human cervical cancer cells. It is noted that while the gFND is not so well suitable for the bioimaging applications because of the high autofluorescence background in the wavelength of 500 nm region [71], time-resolved techniques [29, 65] can be fruitfully exploited to enhance the image contrast.

6.3.4.2 Single Particle Imaging and Tracking

The essence of using fluorescent nanoparticles as cellular markers is to probe the complex cellular machinery at the single particle level without interfering much with the associated biochemical processes. For the rFNDs, we have been able to detect the particles as small as 35 nm individually in live cells. Shown in Fig. 6.13a and b are the images of 35-nm rFNDs internalized by the HeLa cell [21]. While many rFNDs are found to form aggregates in the cytoplasm, some isolated diamond nanoparticles can be detected. These particles, marked in yellow boxes in the figures, are identified as single rFNDs because the spot sizes of their images are diffraction limited (Fig. 6.13c) and, moreover, their fluorescence intensities are

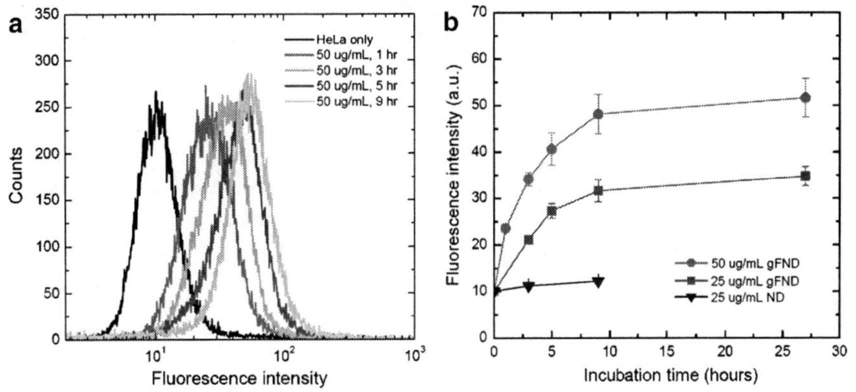

Fig. 6.12 Flow cytometry analysis of 70-nm gFNDs uptaken by HeLa cells as a function of incubation time. (**a**) Fluorescence intensity histograms of the uptaken gFNDs and (**b**) arithmetic means of the fluorescence intensity measurements for the uptaken gFNDs at two different particle concentrations. *ND* stands for nanodiamond particles without the irradiation/annealing treatment. Adapted from [23]

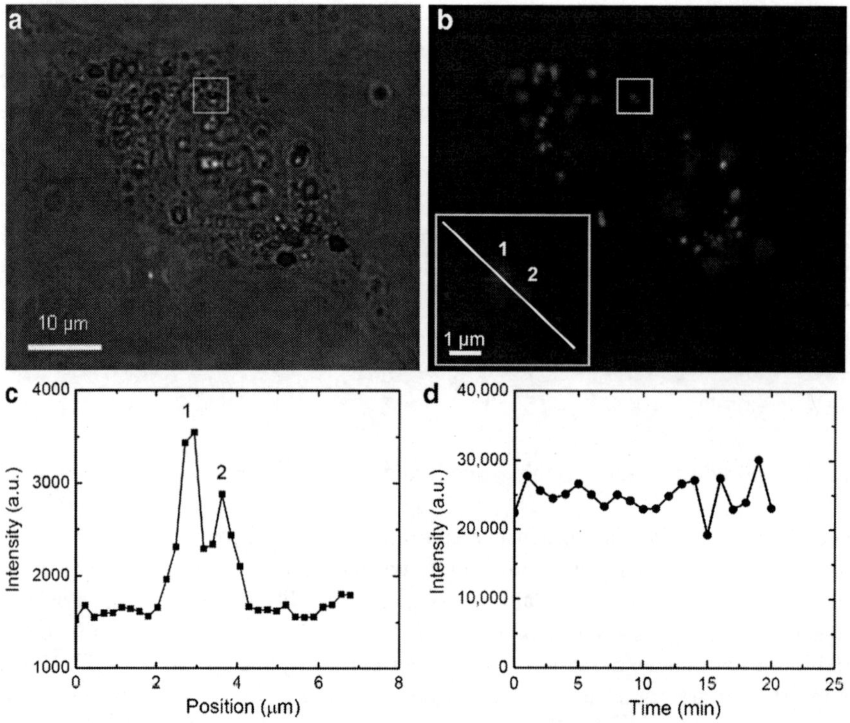

Fig. 6.13 Observation of single 35-nm rFNDs in HeLa cells. (**a**) Bright-field and epifluorescence images of a HeLa cell after uptake of 35-nm rFNDs. Most of the uptaken rFNDs are seen to distribute in the cytoplasm. (**b**) Epifluorescence image of a single HeLa cell after the rFND uptake. Inset: An enlarged view of two fluorescence spots (denoted by "1" and "2") with diffraction-limited sizes (FWHM ~ 500 nm). The separation between these two particles is ~1 μm. (**c**) Intensity profile of the fluorescence image along the line drawn in the inset of (**b**). (**d**) Integrated fluorescence intensity as a function of time for particle "1". The signal integration time was 0.1 s. Adapted from [21]

comparable to that of single rFNDs spin-coated on the glass plate. It is of importance to note that these internalized rFNDs are photostable even after continuous excitation of the sample for 20 min at a laser power density of 100 W/cm^2 (Fig. 6.13d). Similar to the earlier findings, neither photobleaching nor photoblinking was observed for these rFNDs within the limits of our detection sensitivity and time window.

The excellent photostability and high brightness of the nanomaterial make it possible to conduct long-term, three-dimensional tracking [72–74] of a single rFND in a live cell. The technique offers great potential to reveal details of cellular dynamics and intracellular activities, such as drug delivery and virus infection [75], in real time. Displayed in the left panel of Fig. 6.14 is an overlap of bright-field and fluorescence images of a live HeLa cell after uptake of 35-nm rFNDs at a low concentration. Following the procedures described earlier, a single rFND (marked with

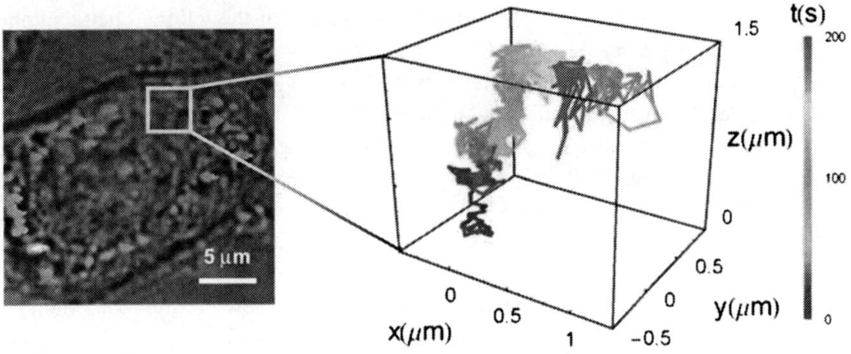

Fig. 6.14 Three-dimensional tracking of a single 35-nm rFND in a live cell. Overlay of bright-field and epifluorescence images (*left*) shows the existence of 35-nm rFNDs in the cytoplasm of a live HeLa cell. The fluorescence intensity is sufficiently high and stable to allow tracking of a single rFND in three dimensions over a time span of 200 s (*right*). Adapted from [22]

a yellow box) can be readily identified in the cytoplasm. By using the homebuilt servo control system and operating the fluorescence microscope in the widefield mode, we were able to track this particular particle over a time span of more than 200 s [22]. From a mean square displacement analysis [76] of the three-dimensional trajectory (the right panel in Fig. 6.14), a diffusion coefficient of $3.1 \times 10^{-3}\,\mu m^2/s$ was determined for the internalized 35-nm rFND particle. This value, in good agreement with the reported diffusion coefficients for quantum dots confined within an endosome [73], indicates that the particle was not moving freely in the cytoplasm.

6.4 Conclusion and Future Work

FND is a new nanomaterial that possesses several unique properties including good biocompatibility, excellent photostability, and facile surface functionalizability. Depending on the type of N-V defect centers embedded in the crystal lattice, the material can emit multicolor fluorescence upon photoexcitation with visible light (cf. Table 6.1). Of particular interest is the $(N\text{-}V)^-$ center, which emits tissue penetrating red photons and is well suited for bioimaging applications [63]. It is demonstrated that the fluorescence intensity of the rFNDs is sufficiently high and stable to allow the three-dimensional tracking of a single particle in cells by widefield fluorescence microscopy. These excellent photophysical characteristics, together with the feasibility of scale-up production of the material, make FND (both red and green) an ideal nanoprobe for long-term, high temporal as well as high spatial resolution imaging applications in biology and nanoscale medicine.

Our future experiments will be focused on the production of smaller and brighter FNDs in a large scale for biomedical use. Particles as small as 5 nm in diameter are highly desirable when used as fluorescent biolabels and, fortunately, the raw materials are now readily available [77]. We will take advantage of the unique chemical and photophysical properties of these FNDs and applying them as a novel carrier of bioactive compounds for targeted delivery and controlled release in live cells and animals. Specifically, we will use medicine loaded FNDs as light beacons in chemotherapy so that these medicines are ensured to be accurately delivered to the target zone. Researches will also be carried out to determine the animal toxicity of FNDs and how long these particles will remain in whole animals before dissipating. Clearly, there are still many things to explore with this novel nanomaterial [78]. It is anticipated that the surface functionalized fluorescent nanodiamonds will meet up with several biological and medical needs in the next few decades to come.

Acknowledgment The author thanks Y.-R. Chang and H.-Y. Chou for their assistance in preparing this manuscript. This research was supported by the Academia Sinica and the National Science Council (Grant No. NSC 96-2120-M-001-008-) of Taiwan, ROC.

References

1. Niemeyer CM (2001) Angew Chem Int Ed Engl 40:4128–4158
2. Sahoo SK, Labhasetwar V (2003) Drug Discov Today 8:1112–1120
3. Ferrari M (2005) Nature Rev Cancer 5:161–171
4. Alivisatos P (2004) Nat Biotechnol 22:47–52
5. Bharali DJ, Klejbor I, Stachowiak EK, Dutta P, Roy I, Kaur N, Bergey EJ, Prasad PN, Stachowiak MK (2005) Proc Natl Acad Sci U S A 102:11539–11544
6. Michalet X, Pinaud FF, Bentolila LA, Tsay JM, Doose S, Li JJ, Sundaresan G, Wu AM, Gambhir SS, Weiss S (2005) Science 307:538–544
7. Vosch T, Antoku Y, Hsiang JC, Richards CI, Gonzalez JI, Dickson RM (2007) Proc Natl Acad Sci U S A 104:12616–12621
8. Medintz IL, Uyeda HT, Goldman ER, Mattoussi H (2005) Nat Mater 4:435–446
9. Derfus AM, Chan WCW, Bhatia SN (2004) Nano Lett 4:11–18
10. Yao J, Larson DR, Vishwasrao HD, Zipfel WR, Webb WW (2005) Proc Natl Acad Sci U S A 102:14284–14289
11. Cui BX, Wu CB, Chen L, Ramirez A, Bearer EL, Li WP, Mobley WC, Chu S (2007) Proc Natl Acad Sci U S A 104:13666–13671
12. Field JE (ed) (1992) Properties of Natural and Synthetic Diamond. Academic, London, UK
13. Nebel CE, Shin DC, Rezek B, Tokuda N, Uetsuka H, Watanabe H (2007) J R Soc Interface 4:439–461
14. Holt KB (2007) Phil Trans R Soc A 365:2845–2861
15. Krueger A (2008) Chem Eur J 14:1382–1390
16. Dion I, Baquey C, Monties JR (1993) Int J Artif Organs 16:623–627
17. Yang W, Auciello O, Butler JE, Cai W, Carlisle JA, Gerbi JE, Gruen DM, Knickerbocker T, Lasseter TL, Russell JN Jr, Smith LM, Hamers RJ (2002) Nat Mater 1:253–257
18. Hartl A, Schmich E, Garrido JA, Hernando J, Catharino SCR, Walter S, Feulner P, Kromka A, Steinmuller D, Stutzmann M (2004) Nat Mater 3:736–742
19. Huang H, Pierstorff E, Osawa E, Ho D (2007) Nano Lett 7:3305–3314
20. Yu S-J, Kang M-W, Chang H-C, Chen K-M, Yu Y-C (2005) J Am Chem Soc 127: 17604–17605
21. Fu C-C, Lee H-Y, Chen K, Lim T-S, Wu H-Y, Lin P-K, Wei P-K, Tsao P-H, Chang H-C, Fann W (2007) Proc Natl Acad Sci U S A 104:727–732

22. Chang Y-R, Lee H-Y, Chen K, Chang C-C, Tsai D-S, Fu C-C, Lim T-S, Tzeng Y-K, Fang C-Y, Han C-C, Chang H-C, Fann W (2008) Nature Nanotech 3:284–288
23. Wee T-L, Mau Y-W, Fang C-Y, Hsu H-L, Han C-C, Chang H-C (2009) Diamond Relat Mater 18:567–573
24. Smith BR, Niebert M, Plakhotnik T, Zvyagin AV (2007) J Lumin 127:260–263
25. Chao J-I, Perevedentseva E, Chung P-H, Liu K-K, Cheng C-Y, Chang C-C, Cheng C-L (2007) Biophys J 93:2199–2208
26. Perevedentseva E, Cheng C-Y, Chung P-H, Tu J-S, Hsieh Y-H, Cheng C-L (2007) Nanotechnology 18:315102
27. Neugart F, Zappe A, Jelezko F, Tietz C, Boudou JP, Krueger A, Wrachtrup J (2007) Nano Lett 7:3588–3591
28. Vial S, Mansuy C, Sagan S, Irinopoulou T, Burlina F, Boudou JP, Chassaing G, Lavielle S (2008) ChemBioChem 9:2113–2119
29. Faklaris O, Garrot D, Joshi V, Druon F, Boudou J-P, Sauvage T, Georges P, Curmi PA, Treussart F (2008) Small 4:2236–2239
30. Kossovsky N, Gelman A, Hnatyszyn HJ, Rajguru A, Garrell RL, Torbati S, Freitas SSF, Chow G-M (1995) Bioconjug Chem 6:507–511
31. Ushizawa K, Sato Y, Mitsumori T, Machinami T, Ueda T, Ando T (2002) Chem Phys Lett 351:105–108
32. Huang L-CL, Chang H-C (2004) Langmuir 20:5879–5884
33. Krueger A, Liang YJ, Jarre G, Stegk J (2006) J Mater Chem 16:2322–2328
34. Chung P-H, Perevedentseva E, Tu J-S, Cheng C-L, Liu K-K, Chao J-I, Chen P-H, Chang C-C (2006) Diamond Relat Mater 15:622–625
35. Cheng C-Y, Perevedentseva E, Tu J-S, Chung P-H, Cheng C-L, Liu K-K, Chao J-I, Chen P-H, Chang C-C (2007) Appl Phys Lett 90:163903
36. Nguyen TTB, Chang H-C, Wu VW-K (2007) Diamond Relat Mater 16:872–876
37. Krueger A, Stegk J, Liang YJ, Lu L, Jarre G (2008) Langmuir 24:4200–4204
38. Yeap WS, Tan YY, Loh KP (2008) Anal Chem 80:4659–4665
39. Liu K-K, Chen M-F, Chen P-Y, Lee TJF, Cheng C-L, Chang C-C, Ho Y-P, Chao J-I (2008) Nanotechnology 19:205102
40. Jia G, Wang H, Yan L, Wang X, Pei R, Yan T, Zhao Y, Guo X (2005) Environ Sci Technol 39:1378–1383
41. Liu K-K, Cheng C-L, Chang C-C, Chao J-I (2007) Nanotechnology 18:325102
42. Schrand AM, Huang HJ, Carlson C, Schlager JJ, Osawa E, Hussain SM, Dai LM (2007) J Phys Chem B 111:2–7
43. Schrand AM, Dai LM, Schlager JJ, Hussain SM, Osawa E (2007) Diamond Relat Mater 16:2118–2123
44. Hall HT (1961) J Chem Educ 38:484–489
45. Davies G (ed) (1994) Properties and growth of diamond, emis datareviews series No. 9, INSPEC. The Institute of Electrical Engineers, London, Chap. 3.
46. Davies G, Hamer MF (1976) Proc R Soc Lond A 348:285–298
47. Jelezko F, Tietz C, Gruber A, Popa I, Nizovtsev A, Kilin S, Wrachtrup J (2001) Single Mol 2:255–260
48. Wee T-L, Tzeng Y-K, Han C-C, Chang H-C, Fann W, Hsu J-H, Chen K-M, Yu Y-C (2007) J Phys Chem A 111:9379–9386
49. Collins AT, Thomaz MF, Jorge MIB (1983) J Phys C 16:2177–2181
50. Rand SC (1994) In: Davies G (ed.), Properties and growth of diamond, emis datareviews series no. 9, INSPEC. The Institute of Electrical Engineers, London, Chap. 7.4.
51. Gruber A, Drabenstedt A, Tietz C, Fleury L, Wrachtrup J, von Borczyskowski C (1997) Science 276:2012–2014
52. Kurtsiefer C, Mayer S, Zarda P, Weinfurter H (2000) Phys Rev Lett 85:290–293
53. Beveratos A, Brouri R, Gacoin T, Villing A, Poizat JP, Grangier P (2002) Phys Rev Lett 89:187901

54. Dutt MVG, Childress L, Jiang L, Togan E, Maze J, Jelezko F, Zibrov AS, Hemmer PR, Lukin MD (2007) Science 316:1312–1316
55. Davies G, Lawson SC, Collins AI, Mainwood A, Sharp S (1992) Phys Rev B 46:13157–13170
56. Davies G, Nazare MH, Hamer MF (1976) Proc R Soc Lond A 351:245–265
57. Rand SC, DeShazer LG (1985) Opt Lett 10:481–483
58. Roberts WT, Rand SC, Redmond S (2005) NASA Tech Briefs NPO-30796.
59. Crossfield MD, Davies G, Collins AT, Lightowlers EC (1974) J Phys C 7:1909–1917
60. Ziegler JF, Biersack JP, Littmark U (1985) The stopping and range of ions in solids. Pergamon: New York. Free SRIM software (version 2003) is available from the website http://www.srim.org/
61. Osswald S, Yushin G, Mochalin V, Kucheyev SO, Gogotsi Y (2006) J Am Chem Soc 128:11635–11642
62. De Weerdt F, Van Royen J (2001) Diamond Relat Mater 10:474–479
63. Lim YT, Kim S, Nakayama A, Stott NE, Bawendi MG, Frangioni JV (2003) Mol Imaging 2:50–64
64. Hui YY, Chang Y-R, Lim T-S, Lee H-Y, Fann W, Chang H-C (2009) Appl Phys Lett 94:013104
65. Lim T-S, Fu C-C, Lee K-C, Lee H-Y, Chen K, Cheng W-F, Pai WW, Chang H-C, Fann W (2009) Phys Chem Chem Phys 11:1508–1514
66. Pante N, Kann M (2002) Mol Biol Cell 13:425–434
67. Muirhead KA, Horan PK, Poste G (1985) Nat Biotechnology 3:337–356
68. Chithrani BD, Ghazani AA, Chan WCW (2006) Nano Lett 6:662–668
69. Chithrani BD, Chan WCW (2007) Nano Lett 7:1542–1550
70. Slowing I, Trewyn BG, Lin VS-Y (2006) J Am Chem Soc 128:14792–14793
71. Billinton N, Knight AW (2001) Anal Biochem 291:175–197
72. Speidel M, Jonas A, Florin E-L (2003) Opt Lett 28:69–71
73. Holtzer L, Meckel T, Schmidt T (2007) Appl Phys Lett 90:053902
74. Cang H, Xu CS, Montiel D, Yang H (2007) Opt Lett 32:2729–2731
75. Greber UF, Way M (2006) Cell 124:741–754
76. Hong QA, Sheetz MP, Elson EL (1991) Biophys J 60:910–921
77. Morita Y, Takimoto T, Yamanaka H, Kumekawa K, Morino S, Aonuma S, Kimura T, Komatsu N (2008) Small 4:2154–2157
78. Vaijayanthimala V, Chang H-C (2009) Nanomed 4:47–55

Chapter 7
Nanodiamond-Mediated Delivery of Therapeutics via Particle and Thin Film Architectures

Houjin Huang, Erik Pierstorff, Karen Liu, Eiji Ōsawa, and Dean Ho

Abstract Due to their integrative properties that are conducive towards applications in nanomedicine, nanodiamonds (NDs) can serve as highly versatile and biocompatible carbon-based platforms for the controlled functionalization and delivery of a wide spectrum of therapeutic compounds (e.g. small molecule, protein/antibody, nucleic acid, etc.). This chapter explores the development, effectiveness, and potential of drug-functionalized ND materials (2–8 nm in diameter) via particle and thin film architectures for chemotherapeutic delivery. In this study, doxorubicin hydrochloride (Dox), an apoptosis-inducing cytotoxic drug widely used in chemotherapy, as well as Dexamethasone (Dex), a clinically applicable anti-inflammatory were successfully applied toward the functionalization of NDs and introduced towards murine macrophages (e.g. RAW 264.7) and human colorectal carcinoma cells (e.g. HT-29) with preserved cytotoxic efficacy. The adsorption of Dox onto the NDs and its reversible release were achieved by regulating Cl^- ion concentration, among other mechanisms and chemical treatment methodologies, and as such, ND particles were found to be able to efficiently ferry the drug inside a broad spectrum of cell lines. Moreover, novel ND thin films were developed by assembling detonation NDs dispersed in aqueous solution into a closely packed ND multilayer nanofilm with positively charged poly-L-lysine (PLL) via the layer by layer (LBL) deposition technique. Comprehensive assays were performed to quantitatively assess and confirm inherent ND biocompatibility in both the particle and thin film architectures via cellular gene expression examination by real time polymerase chain reaction (RT-PCR), DNA fragmentation assays, and mitochondrial function (MTT) analysis, confirming the functional apoptosis-inducing and inflammation-suppressing mechanisms driven by the Dox-functionalized NDs and Dex-functionalized NDs, respectively. The relevance of the Dox–ND composites has been extended toward a translational context, where MTT assays were performed on the HT-29 colon cancer cell line to examine Dox–ND-induced cell death and ND-mediated chemotherapeutic surface sequestering for potential slow- and sustained-release capabilities. Additionally, the

Dean Ho (✉)
Northwestern University, Evanston, IL, USA
d-ho@northwestern.edu

D. Ho (ed.), *Nanodiamonds: Applications in Biology and Nanoscale Medicine*,
DOI 10.1007/978-1-4419-0531-4_7, © Springer Science+Business Media, LLC 2010

functionality of the Dex-ND films was assessed via interrogation of the suppression of inflammatory cytokine release. Suppression of lipopolysaccharide-mediated inflammation was observed through the potent attenuation of tumor necrosis factor alpha (TNFα), interleukin-6 (IL-6), and inducible nitric oxide synthase (iNOS) levels following ND thin film interfacing with RAW 264.7 murine macrophages. Furthermore, basal cytokine secretion levels examined innate material compatibility, revealing unchanged quantitative cellular inflammatory responses, which strongly support the non-toxicity and relevance of the NDs as effective treatment platforms for nanoscale medicine. In addition to their straightforward/facile preparation, robustness/resistance to delamination, and fine controllability of the film structures, these hybrid materials possess enormous potential towards biomedical applications such as localized drug delivery and anti-inflammatory implant coatings and devices, as demonstrated in vitro in this chapter.

7.1 Introduction

The concept of applying nanomaterials towards a broad array of applications including therapeutic release and the diagnosis and treatment of a broad range of physiological disorders has attracted much attention [1–37]. These novel technologies are often based upon the use of natural or artificial nanostructures for biological applications and the fabrication of biofunctional devices from the integration of these nanostructures [2, 3]. Among the many classes of nanoscale building blocks that are being explored for biological applications, carbon-based nanomaterials (e.g., fullerene, carbon nanotubes and NDs) are receiving significant attention due to their remarkable physical, chemical and biological properties [4–6].

Regarding clinical applications, the controlled delivery and elution of therapeutic agents are often important as their tailored dosing is vital toward the reduction of major side effects and complications (including infection, extravasation, and even patient mortality), given certain characteristics such as nonspecific mechanisms of action [7, 8]. The application of nanoparticle-based vehicles as multifunctional, versatile, and biocompatible drug carriers would then serve as a potentially ideal approach toward enhancing the specificity of therapy. Their many advantages include, but are not limited to the ability to target specific locations in the body to increase treatment specificity, the reduction of the overall quantity of drug used at the active dosing site which can significantly reduce patient toxicity and impact upon the immune system, and the potential to reduce the concentration of the drug at healthy and unaffected sites, resulting in fewer side effects that can complicate the efficacy of treatment [1, 9]. Due to the advantages of nanoparticle-based chemotherapeutic delivery, the prevention of generalized drug release, which is also a key mechanism for drug-induced medical treatment complications, may then be realized, dramatically improving the prospects for high-efficacy/specificity and high biocompatibility treatment that is clinically significant. To bring this concept to fruition, efforts have been devoted to the intelligent/rational design and synthesis of novel nanostructured materials [10–13]. With respect to the application of nanocarbon platforms in a biological or medical context, the scientific community has been primarily focused on fullerene and carbon nanotube-based modalities, while the activation of sub-cellular processes following cell–nanocarbon interaction receives continued evaluation [14, 15]. In particular, single-walled carbon nanotubes (SWNTs) have been recently explored as drug delivery agents [16–20], as well as potential cancer cell-targeting therapeutic and imaging agents via infrared therapy [21]. However, they have also been previously shown to induce oxidative stress in human keratinocytes via the nuclear factor-kappa B (NF-κB) transcription factor [22].

Because of their integrative physical properties and biocompatibility, diamond-based nanomaterial platforms have emerged as promising materials for biomedical applications. These include the use of diamond films for robust implant coatings and fluorescent NDs (NDs, 35 or 100 nm) as stable biomarkers [23–25]. Several of these approaches have been based on polycrystalline diamond films or diamond particles with a relatively large grain size produced by chemical vapor deposition (CVD). To generate ND powder-based particles with a much smaller size (<10 nm in diameter), the detonation technique has also been applied [26]. Recent progress in the dispersion of detonation NDs (2–8 nm) in aqueous media, developed by Osawa and colleagues, has facilitated the use of NDs in physiological solutions [27]. Furthermore, the biocompatibility of detonation NDs has previously been assessed, where MTT assays and luminescent ATP production showed that the NDs are not toxic to a broad spectrum of cell types [28]. Compared to other nanocarbon materials such as carbon nanotubes (CNTs), which are toxic in some studies and naturally not water-soluble, it is thus observed that NDs can serve as an important, scalably processed material for biomedical applications [29]. While detonation NDs have been used to adsorb proteins [29–31], these studies do not address the broad range of drug carrying and releasing applications of which this material can serve as a foundation. Furthermore, critical testing of the innate biocompatibility of NDs through gene expression studies, which provide quantitative insight into ND–cell interactions, was not performed in these earlier studies, even though preliminary studies of cell viability were examined.

In this context, this chapter discusses the efficiency of ND particles as foundations for chemotherapeutic drug carriers and anti-inflammatory interfaces. Doxorubicin hydrochloride (Dox), an apoptosis-inducing drug widely used in chemotherapy, was successfully coated on to ND surfaces and/or embedded into the crevices of ND aggregates, and subsequently introduced into living cells. NDs were internalized quickly, transferring the drug inside living cells efficiently. In addition, quantitative gene expression assays were performed, confirming the innate biocompatibility of the NDs via real time polymerase chain reaction (RT-PCR) and electrophoretic studies of DNA fragmentation to provide deeper insight into cellular responses to ND interaction.

With regards to nanofilm architectures, current techniques of producing ultra-nanocrystalline diamond thin films via CVD processing have also yielded important fundamental materials studies as well as application-relevant coatings. CVD diamond deposition typically involves a gas-phase chemical reaction using specialized instrumentation (reactors, vacuum furnaces, heaters and plasma generators, etc.) and specific reaction conditions such as precisely controlled gas flow, high vacuum parameters and high temperatures (>1,000 K) [32]. Alternatively, diamond thin films can also be assembled from a fluorine-functionalized detonation ND powder, followed by generating an ND coating on a glass surface through the reaction of a fluorinated ND powder with an amino-functionalized glass surface [33, 34]. For biomedical applications, an additional consideration pertains to whether or not the generated diamond film, which is typically very hard, is compatible with soft tissues in the living body, and if the material-biology interface induces an inflammatory response Addressing this element of material development is important in that adverse immune response can reduce the lifetime of an implanted device, resulting in additional medical procedures (e.g. replacement, etc.). As such, the development of additional approaches that utilize mild coating conditions, a low-cost and scalable fabrication methodology, and the engineering of bio-functionality (e.g. drug carrying) into the composite nanofilm system is of significant importance.

To address this need, this chapter also describes how single and multilayer diamond films were assembled from dispersible nontoxic NDs with controllable thickness via a soft coating layer-by-layer (LBL) deposition technique (Fig. 7.1). The LBL method introduced by Decher and co-workers in the early 1990s has attracted extensive attention [35, 36]. The advantages of this technique for biomedical applications include ease of preparation/scalability, versatility with respect to controlling substrate size and topology. Deposition upon a broad array of surfaces, the capability of incorporating a high loading capacity of different types of biomolecules in the films, fine control over the material structure, and robustness of the products under ambient and physiological conditions for sustained activity [37]. This chapter also examines the biocompatibility of NDs in both free-floating and LBL thin film

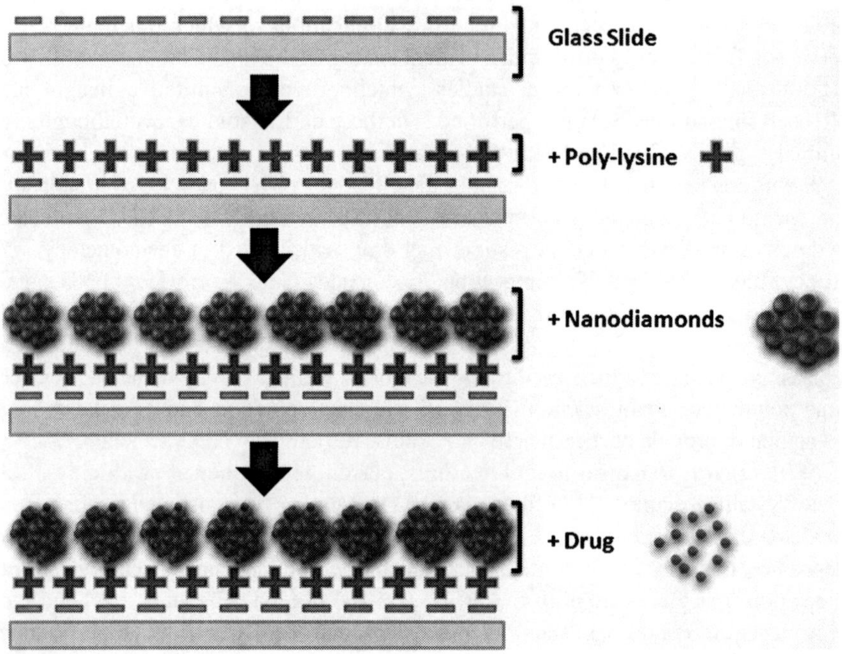

Fig. 7.1 Schematic drawing of ND nanofilm formation and drug incorporation into the film. Poly-L-lysine is used to attract NDs onto the glass surface

structures at the gene expression level via RT-PCR, in which pro-inflammatory cytokine levels were examined. This served as an important interrogation of the ND bio-interfacial phenomena generated by inherent material properties in both the particle and thin film approaches, with the results indicating their potential for medical or clinical applicability. LBL ND film activity as an anti-inflammation drug matrix was also demonstrated via observed suppression of a broad spectrum of cytokines and signaling molecules. Results discussed in this chapter, thus, also demonstrate the potential that NDs possess as a broad drug-functionalization thin film platform technology for nanoscale medicine.

7.2 ND Preparation, Characterization, and Dispersion

ND powder (NanoCarbon Research Institute Ltd.) was synthesized according to previously reported detonation techniques [26]. Multiple or combinatorial methods of impurity removal were employed that include both mechanically and chemically-based methodologies, as this addresses the removal of the spectrum of impurities (e.g., nondiamond carbon, metals, etc.) associated with detonation ND synthesis that can potentially interfere with drug adsorption or film deposition. For example, from a mechanical context, the NDs were subjected to stirred-media milling with micron-sized ceramic beads, leading to disintegration into ND primary particles [27]. Chemical methodologies for purification were also employed, such as nitric acid treatment which could have also introduced surface-bound functional groups to further promote dispersibility and drug adsorption. The NDs typically formed clusters or aggregates with a size distribution in the range of a few nanometers to hundreds of nanometers, depending on sonication conditions, the solvent used and settling time, and were able to easily disperse and remain dispersed in water.

Prior to ND film assembly, the chemical properties/purity of the NDs were thoroughly examined with various characterization techniques, including X-ray photoelectron spectroscopy (Omicron, ESCA probe) using monochromatic Al Kα radiation at a power of 300 W, Fourier transformed infrared spectroscopy (FTIR, Thermal Nicolet, Nexus 870), Raman spectroscopy (Renishaw, inVia reflex microRaman, 514.5 nm laser), thermogravimetric analysis (TGA, TA instruments, SDT 2960), and transmission electron microscopy (TEM, Hitachi H-8100). From this data, the size of the NDs from this sample was found to be in the range of 2–8 nm, the purity of the NDs was over 90%wt, and the NDs were heavily functionalized with hydrophilic functional groups such as –OH and –COOH on the surface.

Integrating the presence of the hydrophilic functional groups and the fact that the chemical linkage between the ND particles in the ND solids were largely disintegrated by stirred-media milling, a stable dispersion of NDs in water ($\leq$100 μg/ml) could be readily achieved via mild ultrasonication (100 W, VWR 150D sonicator) for 30 min. Some polar aprotic solvents, such as dimethyl sulfoxide (DMSO) and *N*-methylpyrrolidone, have been confirmed to be favorable solvents for dispersion of the NDs [31]. However, for cellular studies, biologically-relevant solvents were utilized.

7.3 ND Thin Film Fabrication via LBL Assembly

Positively charged poly-L-lysine (Sigma) backbones could be easily coated onto naturally negatively charged glass and quartz surfaces (Fig. 7.2). Following the coating of the surfaces with poly-L-lysine (PLL) and a thorough rinsing with pure water using multiple steps, negatively charged ND clusters were then self-assembled on top of the charged PLL polymer by simply dipping the glass slide into an ND aqueous dispersion (0.1 mg/ml). LBL assembly thus enabled fine control of ND film thickness. ND–PLL multilayer deposition was monitored by measuring UV-Vis film absorption on a quartz substrate which was pre-treated with concentrated NaOH to enhance the negative charge density atop the quartz surface. Drug molecule incorporation into the ND thin film was accomplished by dipping-induced adsorption followed by a drying process under steady air flow, or by coating the drugs on the NDs prior to thin film self-assembly.

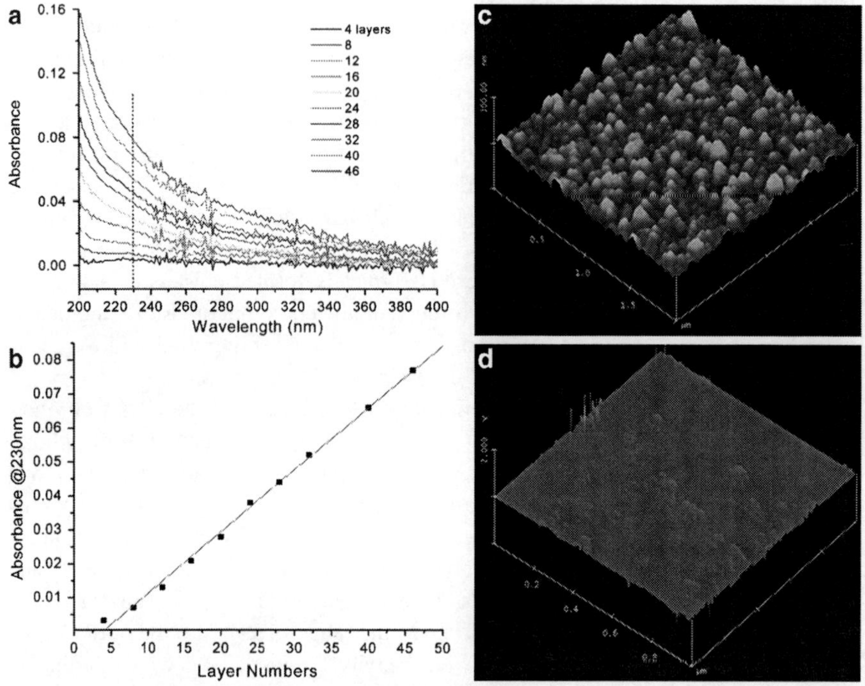

Fig. 7.2 (**a**) Absorbance readings demonstrated the ability to gradually assemble more layers of ND films which was mediated by alternating layers of poly-L-lysine (PLL). The process could be performed in a very controlled fashion, generating uniform films with good layer by layer control. (**b**) Utilizing the layer-by-layer methodology, fine control over film architecture could be achieved as shown via absorbance readings of the multi-layered coatings. (**c**) An atomic force microscope (AFM) image reveals the comprehensive coverage that can be obtained via PLL-mediated ND assembly into multi-layered thin films. This may result in the generation of tunable dosing parameters for sustained localized drug delivery. Solvent-dependent deposition of the NDs can determine the extent of ND coverage atop the substrate surface. Reduced coverage, shown here, also further confirms the high density ND coverage shown in figure 7.2c

To illustrate the chemical nature and purity of the materials used in fabricating ND thin films, methods such as thermogravimetric analysis (TGA) of the NDs were previously applied. The temperature observed at the maximum weight-loss rate of the NDs was shown to be ~570°C, and the majority of the NDs (>95%) were oxidized between 495 and 605°C. Since amorphous carbons (<400°C) and C_{60} (425°C) are oxidized at much lower temperatures [38, 39], the TGA studies clearly indicated that the NDs contained no amorphous carbons and C_{60}. As a result of these side effects, the temperature at the maximum weight-loss rate (570°C) of the NDs was observed to be ~60°C lower than the temperature observed with micrometer-sized diamonds [39]. Another observed feature in the TGA curve was that the oxidation-induced weight gain around 450°C, which was believed to be due to the fullerene-like conjugated sp^2 carbons on the ND surface, was about 2.2% by weight, indicating that sp^2 carbons comprised 1.7% by atom, in terms of one carbon atom reacting with one oxygen atom.

With regards to ND composition, sp^2 carbon presence was also previously confirmed by UV-Vis, Raman and FTIR studies. Since diamond is a wide bandgap (5.45 eV) semiconductor material, pure diamond should not have any optical absorption beyond 300 nm (4.13 eV). The observation of optical absorption extension towards the near IR region could be attributed to conjugated sp^2 carbons present on the ND surfaces. Raman spectra of the NDs also revealed a peak at around 1,327 cm^{-1} which represented clear evidence of the presence of a diamond phase in the sample [40]. As previously noted by Osawa and colleagues, the isolation of detonation NDs in pure form has never before been achieved [27]. Similar to the formation of fullerols [41], the fullerene-like surface of NDs [42], as shown by the dark-brown color of the as-received NDs, can be oxidized by strong acids during the purification process to produce an abundance of surface-bound –OH and –COOH functional groups, as found in the FTIR spectrum. These groups enable opportunities for broadened conjugation methods to the ND surface for applications in therapeutics and diagnostics. The rich presence of the functional groups confers to the NDs two additional important functions: (1) their dispersibility in water which is critical for biomedical and other applications, and (2) their charged surfaces, thus enabling efficient and ordered self-assembly onto a substrate when templated by other charged polymeric species or via covalent conjugation methods, among others.

Dispersed NDs can be readily adsorbed onto PLL-coated glass slides by dipping the slide into the ND aqueous solution. Atomic Force Microscopy (AFM) studies of bare glass, a PLL-coated slide and an ND-coated slide demonstrated this capability. Due to the uniform distribution of the ND particles via aforementioned processing methodologies (including nitric acid surface treatment, centrifugal purification, milling, and sonication, among others) coupled with PLL-mediated templating, the ND particles were uniformly distributed on the glass slide and no large ND aggregates (>200 nm) were observed, resulting from the thoroughly-dispersed NDs in water which further contributes to the scalability of this process. While other solvents such as DMSO generate a much higher dispersibility for NDs (compared to water), the NDs did not readily adsorb to PLL-coated glass under these conditions, as shown in Fig. 7.2d [43]. This interesting observation may have been due to the aprotic characteristics of the solvent as well as the innate charge properties of the surfaces of the slide as well as the ND particles. Further investigations are moving towards optimized and scalable/high yield device coating with bio-functionalized NDs.

Taking advantage of UV-Vis optical absorption of conjugated sp[2] carbons in the NDs, the construction of ND–PLL multilayers by the LBL technique was monitored, and the optical absorbance increased linearly with the subsequent deposition of additional layers. Colorful reflected interference patterns could be observed under light when the number of layers reached about 10, and disappeared after depositing approximately 20 layers. From AFM imaging, the surface roughness and morphology of a multilayer ND film of up to 50 layers were essentially the same as that found in a single-layer film. The NDs in the LBL film could not be washed away by vigorous water flow, and thus the film was attached in a robust fashion onto the glass substrate via electrostatic and Van der Waals interactions. This observation served as an important characteristic contributing to translational relevance and overall film robustness. Additionally, no film detachment was observed in the salt-containing solutions, which was confirmed by comparing the AFM images before and after cell culture. The ability to construct multilayer structures is particularly important for drug delivery in a medical context because the tuned release of a biological compound (e.g., therapeutic small molecules, proteins/antibodies, RNA/DNA, etc.) can be achieved for basic cellular interfacing and controlled release studies, whereby the LBL method can be used to controllably create multilayers of drug-ND hybrids. In a translational context, novel medical implant coatings can be developed for layer-dependent sequential dosing and tunable dosing-rate applications.

7.4 ND-Mediated Loading and Release of Doxorubicin Hydrochloride (Dox)

Because ND surfaces are negatively charged (e.g., ND–COO^-) and Dox ions (D-NH_3^+) are cationic, the interaction between NDs and Dox appears to be straightforward. However, Dox ions are not easily adsorbed by NDs due to the high aqueous dispersibility

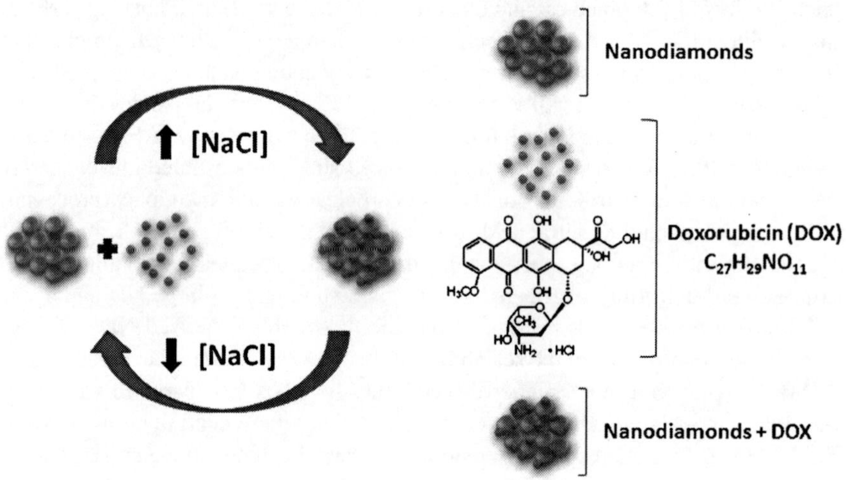

Fig. 7.3 Schematic drawing of NaCl-mediated loading and release of doxorubicin hydrochloride (Dox). The addition of the salt induces functionalization of the drug onto the ND aggregate surface. Salt removal drives drug release. This mild switching process is amenable towards medically relevant processes

of both cations and anions. In order to observe the precipitation of Dox–ND composites from their aqueous suspension upon centrifuging, the addition of salt, such as NaCl, is a necessary component of the loading process (Figs. 7.3 and 7.4). Therefore, these studies demonstrated that salt can promote the adsorption of Dox onto the NDs, revealing solvent-mediated effects upon drug-ND binding. Further studies have revealed a host of other chemical treatment methodologies applicable towards promoting robust drug adsorption onto the ND surfaces. It was also important to note that this binding mechanism generated an extraordinary slow release effect with preserved drug efficacy that could be harnessed for significant enhancements to treatment efficacy. Without the addition of NaCl, less than 0.5%wt of Dox could be adsorbed onto the NDs. In contrast, 10%wt adsorption of Dox on NDs was achieved with the addition of NaCl (10 mg/ml). The mechanism of salt-mediated adsorption of Dox onto the NDs can be explained by the observation that in aqueous solution, without the addition of

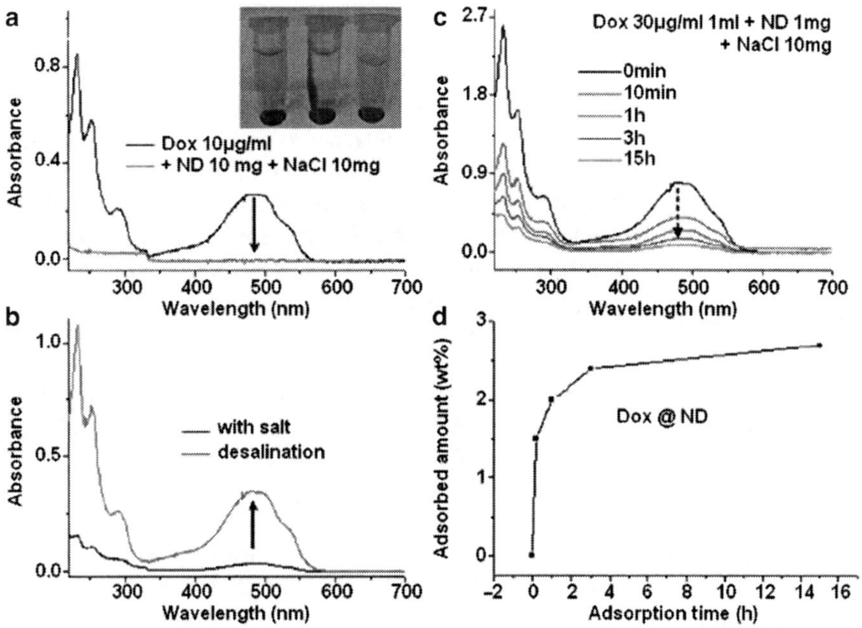

Fig. 7.4 Dox-ND interactions can be modulated via chemical treatment of the NDs as well as solvent conditions. (**a**) UV-Vis spectra of Dox before and after addition of NDs and NaCl. Optimal binding between Dox and the NDs was observed via changes in solvent conditions during the mixing process. Dox (10 mg/ml) was completely removed from solution following the addition of NDs (10 mg) and NaCl (10 mg) with centrifuging at 10 kg. (Inset): Aqueous solution of Dox (10 mg/ml), centrifuged ND/ Dox/NaCl solution and re-dispersed ND/Dox composite drug in water. (**b**) UV-Vis reading shows that desalination results in Dox release from the NDs in which ~1.0% weight of Dox has already been adsorbed. Desalination was performed by partial exchange of the NaCl-containing supernatant with pure water. To further confirm the effect of NaCl on Dox-ND binding, the gradual adsorption of the Dox onto the NDs was monitored using UV-vis spectrophotometry. (**c**) The gradual increase in Dox adsorption to the ND surface over time mediated by NaCl is revealed by the disappearance of free Dox from solution and interaction with the NDs. (**d**) Represents the gradual increase in weight % of the adsorbed Dox onto the ND surface

any other perturbation ions, the repulsive interaction of Dox–Dox and ND–ND is greater than the cohesive interaction of ND–Dox. As such, this was believed to contribute to an insignificant amount of Dox being adsorbed onto the NDs. With the addition of NaCl, the increase of Cl^- ions may have shifted the balanced interactions toward the formation of Dox–ND complexes, since cationic Dox is also balanced with anionic Cl^- ions. Release of Dox from NDs could be easily achieved by desalination. In addition to salt-mediated drug adsorption, other methods are currently being explored to enhance loading capacity and release efficacy via high throughput methods.

Optimizing the manufacturing processes of drug carriers would enable the fabrication of nanoparticles that support the integration of a drug that is both shielded as well as adsorbed to the interior of the carrier to prevent both systemic cytotoxicity or over-elution through both drug sequestering from unintended tissue exposure. This may also minimize generalized and non-specific activity through drug leakage, where the drug adsorption would serve as a potent silencing mechanism. This demonstrates the Dox adsorption/desorption from ND aggregates as an elucidative way to characterize novel material development routes for therapeutic delivery. Therefore, these observed phenomena may then have enabled a slow-release and delamination element of the ND carrier system. Drug adsorption to the NDs can be also be demonstrated by the observations provided in the proceeding sections when comparing the activity of the cell cultures in media only, free Dox upon cell behavior, and Dox–ND composites. While the cells cultured in media alone and drug alone generated a reference (demonstrated quantitatively via MTT assays and DNA fragmentation assays) for drug-ND effects, it could be seen that Dox–ND composites generated an attenuated cell death response which was expected due to the demonstrated potent drug adsorption/binding capabilities of the NDs, confirmed through both the NaCl-mediated binding mechanism, TEM imagery, FTIR, and quantitative MTT and DNA fragmentation analyses. Non-binding between the Dox and NDs would have thus generated an identical effect of cell death (nonattenuated) for the free Dox due to innate cytotoxicity and Dox–ND conditions. As such, through the several methodologies of verification conducted as detailed above, it could be determined that the Dox–ND interaction as well as Dox–ND driven cell death was occurring. Furthermore, many chemotherapeutic molecules possess a generalized mechanism of activity as they can act on both healthy and cancerous cells. This can result in severe complications and patient mortality. The ability to shield non-utilized drug molecules from the surrounding environment would then protect against cytotoxic therapeutic activity against unaffected cells. Subsequent studies of functionalizing the ND surfaces with cellular targeting elements (e.g., antibodies, aptamers, etc.) would add another level of intelligence to this carrier technology to drive targeted and slow drug release [44–47].

TEM imagery before and after Dox-ND mixing, respectively, showed that the NDs were coated with another material which was different from the pristine NDs. The thickness of the coated materials on the ND surface was in the range of 2–10 nm. Because no other solid materials except the drug were used, the composition of the coated layer could then be based upon the presence of the drug. This was also confirmed by FTIR analysis of the drug-coated NDs. The amenability of the ND carrier technology with water as a solvent makes it a suitable system with facile dispersal parameters for future in vivo testing and facile injection protocols. Furthermore, given the potent binding properties of the ND surfaces, As such, the addition and release of additional therapeutic compounds such as therapeutic proteins and nucleic acids has been realized. These collective capabilities of the NDs thus further contribute to their relevance as both a localized and systemic drug delivery strategy.

7.5 Therapeutic Drug Functionalization of NDs

Due to surface charge properties, NDs can aggregate into clusters with numerous interparticle voids. As such, therapeutic elements, such as dexamethasone (Dex), a clinically-relevant anti-inflammatory drug which is demonstrated and discussed here, can be easily incorporated into the ND particles and thin films. Dexamethasone is a potent synthetic member of the glucocorticoid class of steroid hormones and acts as an anti-inflammatory and immunosuppressant. Here, the application of a single-layer ND thin film for Dex incorporation was explored as a potential localized release platform. The drug release was believed to be driven by a diffusion process because of the concentration difference that exists between the film region and solution body. No film degradation or delamination was observed after drug release by intensive washing using water, further demonstrating film robustness. Because the film layers can be controllably constructed by the LBL process, the drug could have been incorporated into the inner layers and thus, the ND-mediated drug desorption process toward a controlled release process could be realized. Further film engineering and surface modification work is ongoing towards optimized drug loading and enhancing the specificity of the release processes for potential ND film clinical translation. This study demonstrated that inflammation in the macrophage cells can be effectively suppressed by the gradually released Dex molecules which are capable of suppressing inflammatory gene induction. The resultant potent attenuation of lipopolysaccharide (LPS)-mediated inflammation demonstrated highly efficient multi-layered ND film-mediated drug release with preserved efficacy upon the interfaced macrophages. While many approaches in the form of polymeric and lipid-based materials have been explored as exciting technologies for drug carrying solutions [48–55], the protein-templated ND thin film potentially serves as a clinically applicable platform strategy for localized therapeutic delivery and drug functionalization that can be rapidly accomplished with specified dimensions.

7.6 Therapeutic Release from ND Thin Films

In order to confirm that the inflammatory suppression observed was indeed due to the Dex therapeutic, Dex desorption/release from the PLL–ND hybrid substrates was examined to ascertain that the anti-inflammatory molecules were indeed capable of being released from the substrate surface for subsequent anti-inflammation activity upon the macrophages. This study was performed to further confirm that the suppression of inflammatory gene expression was not due to cellular necrosis or adverse reactivity towards the ND films. Dex elution was examined using two substrate conditions, with one based upon a single PLL–ND hybrid layer and the other being based upon 20 layers of the PLL–ND surfaces to illustrate in a more pronounced way the detachment and delivery of Dex between the cellular interface and ND surface. The release characteristics of fluorescent Dex from 20 layers of the PLL–ND films displayed a sustained release behavior. The time scale of the release properties (on the order of hours) were consistent with the common time scales utilized for the incubation of Dex with the macrophages, were envisioned to result in the potent suppression of inflammatory cytokines, which will subsequently be examined in detail. A similar release effect as well as time scale of elution was observed with the single PLL–ND layer. Coupled with the RT-PCR as well as proceeding MTT and DNA fragmentation assays, it was demonstrated that the material-mediated suppression of inflammatory gene expression was in fact due to the release of Dex and the preservation of its activity and not associated with innate or material-induced cell death.

7.7 Visualization of ND–Cell Interaction

In order to understand the rate of ND particle movement into living cells, which is a vital property for assessing the NDs as efficient drug carriers, NDs were coated with fluorescent FITC-linked PLL (PLL–ND = 10%wt) via a physical adsorption mechanism analogous to the adsorptive mechanism employed for Dox–ND interfacing, and cells were subsequently incubated with these fluorescent agent modified NDs. It should be noted that PLL served as a useful in vitro imaging agent, but due to its previously observed toxicity, additional strategies for imagery in vivo are being developed. The living cells were fixed in a series of growth periods (0, 1, 3, 5, 10, 24 h) using freshly depolymerized paraformaldehyde (4.0%wt) in a phosphate buffer (pH = 7.2), and then examined using confocal microscopy. In the confocal images, the NDs were observed to attach to the cell membranes instantly, as seen in Fig. 7.5a, in which the fluorescent rings are from PLL-coated NDs. Small fluorescent dots (instead of ring-like structures) were visible in the cells fixed at a growth period of 1 h or longer (Fig. 7.5b), indicating that the living cells interacted with the NDs dynamically. While it is difficult to determine the exact internalization speed in this dynamic process, a large portion of the fluorescent ND clusters were observed to be inside the cells within 10 h by z-sectional confocal imaging. To visualize the nuclei of the cells grown with the addition of the NDs, half of the cells were stained with the DNA-binding dye TOTO-1 (T3600, Invitrogen). As shown in Fig. 7.5c, while the NDs were not visible in this case, the clear profile and integrity of the cell nuclei implied that the cell nuclei were not affected by the internalized NDs.

The internalization of the NDs and Dox–ND composites was also examined via TEM. The efficiency of the NDs as drug carriers was evidenced by the rich presence of ND aggregates inside the cells (Fig. 7.5). Although this image doesn't specify

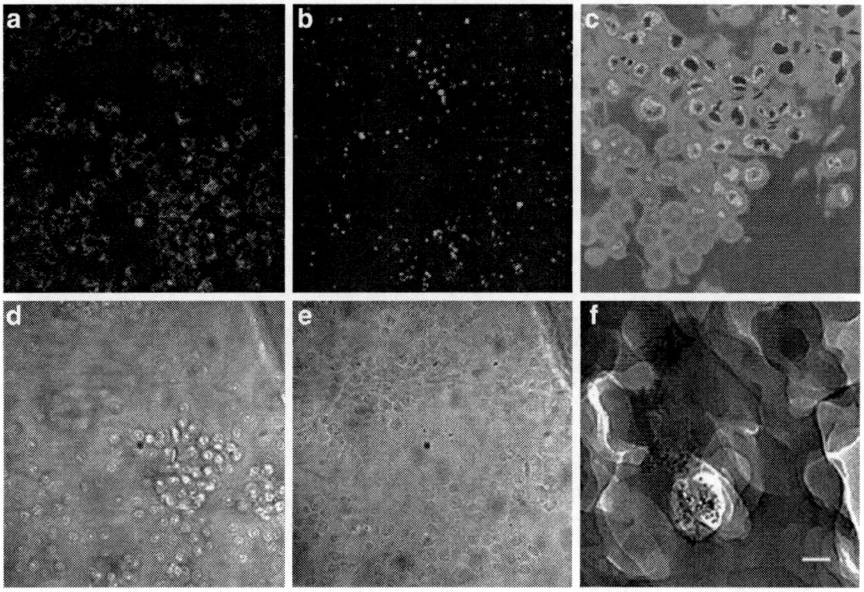

Fig. 7.5 (**a–e**) Confocal images of ND internalization into macrophage RAW cells is shown here. Results indicate the appearance of ND non-toxicity. NDs (20 μg/ml) coated with fluorescent poly-L-lysine (FITC-PL/ND ~10 wt. %) were used. (**a**, **b**) These images were taken from the cells after the addition of NDs without and with incubation (37°C for 10 h), respectively. The excitation wavelength utilized was 488 nm. (**d**, **e**) Bright-field microscopy images corresponding to (**a**) and (**b**), respectively. (**c**) This image shows the same cells as in (**b**) but the cells were stained with TOTO-1 dye and excited with a 514 nm laser. The nucleus of the cells can be clearly identified. (**f**) A TEM image showing NDs in macrophage cytoplasm reveals cellular internalization. The image was taken after a 3h incubation of the cells with the addition of Dox coated NDs (10 wt. %). The cells were dehydrated, fixed and sliced for TEM observation. The scale bar represents 20 nm

the exact location of the ND particles within the cells, most of them appeared to situate themselves in the cytoplasm in agglomerates with size ranges of 50–500 nm. It is possible, given the loading capabilities of the NDs and the aforementioned concept of Dox being integrated within agglomerate interiors due to ND–Dox particle clustering observed via TEM, that drug molecules were buried within the ND agglomerates such that, in addition to the clearly shown surface bound Dox molecules, the agglomerates contained excess and shielded surface-bound Dox molecules not exposed to the ambient solution which were subsequently introduced quickly into the cells. This was demonstrated by the rapid transmembrane transport capabilities of the NDs and the preservation of fluorescent PLL molecules which were interfaced with the NDs through a parallel physical binding mechanism used for Dox binding. In addition, some of the drug could also have been released after internalization and thus, the reduction of overall quantity of the drug to achieve the same efficacy could have been potentially anticipated. Compared to free drug administration, if any embedded drug within the ND agglomerates was released, a potential elution mechanism could have been based on gradual desorption and subsequent escape from the clusters. Furthermore, the preservation of PLL adsorption to the NDs during intracellular trafficking and observation of ND-mediated drug adsorption capabilities (verified both optically and quantitatively), coupled with demonstrated apoptosis and DNA fragmentation through Dox activity, pointed to a desorption-mediated activity. Therefore, the ability to maintain low drug concentrations coupled with high efficacy over a longer period of time would thus be an important parameter of advanced drug delivery systems currently in development.

7.8 Examination of Innate Biocompatibility of ND Delivery Systems

Measurements of cellular inflammation were chosen as readouts for biocompatibility for several reasons including the established knowledge that inflammation is a key process in the body that confers resistance to infection and for the protection from foreign bodies [56]. Chronic inflammation has been shown to induce apoptosis in various tissues and even the onset of prostate cancer [57, 58], thus adding to challenges associated with the prolonged implantation or administration of synthetic materials within the body. The translational potential of novel nanomaterials being developed for medical applications is, in a significant way, determined by the ability for the material alone to favorably interface with surrounding biological tissue to preclude adverse secondary complications (e.g., stress, infection, etc.) that would innately counteract any benefits of the initial delivery of a therapeutic material. Examples of bio-amenable interfaces include the absence of basal inflammatory cellular responses, as well as protection against toxicity (e.g., necrosis) and apoptosis (e.g., programmed cell death) as well as other unfavorable biological reactions induced by the foreign material as cytokine release can inherently act as a complication-inducing signaling factor. For example, tissue specific responses to increased expression of some inflammation genes can have a broad range of physiological effects to complicate anti-inflammatory and cancer treatment cases. IL-6 expression has been shown to be higher in hearts with advanced heart failure [59, 60]. High levels of TNFα can directly damage or sensitize liver tissue to damage [61]. A link between iNOS expression and cholangiocyte carcinogenesis, possibly through an increase in DNA damage has also been shown [62]. In terms of potentially applying the ND nanofilms as implant coatings or ND particles as systemic drug delivery vehicles, the indication that they do not lead to increased inflammatory cytokine expression may serve as an important benefit toward its application as a medical technology, as the ND biocompatibility reduces the potential for secondary physiological disorders associated with material-induced cell stress.

While the assessment of IL-6, TNFα, iNOS, and Bcl-x explains certain elements of cellular regulatory behavior in response to nanomaterial exposure, the four specific genes served as a representative component of the collective cellular gene response as they provide important insight into cell-material interactions from two perspectives. One perspective serves as an indicator of basal cytokine secretion in terms of examining the translational potential of the material through confirming the lack of evident increases in inflammatory gene expression, which contributes strongly to strengthening the clinical potential of the material. The second perspective addresses the complications associated with increased basal inflammation and the predisposition, onset, and proliferation of cancer, all of which can be promoted through increased endogenous levels of IL-6, TNFα, and iNOS, for example. More specifically, the absence of ND-mediated upregulation of cytokine production served as a positive indicator of ND biocompatibility, as increases in endogenous cytokine levels have been shown to play a role in the onset as well as progression of multiple disorders including silencing of chemotherapeutic activity as well as tumor angiogenesis, all of which would serve as counteracting events to the benefits of nanomaterial-mediated drug delivery.

To examine the quantitative biocompatibility of NDs for their potential use as a material towards medical and clinical applications (e.g., implant coatings and drug delivery in patients), the internal cellular response towards incubation with both free NDs and the ND thin films in culture was investigated. As a first step to studying bio-interfacial ND properties, cell viability during growth in culture with the NDs was monitored at 24 and 72 h. Genetic responses involved in inflammation, which are not examined in conventional cell viability tests, were also assessed [28]. For both basal inflammation as well as biological functionality experiments, RAW 264.7 (ATCC) murine macrophage cells were cultured at 37°C in DMEM supplemented with 10% FBS and 5% Penicillin/Streptomycin. Cell growth, morphology, and the response of genes involved in inflammation were initially monitored using conventional microscopy (Leica Lambda DG-4, Magnification × Objective: (40 × 10)). Specifically, IL-6, TNFα, and iNOS were evaluated using real time polymerase chain reaction (RT-PCR). Biocompatibility studies were performed without LPS stimulation to assess basal inflammation and inherent cellular response. To examine the bio-functionality of the therapeutic-activated ND thin films, Dexamethasone (1 mg/ml in ethanol), a glucocorticoid anti-inflammatory drug, was incorporated into ND clusters in the thin film (single-layer, 30 min dipping). For these experiments, lipopolysaccharide (LPS-4 h stimulation) mediated upregulation of inflammatory cytokine expression in vitro was performed to examine the suppression of inflammatory cytokine gene expression by the Dex-incorporated ND thin film which was also determined using RT-PCR. Isolation of the genetic material was accomplished by adding 1 ml of TRIzol cell lysis solution to wash the cell-coated slides. RNA isolation was performed according to the manufacturer's protocol [63]. Subsequent RNA to cDNA conversion was performed using the I-script enzyme (Bio-Rad). Following conversion of the isolated mRNA to cDNA, RT-PCR analysis was performed.

The following genes involved in cellular inflammation were then monitored by RT-PCR using the MyiQ Single Color Real-Time PCR Detection System (BioRad): TNFα, IL-6, and iNOS. Beta-actin was used as a housekeeping gene to normalize RNA levels across all samples. With regards to cancer, IL-6 is a secreted protein and plays a significant role in several physiological disorders including tumor angiogenesis, the protection of tumor cells from the host immune system, and inflammation.

Its signaling mechanism involves Jak/STAT, Ras/MAP Kinase (MAPK), and PI-3 Kinase (PI3-K)/Akt pathways which can be correlated with multiple cellular processes including mechanosensation and cancer [64–68]. It has also been observed that the role of IL-6 in cancer varies depending on the cell type upon which it acts. For example, IL-6 has been shown to promote the growth of multiple myelomas as well as prostate cancer cells. However, IL-6 can also impact breast cancer as well as lung cancer cell growth in vivo [69–71]. Regarding tumor angiogenesis, IL-6 can also enhance the angiogenic capabilities of cervical tumor cells via up-regulation of vascular endothelial growth factor (VEGF) [72].

IL-6 has also been implicated in the progression of prostate cancer via transforming growth factor (TGF)-β1-mediated secretion through the NF-κB, JNK, and Ras signaling pathways [73]. Elevated serum IL-6 levels have also been linked to hormone-refractory disease as well as cancer morbidity [74–77]. Also, IL-6 is a potent autocrine growth factor in vivo for primary, androgen-dependent cancers [72, 78, 79]. The role of IL-6 in tumor progression was observed through a preference for prostate cancer cell spreading towards organs that exhibited elevated levels of IL-6 secretion including bone, lymph nodes, and the lungs [80]. With regards to chemotherapeutic delivery, endogenous IL-6 can induce prostate carcinoma cell resistance toward therapeutic agents including cis-diamminedichloroplatinum and etoposide. Therefore, the observation that tissue-ND incubation does not induce IL-6 upregulation is important in that favorable nanomaterial-tissue interfacing at the genetic level is critical for the preserved activity of therapeutic systems, for which the novel delivery strategy was initially developed.

The significance of maintained TNFα expression following ND exposure can be explained by its relationship to cancer progression at elevated levels. For example, TNFα can induce cellular DNA damage via reactive ion species [81]. In murine in vivo testing, elevated TNFα levels induced oxidative stress, damage to nucleotides, as well as mouse embryo fibroblast malignant transformations TNFα can also play a role in promoting chronic inflammatory diseases including skin ulcers and chronic hepatitis in addition to other precursors to cancer development. Blockage of the TNFα signaling pathway suppressed the formation of chemically induced skin cancers [82, 83]. TNFα is also involved in inflammation-based cancer development through NF-κB activation which in turn can support the viability of, and inhibit apoptosis of precancerous and transformed cells [84–87]. Therefore, ND incubation with RAW 264.7 macrophages demonstrated a significant property of the material, in that basal TNFα levels were not elevated, indicating a lack of cellular inflammatory response in vitro using an additional cytokine. This observation may also provide additional support to the ND as a medically/clinically relevant technology, given the established adverse downstream effects that elevated TNFα levels would have activated.

Inducible Nitric Oxide Synthase (iNOS) plays a role in tumor progression, angiogenesis, and suppression of apoptosis mechanisms by engaging in crosstalk behavior with the COX-2 pathway [88]. Therefore, a non-elevated iNOS response following cellular-ND interaction served as another positive aspect of the non-functionalized ND as a biocompatible drug carrier system.

In this study, the IL-6, TNFα, and iNOS genes were examined for increased expression as indicators of inflammation while Bcl-x was examined for decreased expression, serving as an indicator for decreased cellular protection against apoptosis and potential cytotoxicity. Cytokine expression levels were not significantly upregulated following ND incubation, and Bcl-x expression was not decreased, indicating that the NDs are biocompatible at the genetic level in vitro. While any nanomaterial-induced enhancement in IL-6, TNFα, iNOS levels, or decreases in Bcl-x expression would clearly not be correlated to the onset of cancer, increased levels of the inflammatory regulatory molecules in the body can certainly play a relevant part in signaling element activation that can in turn generate adverse physiological reactions. Furthermore, an important aspect of advanced chemotherapy mediated by these novel ND systems is their ability to maintain steady levels of cytokines such that secondary effects due to ND presence are not activated. As cytokine upregulation has been implicated in physiological disorders that extend well beyond cancer progression, the absence of IL-6, TNFα, and iNOS upregulation or Bcl-x expression decreases following cellular incubation with NDs provides a quantitative biocompatibility readout and demonstrates their innate potential for clinically-significant drug delivery.

Following studies performed upon free NDs, IL-6, TNFα, and iNOS expression was again evaluated using RT-PCR to assess the biocompatibility of the protein-templated ND films. Genetic analysis of inflammation of RAW 264.7 murine macrophages grown for 24 h without and with addition of free floating NDs (30 μg/ml) was explored. No significant change of relative mRNA levels was found for all three genes (IL-6, TNFα, and iNOS) studied. In addition, cells were introduced to ND-only solutions, and the morphological as well as the architectural characteristics of the cells were monitored. ND solutions were used to introduce a pervasive presence of the NDs toward the cells, which was envisioned to be a more comprehensive test of biocompatibility. Both cells with and without the addition of the ND solutions grew readily with no noticeable cell morphology difference or necrotic activity. The same cell population was utilized to demonstrate the lack of adverse effects generated by ND solutions in a quantitative fashion. Additionally, subsequent RT-PCR analysis of cells cultured atop ND films also showed that at an intrinsic genetic level, the cells were unaffected by the ND substrates shown by the absence of ND substrate-mediated up-regulation of inflammatory gene expression.

Therefore, in addition to the conventional application of the ND thin films as inflammation-suppressing coatings on implants, the NDs may also serve as a foundational technology for localized chemotherapeutic elution applications. The demonstrated efficient attenuation of cytokine release both actively via ND-mediated Dex elution, as well as the absence of basal cytokine secretion following cellular interaction with the NDs demonstrates significant implications of ND-based thin films as interfacial materials for the inhibition of malignant tumor growth and localized post-operative chemotherapeutic delivery for cancer applications as a combinatorial delivery strategy, given the potent impact that cytokine secretion possesses upon a multitude of cancer-relevant signaling pathways.

7.9 Cytotoxicity Analysis of ND–Cell Interfacing

In addition to quantitative gene expression analysis, ND biocompatibility was analyzed from the standpoint of cytotoxicity using MTT assay. DNA fragmentation was also used to examine any presence of apoptotic activity due to cellular–ND interactions. For these studies, RAW 264.7 murine macrophage cells were cultured in DMEM media supplemented with 10% Fetal Bovine Serum and 1% Penicillin/Streptomycin. Cells were grown in the presence or absence of NDs and monitored for the onset of apoptosis and/or cell death due to the presence of NDs. Dox was utilized as a positive control for apoptosis and cell death. To first monitor the onset of apoptosis, a DNA fragmentation assay that visualizes the patterned degradation of DNA due to apoptosis was used. Cells were seeded at ~40% confluency with no additives or in the presence of 25 μg/ml NDs or 2.5 μg/ml dox and grown for 24 h. DNA was collected via a standard harvesting protocol [89]. Cells were lysed in lysis buffer (10 mM Tris–HCl (pH 8.0), 10 mM EDTA, 1% TritonX-100) for 10 min at room temperature. Samples were collected and treated with 40 μg/ml RNase for 30 min at 37°C and 100 μg/ml Proteinase K for 30 min at 37°C followed by phenol chloroform isolation and isopropanol precipitation. The DNA pellet was washed in 70% ethanol and re-suspended in water. Samples were loaded onto a 0.8% agarose gel in sodium borate buffer, run and stained with ethidium bromide [90]. The gel was visualized on a short wave UV box with a CCD camera.

ND bio-amenability was monitored via the MTT-based cell viability using the following procedure. Again, cells seeded at ~40% confluency in media containing or lacking 25 μg/ml NDs with 2.5 μg/ml dox were used as a positive control for cell death. Cells were grown for 24 h and the cell viability assay was performed using the manufacturer's protocol (Sigma-Aldrich) and analyzed in a Safire multiwell plate reader (Tecan) using Magellan software (Tecan). Samples were examined in triplicate.

These observed results were consistent with previous cell viability studies conducted by examining MTT and luminescent ATP production, indicating that the NDs are biocompatible with the cell lines that were studied and thus, appear to be suitable for implant coatings or as the foundational material for drug delivery devices or systemic therapy. Cell viability and cytokine expression up to a concentration of 100 μg/ml was also examined. There was no evidence of cell death in the form of cellular delamination from the ND substrate and inflammation in the form of upregulated cytokine gene expression, or cytoxicity and apoptotic induction. A concentration of 30 μg/ml was chosen, instead of 100 μg/ml to ensure that the NDs did not agglomerate during cell growth in culture media. If the concentration was too low, the biocompatibility of NDs may not have been thoroughly addressed. As such, because the specified concentration enabled complete/high density film formation while resisting agglomeration and served as a favorable fabrication parameter, it was also selected as a concentration for biocompatibility studies. RT-PCR studies showed that ND-cellular incubation resulted in the maintenance of IL-6, TNFα, and iNOS levels when comparing glass-only and ND-only samples. This served as one of the many additional indicators of ND favorability in a biological environment.

7.10 Dox–ND-Induced Cellular Apoptosis

Interactions between pristine NDs and murine macrophages, as well as the Dox–ND composites and macrophages were imaged by optical microscopy, and morphological analysis was performed to assess Dox–ND effects and impact upon cellular health. This visual and morphological assessment was then supported by substantial further testing that provided quantitative insight into several factors including the innate biocompatibility of the NDs, Dox–ND efficacy in multiple cell lines (RAW 264.7 and HT-29), free Dox efficacy, and LPS-inhibited Dox activity (conditions assessed via RT-PCR, MTT assays, DNA fragmentation assays). Cell imagery and visual analyses were conducted 18 h post incubation under the given parameters.

On a qualitative level, the macrophages cultured in media-only environments resulted in unaltered morphologies, density, and viability which was previously confirmed via MTT and DNA fragmentation assays. Altered cellular morphology followed the addition of pure Dox, which was also confirmed from the MTT and DNA fragmentation assays (3 μg/ml Dox). ND-only incubations also showed a clear difference from Dox addition trials, as they were indicative of healthy growth, further confirming ND bio-amenability via MTT and DNA fragmentation assays (30 μg/ml NDs). Dox–ND delivery upon the cells resulted in an expected attenuated but clearly observed cell death due to the adsorptive effects of the NDs upon the Dox and subsequent reduction in Dox-mediated effects upon cellular morphology, and cytotoxicity with impact upon cell death also confirmed through the MTT and DNA fragmentation assays (30 μg/ml NDs + 3 μg/ml Dox).

To further examine and confirm that cell death, upon addition of the Dox–ND composites, was indeed Dox-induced activation of apoptosis, studies of LPS-mediated effects upon Dox activity were also conducted, where cells were pretreated with LPS (10 ng/ml), an established inhibitor of Dox-mediated apoptosis, followed by Dox-ND addition [89]. Cell death in the presence of the same additives was significantly decreased. Apoptotic cells exhibit a number of characteristic phenotypes, including damage and fragmentation of cellular DNA, activation of caspase-3 (a mediator of apoptosis), and the cleavage of poly-(ADP-ribose) polymerase. Previous studies have revealed the progression of these three apoptosis signals upon the treatment of murine macrophage cells with Dox [89]. Inflammatory pathway activation has previously been shown to be a potent blocker of Dox-mediated cell death. The stimulation of inflammation involves the induction of gene expression from a number of loci, including IL-6. Overexpression of IL-6 has also been shown to inhibit p53 mediated apoptosis [91, 92]. Therefore, LPS pretreatment of macrophage cells served as an effective inhibitor of Dox (released from NDs) induced apoptosis.

MTT assays were conducted utilizing Dox to serve as a reference for ND-induced apoptosis and cytotoxicity levels. Whereas, Dox was capable of rapidly causing cell death which was visible via conventional brightfield microscopy coupled with apparent decreases in cellular viability, comparisons of cell viability between cultures grown in Petri dishes and media, and in ND-containing solutions showed a negligible difference, further revealing that the cells were capable of remaining viable and proliferating in ND cultures. Furthermore, this observation further confirmed that inflammatory attenuation was not due to ND-induced cell death.

The presence of NDs also did not lead to the characteristic DNA fragmentation ladder pattern associated with apoptosis. However, the addition of Doxorubicin alone induced DNA fragmentation, serving as a reference and indicator of cell death. Thus, it was clearly apparent that apoptosis was not induced by the addition of NDs to the RAW 264.7 macrophage cells.

7.11 DNA Assay for Dox–ND Composite Driven Cell Death

Because the fragmentation of cellular DNA is indicative of apoptotic cell death, DNA laddering assays were also used as efficacy readouts for the Dox-ND complexes. The different magnitudes of cell death induced by free Dox and Dox–ND composites can also be examined via gel electrophoresis of fragmented DNA. Control assays and ND-only assays revealed no DNA fragmentation. Positive Dox-only controls showed an expected fragmentation pattern. As previously stated, the reduced attenuated effect of the Dox–ND composites and the resultant onset of cell death as revealed by the MTT assays was again observed and expected due to the potent adsorptive capability of the NDs, and further confirmed through the NaCl-mediated mechanism, TEM imagery, FTIR, and quantitative MTT and DNA fragmentation analyses. Non-binding between the Dox and NDs would have resulted in the identical potency of cell death in the form of observed fragmentation (e.g., non-attenuated) for the free Dox and Dox–ND conditions because the same amount of drug would have been acting on the cells due to the equal amounts added to the free Dox and Dox–ND solutions (3 μg/ml). On the contrary, however, to further confirm the TEM imagery and MTT data observed, the DNA fragmentation pattern for the Dox–ND composites were then expectedly shown through the ND modulation of drug sequestering and reduced effect upon inducing cell death. As such, through the several methodologies of verification conducted as detailed above, it could be determined that the Dox–ND interaction as well as Dox–ND driven cell death was occurring. In addition, longer cellular incubation with Dox–ND composites led to a more pronounced progression of apoptotic cell death, which also demonstrated the functionality of the Dox–ND composites. The relative cell viability with the three different added components, which was reflected in the fragmented DNA patterns, was as follows: NDs > Dox–ND > Dox. This result was consistent with the qualitatively observed cell morphology changes observed previously via optical microscopy.

7.12 MTT Assay of Dox–ND Activity in HT-29 Colorectal Cancer Cell Line

To also explore the platform capabilities of the ND drug carrying particle technology, Dox–ND composites were also applied towards the HT-29 human colorectal cancer cell line. MTT-based cell viability assays were done using HT-29 human colorectal adenocarcinoma cells (ATCC). Cells were seeded at ~40% confluency in McCoy's 5A media supplemented with 10% fetal bovine serum. NDs were subsequently added to the cells at a final concentration of 25 μg/ml, Dox at a concentration of 2.5 μg/ml, or a mixture of NDs and Dox at concentrations of 25 and 2.5 μg/ml respectively. Cells were then grown for 41 h for ND particle architectures, and 24 h for ND thin film architectures, at 37°C with 5% CO_2. The cell viability assay was performed in triplicate following the manufacturer's protocol (Sigma-Aldrich) and read in a Safire multiwell plate reader (Tecan) using Magellan software (Tecan) for analysis.

Results yielded no significant difference in cell viability when cells were grown in the presence of NDs, which further confirmed the biocompatibility of the NDs in an additional cell line. Doxorubicin alone significantly decreased cell viability, as expected, compared with the Dox–ND composite effects upon HT-29 viability. Following the addition of the Dox–ND composites, the effect of the Dox cytotoxicity was attenuated through ND-mediated Dox sequestering. As such, this translational study further extended the applicability of the Dox–NDs from the model RAW 264.7 murine macrophage, serving as a testing platform for both drug-ND efficacy as well as innate biocompatibility, towards a more clinically relevant colorectal cancer cell line to demonstrate the translational potential of this technology.

7.13 Conclusion

NDs serve as concentration-tuning substrates that can be interfaced with a chemotherapeutic drug while further serving as stabilization matrices to preserve drug functionality for both localized and systemic drug elution. Doxorubicin hydrochloride (Dox), an apoptosis-inducing drug widely used in chemotherapy, was successfully and reversibly coated on the ND platforms (NDs, 2–8 nm) following chemical treatment of the NDs and introduced into living cells. The adsorption of Dox onto the NDs and its reversible release were achieved by regulating Cl^- ion concentration to elucidate the potential of switchable drug elution mediated at the ND surface using ND particles. Following the observation of efficient ND-driven drug internalization, comprehensive bioassays were performed to assess and confirm the innate biocompatibility of the NDs and stable gene program activity via RT-PCR. Genetic interrogation was utilized to observe functional apoptosis-inducing mechanisms enabled by the Dox-functionalized ND composite materials, confirming its applicability as a platform transport technology for a multitude of therapeutic molecules and broad nanoscale medicine modality.

This chapter also discussed the novel fabrication of ND-based thin films using a protein-mediated LBL deposition procedure. NDs used for the film formation were characterized by various analytical metholodogies including TEM, FTIR, and TGA. Also, both free floating and nanofilm-assembled ND biocompatibility was evaluated at the gene expression level to confirm the innate benefits of the ND material due to its favorable biological properties and lack of observed inflammatory cytokine up-regulated release. As increases in the endogenous levels of inflammatory cytokines, such as IL-6, have been shown to promote tumor progression or directly silence therapeutic activity, the observation that the NDs do not elicit adverse cellular reactions (as shown through unchanged basal cytokine release levels in vitro for IL-6, TNFα, and iNOS, as well as MTT and DNA fragmentation assays which confirmed the nonapoptotic and noncytotoxic ND properties) served as a strong indicator for the nanofilms and ND particles as potential scalable drug vehicle platforms with translational significance. Bio-functionality and the preserved drug efficacy of the ND thin film as an anti-inflammation drug matrix was also demonstrated by Dex anti-inflammatory elution to the interfaced RAW 264.7 murine macrophages. Cytokine expression levels were significantly reduced as confirmed by RT-PCR, indicating efficient drug elution characteristics and stable drug binding with the ND nanofilm multilayers. The straightforward fabrication metholodogies and biologically amenable ND processing conditions that were employed towards developing nanofilms with rapidly tailorable thicknesses open up new opportunities in tuning drug concentration and dosage control capabilities in a bioinert material with potential applications in combinatorial therapy and sequential drug delivery using a novel nanomaterial platform. As such, the ND particle and thin film architectures demonstrate enormous potential toward biomedical applications such as implant coatings, stand-alone therapeutic drug-elution technologies, and injectable vehicles for efficient cancer therapy.

References

1. Niemeyer CM (2001) Angew Chem Int Ed 40:4128–4158
2. Michalet X, Pinaud FF, Bentolila LA, Tsay M, Doose S, Li JJ, Sundaresan G, Wu AM, Gambhir SS, Weiss S (2005) Science 307:538–544
3. Cui Y, Wei QQ, Park HK, Lieber CM (2001) Science 293:1289–1292
4. Baughman RH, Zakhidov AA, de Heer WA (2002) Science 297:787–792
5. Bianco A, Prato M (2003) Adv Mater 15:1765–1768
6. Krüger A (2006) Angew Chem Int Ed 45:6426–6427
7. Langer R, Tirrell DA (2004) Nature 428:487–492
8. Langer R (1998) Nature 392:5–10
9. Allen TM, Cullis PR (2004) Science 303:1818–1822
10. Rao CNR, Cheetham AK (2001) J Mater Chem 11:2887–2894
11. Caruso F (2001) Adv Mater 13:11–22
12. Gao XH, Cui YY, Levenson RM, Chung LWK, Nie SM (2004) Nat Biotechnol 22:969–976
13. Moghimi SM, Hunter AC, Murray JC (2001) Pharma Rev 53:283–318
14. Lacerda L, Bianco A, Prato M, Kostarelos K (2006) Adv Drug Deliv Rev 58:1460–1470

15. Smart SK, Cassady AI, Lu GQ, Martin DJ (2006) Carbon 44:1034–1047
16. Kam NWS, Jessop TC, Wender PA, Dai HJ (2004) J Am Chem Soc 126:6850–6851
17. Kam NWS, Dai HJ (2005) J Am Chem Soc 127:6021–6026
18. Pantarotto D, Briand JP, Prato M, Bianco A (2004) Chem Commun 1:16–17
19. Pantarotto D, Singh R, McCarthy D, Erhardt M, Briand JP, Prato M, Kostarelos K, Bianco A (2004) Angew Chem Int Ed 43:5242–5246
20. Cai D, Mataraza JM, Qin ZH, Huang ZP, Huang JY, Chiles TC, Carnahan D, Kempa K, Ren ZF (2005) Nat Methods 2:449–454
21. Kam NWS, O'Connell M, Wisdom JA, Dai HJ (2005) Proc Natl Acad Sci U S A 102:11600–11605
22. Manna SK, Sarkar S, Barr J, Wise K, Barrera EV, Jejelowo O, Rice-Ficht AC, Ramesh GT (2005) Nano Lett 5:1676–1684
23. Narayan RJ, Wei W, Jin C, Andara M, Agarwal A, Gerhardt RA, Shih CC, Shih CM, Lin SJ, Su YY, Ramamurti Y, Singh RN (2006) Diam Rel Mater 15:1935–1940
24. Yu SJ, Kang MW, Chang HC, Chen KM, Yu YC (2005) J Am Chem Soc 127:17604–17605
25. Fu CC, Lee HY, Chen K, Lim TS, Wu HY, Lin PK, Wei PK, Tsao PH, Chang HC, Fann W (2007) Proc Natl Acad Sci U S A 104:727–732
26. Greiner NR, Phillips DS, Johnson JD, Volk F (1988) Nature 333:440–442
27. Kruger A, Kataoka F, Ozawa M, Fujino T, Suzuki Y, Alesenskii AE, Vul AY, Osawa E (2005) Carbon 43:1722–1730
28. Schrand AM, Huang HJ, Carlson C, Schlager JJ, Osawa E, Hussain SM, Dai L (2007) J Phys Chem B 111:2–7
29. Huang L-CL, Chang H-C (2004) Langmuir 20:5879–5884
30. Chung PH, Perevedentseva E, Tu JS, Chang CC, Cheng CL (2006) Diam Rel Mater 15:622–625
31. Nguyen TTB, Chang HC, Wu VWK (2007) Diam Rel Mater 16:872–876
32. Gruen DM, Shenderova OA, Vul' AY (eds) (2005) Synthesis. Properties and Applications of Ultrananocrystalline Diamond, Springer, New York, USA
33. Liu Y, Khabashesku VN, Halas NJ (2005) J Am Chem Soc 127:3712–3713
34. Liu Y, Gu ZN, Margrave JL, Khabashesku VN (2004) Chem Mater 16:3924–3930
35. Decher G, Hong JD (1991) Macromol Chem Macromol Symp 46:321–327
36. Decher G (1997) Science 277:1232–1237
37. Tang ZY, Wang Y, Podsiladlo P, Kotov NA (2006) Adv Mater 18:3203–3224
38. Huang HJ, Marie J, Kajiuar H, Ata M (2002) Nano Lett 2:1117–1119
39. Pang LSK, Saxby JD, Chatfield SP (1993) J Phys Chem 97:6941–6942
40. Prawer S, Nugent KW, Jamieson DN, Orwa JO, Bursill LA, Peng JL (2000) Chem Phys Lett 332:93–97
41. Chiang LY, Upasani RB, Swirczewski JW (1992) J Am Chem Soc 114:10154–10157
42. Raty JY, Galli G (2003) Nat Mater 2:792–795
43. Ozawa M, Inaguma M, Takahashi M, Kataoka F, Krüger A, Osawa E (2007) Adv Mater 19:1201–1206
44. Dachs GU, Dougherty GJ, Stratford IJ, Chaplin DJ (1997) Oncol Res 9:313–325
45. Portney NG, Ozkan M (2006) Anal Bioanal Chem 384:620–630
46. Farokhzad OC, Jon S, Khademhosseini A, Tran TT, LaVan DA, Langer R (2004) Cancer Res 64:7668–7672
47. Farokhzad OC, Cheng JJ, Teply BA, Sherifi I, Jon S, Kantoff PW, Richie JP, Langer R (2006) Proc Natl Acad Sci U S A 103:6315–6320
48. Nardin C, Hirt T, Leukel J, Meier W (2000) Langmuir 16:1035–1041
49. Discher BM, Won YY, Ege DS, Lee JC, Bates FS, Discher DE, Hammer DA (1999) Science 284:1143–1146
50. Nardin C, Widmer J, Winterhalter M, Meier W (2001) Eur Phys J E 4:403–410
51. Nardin C, Thoeni S, Widmer J, Winterhalter M, Meier W (2000) Chem Commun 15:1433–1434
52. Discher DE, Ahmed FP (2006) Annu Rev Biomed Eng 8:323–341
53. Geng Y, Discher DE (2005) J Am Chem Soc 127:12780–12781
54. Ahmed F, Discher DE (2004) J Control Rel 96:37–53
55. Discher DE, Eisenberg A (2002) Science 297:967–973

56. Boehm U, Klamp T, Groot M, Howard JC (1997) Annu Rev Immun 15:749–795
57. Hjelmstrom P (2001) J Leukocyte Biol 69:331–339
58. Palapattu GS, Sutcliffe S, Bastian PJ, Platz EA, De Martzo AM, Isaacs WB, Nelson WG (2005) Carcinogenesis 26:1170–1181
59. Zolk O, Ng LL, O'Brien RJ, Weyand M, Eschenhagen T (2002) Circulation 106:1442–1446
60. Corbi P, Rahmati M, Delwail A, Potreau D, Menu P, Wijdenes J, Lecron J-C (2000) Eur J Cardiothorac Surg 18:98–103
61. Hoeflich KP, Luo J, Rubie EA, Tsao M-S, Jin O, Woodgett JR (2000) Nature 406:86–90
62. Ishimura N, Bronk SF, Gores GJ (2005) Gastroenterology 128:1354–1368
63. Perry AK, Chow EK, Goodnough JB (2004) J Exp Med 199:1651–1658
64. Taga T, Kishimoto T (1997) Annu Rev Immunol 15:797–819
65. Willenberg HS, Päth G, Vögeli TA, Scherbaum WA, Bornstein SR (2002) Ann N Y Acad Sci 966:304–314
66. Kishimoto T, Akira S, Narazaki M, Taga T (1995) Blood 86:1243–1254
67. Murakami M, Hibi M, Nakagawa N, Nakagawa T, Yasukawa K, Yamanishi K, Taga T, Kishimoto T (1993) Science 260:1808–1810
68. Hirano T, Nakajima K, Hibi M (1997) Cytokine Growth Factor Rev 8:241–252
69. Danforth JDN, Sgagias MK (1993) Cancer Res 53:1538–1545
70. Klein B, Zhang XG, Lu ZY, Bataille R (1995) Blood 85:863–872
71. Okamoto M, Lee C, Oyasu R (1997) Cancer Res 57:141–146
72. Wei LH, Kuo ML, Chen CA, Chou CH, Lai KB, Lee CN, Hsieh CY (2003) Oncogene 22:1517–1527
73. Park JI, Lee MG, Cho K, Park BJ, Chae KS, Byun DS, Ryu BK, Park YK, Chi SG (2003) Oncogene 22:4314–4332
74. Adler HL, McCurdy MA, Kattan MW, Timme TL, Scardino PT, Thompson TC (1999) J Urol 161:182–187
75. Drachenberg DE, Elgamal AA, Rowbotham R, Peterson M, Murphy GP (1999) Prostate 41:127–133
76. Twillie DA, Eisenberger MA, Carducci MA, Hsieh WS, Kim WY, Simons JW (1995) Urology 45:542–549
77. Siegsmund MJ, Yamazaki H, Pastan I (1994) J Urol 151:1396–1399
78. Giri D, Ozen M, Ittmann M (2001) Am J Pathol 159:2159–2165
79. Deeble PD, Murphy DJ, Parsons SJ, Cox ME (2001) Mol Cell Biol 21:8471–8482
80. Siegall CB, Schwab G, Nordan RP, FitzGerald DJ, Pastan I (1990) Cancer Res 50:7786–7788
81. Yan B, Wang H, Rabbani ZN, Zhao Y, Li W, Yuan Y, Li F, Dewhirst M, Li W (2006) Cancer Res 66:11565–11570
82. Hendrickse CW, Kelly RW, Radley S, Donovan IA, Keighley B, Neoptolemos JP (1994) Br J Surg 81:1219–1223
83. Cianchi F, Cortesini C, Bechi P, Fantappie O, Messerini L, Vannacci A, Sardi I, Baroni G, Boddi V, Mazzanti R, Masini E (2001) Gastroenterology 121:1339–1347
84. Gullino PM (1995) Acta Oncol 34:439–441
85. Pai R, Szabo IL, Soreghan BA, Atay S, Kawanaka H, Tarnawski AS (2001) Biochem Biophys Res Commun 286:923–928
86. Ziche M, Morbidelli L, Choudhuri R (1997) J Clin Invest 99:2625–2634
87. Sugimoto Y, Narumiya S, Ichikawa A (2000) Prog Lipid Res 39:289–314
88. Cianchi F, Cortesini C, Fantappie O, Messerini L, Sardi I, Lasagna N, Perna F, Fabbroni V, Di Felice A, Perigli G, Mazzanti R, Masini E (2004) Clin Cancer Res 10:2694–2704
89. Hassan F, Islam S, Mu MM, Ito H, Koide N, Mori I, Yoshida T, Yokochi T (2005) Mol Cancer Res 3:373–379
90. Brody JR, Kern SE (2004) Biotechniques 36:214–216
91. Lotem J, Gal H, Kama R, Amariglio N, Rechavi N, Domany E, Sachs L, Givol D (2003) Proc Natl Acad Sci U S A 100:6718–6723
92. Duan Z, Lamendola DE, Penson RT, Kronish KM, Seide MV (2002) Cytokine 17:234–242

Chapter 8
Polymeric Encapsulation of Nanodiamond –Chemotherapeutic Complexes for Localized Cancer Treatment

Robert Lam, Mark Chen, Houjin Huang, Eiji Ōsawa, and Dean Ho

Abstract A diverse range of synthetic and natural nanoscale carriers in both particle and film/scaffold formats have been developed to enable controlled therapeutic release. Examples of these systems include metallic nanoparticles, polymer–protein conjugates, liposomes, micelles, dendrimers, polyelectrolyte films, copolypeptides, carbon nanotubes, etc. Nanodiamonds (NDs), in particular, possess several advantageous properties that make them suitable for advanced drug delivery while also remaining biocompatible. We have previously developed a method of functionalizing aqueous solubilized NDs of diameter 2–8 nm with doxorubicin (DOX), a clinically relevant chemotherapeutic capable of inducing potent DNA fragmentation and cellular apoptosis. This work has realized a scalable approach toward the fabrication of ND-embedded polymer microfilms for localized and sustained drug elution for post-operative chemotherapy. Due to their high surface-area-to-volume ratio and noninvasive dimensions, NDs are capable of extremely high loading capacities of therapeutic compounds. In addition, we have demonstrated the capability of ND binding with a broad range of charged therapeutic molecules via physical interactions due to their inherent surface charge properties. NDs are also biologically stable and appear to be non-toxic, which prevents adverse stressful/inflammation-inducing cellular reactions in the event that they are dispersed throughout the body for either systemic or more localized release activity. The combination of these properties in one system makes the NDs promising platforms

R. Lam, M. Chen, H. Huang, and D. Ho(✉)
Departments of Biomedical and Mechanical Engineering, Robert R. McCormick School of Engineering and Applied Science, Northwestern University, 2145 Sheridan Road, Evanston, IL, 60208, USA

E. Osawa
Nanocarbon Research Institute, Ltd., Asama Research Extension Center, Shinshu University, 3-15-1 Tokita, Ueda, Nagano, 386-8567, Japan

D. Ho(✉)
Robert H. Lurie Comprehensive Cancer Center of Northwestern University, Galter Pavilion, 675 N. St. Clair, 21st Floor, Chicago, IL, 60611, USA
e-mail: d-ho@northwestern.edu

D. Ho (ed.), *Nanodiamonds: Applications in Biology and Nanoscale Medicine*,
DOI 10.1007/978-1-4419-0531-4_8,

for slow-release therapeutics to treat a broad array of physiological disorders (e.g. cancer, heart disease, wound healing, etc.).

8.1 Introduction

Nanoparticles serve as exciting injectable platforms for advanced applications in targeted and sustained therapeutic release [1–7]. In particular, nanodiamonds (NDs) possess several integrative properties that make them favorable for the enhanced duration and specificity of treatment. We have previously developed a novel ND technology possessing individual diameters of 2–8 nm functionalized with the doxorubicin (DOX) chemotherapeutic [8]. Due to their high surface-area-to-volume ratio and charged surfaces that are amenable toward physisorption or conjugation-based drug binding, extremely high loading/binding capacities of the therapeutic compounds are achievable. As such, we have demonstrated the capability of ND interfacing with virtually any therapeutic molecule via physical interactions owing to tailorable surface properties and compositions [9]. With highly ordered aspect ratios near unity, NDs have been shown to be biologically stable, allowing them to preclude adverse cellular stress/inflammatory and apoptotic reactions [10, 11]. Several studies have confirmed the inherently amenable biological performance of suspended NDs when they are incubated with cells [8–12]. More specifically, the general cellular viability, morphology and mitochondrial membrane are maintained among various cell types when incubated with suspended NDs [10, 12]. To realize the potential of the benefits offered by NDs, there are several current needs in medicine that would benefit from sustained release from a patch-like platform where nanoparticle properties can be harnessed for localized elution in lieu of, or in parallel with systemic administration. For example, post-operative cardiac surgery complications include the onset of arrhythmia and congestive heart failure. Post-operative implantation of long-term release devices that can elute anti-scarring and anti-inflammatory agents may decrease the severity of frequency of these complications. As such, with the appropriate tailoring of drug elution parameters in conjunction with the proper selection of material matrices to package ND-mediated sustained elution, the reported microfilm devices serve as promising platforms for sustained and localized therapeutic release.

Previous studies functionalizing ND films fabricated via chemical vapor deposition (CVD) with various biological compounds have provided exciting prospects for biosensing applications but are nonetheless difficult to implant due to their rigidity [13, 14]. In this work, we describe an easily fabricated flexible microfilm patch device that is capable of extended localized chemotherapeutic drug delivery with highly ordered release kinetics. Through concentrated slow release, continuously administered small dosages can replicate the efficacy of normally larger prescribed amounts, reducing side effects generally associated with systemic chemotherapeutic treatments [15–17]. The aim of this approach is to realize a scalable device that combines near zero-order release kinetics, biocompatibility, and a platform approach that can be linked to nearly any type of therapeutic to impact several areas of medicine that require localized drug therapy.

Parylene C, a material with well-documented biocompatibility and FDA approval for the coating of approved medical devices, is used as a flexible and robust nanodiamond packaging agent for microfilm fabrication in this study [18, 19]. Parylene coatings have been utilized in several medical applications due to their highly conformal nature of deposition, biostability, and inertness under physiological conditions with no known adverse biological degradation events [18, 20–22]. Therefore, these properties facilitate sustained and localized delivery that is vital for the release of potent therapeutic agents (e.g., chemotherapy) toward specific regions, which when used as a complimentary drug delivery strategy, may minimize the severe side effects associated with continuously repeated as well as nonspecific drug administration.

8.2 Materials and Methods

8.2.1 ND Suspension and Functionalization with DOX

NDs were surface functionalized and dispersed according to protocols described in [8]. Upon ND ultrasonication (100 W, VWR 150D sonicator) for 30 min, DOX and ND solutions were centrifuged together at appropriate concentrations at a 4:1 ratio. Addition of NaCl helped facilitate the binding process in a more efficient manner.

8.2.2 Materials and Device Fabrication

Serving as the backbone of the microfilm device, a conformal base layer of parylene C (3 g) was deposited on pre cut 2.5 cm×2.5 cm glass slides with a Specialty Coating Systems (SCS) PDS 2010 Labcoater (SCS, Indianapolis, IN). The parylene layer was oxidized via oxygen plasma treatment in a Harrick Plasma Cleaner/Sterilizer (Ithaca, NY) at 100 W for 1 min. DOX–ND solution, the functional element of the device, was then added to the base layer so that the final DOX–ND concentration in solution was 6.6 μg/mL. Subsequently, solvent evaporation occurred in isolation at room temperature. Following DOX–ND deposition, an ultrathin parylene C (0.15 g) layer was deposited as an elution-limiting and control component of the microfilm. The final layer of parylene C was treated with oxygen plasma at 100 W for 1 min. Parylene C was pyrolized into a gaseous monomer at 690°C, and deposited at room temperature under vacuum conditions for all deposition processes. The preceding parameters apply to devices fabricated for spectroscopy studies. The DOX–ND concentration was adjusted for devices used in the DNA fragmentation assay so that the final DOX–ND concentration in solution was 33 μg/mL.

8.2.3 Atomic Force Microscopy Characterization

To examine the surface properties and characteristics of the microfilm devices, Atomic Force Microscope (AFM) (Santa Barbara, CA) images of the samples were taken using an Asylum MFP3D. Image dimensions were 20 µm × 20 µm. Contact mode imaging at line scan rates of 0.3–0.5 Hz was performed at room temperature using Olympus TR800PSA 200-µm length silicon nitride cantilevers (Melville, NY).

8.2.4 Spectroscopic Analysis

Microfilm device samples were immersed in nanopure water (2 mL) in six-well plates and placed in an incubator at 37°C and 5% CO_2. At every 24-h interval, samples were transferred to a new well to avoid residue contamination while the remaining eluate (2 mL) was collected. A full wavelength scan from 350 to 700 nm was performed on a portion of the eluate (100 µl) with a Beckman Coulter DU730 Life Science UV/Vis Spectrophotometer (Fullerton, CA).

8.2.5 Contact Angle Measurements

To examine the modification of the device surfaces, static contact angles were measured with DI water (10 µl) with a Ramé-Hart, Inc Imaging System and Auto Pipetting System (Mountain Lakes, NJ).

8.2.6 DNA Fragmentation Assay

Serving as readouts for device activity, RAW 264.7 murine macrophages (ATCC, Manassas, VA) were cultured in Dulbecco's modification of Eagle's medium (Cellgro, Herndon, VA) supplemented with 10% Fetal Bovine Serum (ATCC) and 1% penicillin/streptomycin (Cambrex, East Rutherford, NJ). Cells were cultured in an incubator at 37°C and 5% CO_2. The cells were plated on two sets of uncovered and covered devices at ~40% confluency for 16 h with one set and 20 h with the other set to contrast progression of apoptosis over time as a result of DOX–ND elution from the native and semi-porous devices. DOX (2.5 µg/mL) served as a positive control for apoptosis, and culture media was used as a negative control. Cell harvest comprised of a PBS wash and subsequent lysis in lysis buffer (500 µL, 10 mM Tris–HCl, pH 8.0, 10 mM EDTA, 1% Triton X-100) for 15 min. Thirty-minute incubations with RNase A (33.3 µg/mL) and proteinase K (83.3 µg/mL) that occurred at 37°C followed the buffer treatment, separately. The collected samples then underwent phenol chloroform extraction, followed by DNA isolation and precipitation in 2-propanol at −80°C for at least 2 h.

After washing with 70% ethanol, the samples were resuspended in water and loaded onto a 0.8% agarose gel in sodium borate buffer, run, and stained with ethidium bromide (Shelton Scientific, Shelton, CT).

8.3 Results and Discussion

The hybrid film drug-eluting microfilm device consists of DOX-functionalized NDs integrated between a thick hermetic base and thin semi-porous layer of parylene C (Fig. 8.1a). NDs efficiently sequester DOX that can be released gradually upon appropriate stimuli, i.e., DOX concentration gradients and acidic pH conditions, which have been associated with the microenvironments of cancerous cells/tumors. While ND release was not intended by this device as it was used as a slow release matrix, it is envisioned that device porosity can be tuned to allow for ND-drug release if needed for more systemic applications. A permeable top layer of parylene C additionally acted as a physical barrier that further modulated the release properties of the device. As such, this device represents a novel, minimally invasive technology with the requisite biostability and biocompatibility for optimized chemotherapeutic delivery. Furthermore, this technology serves as a potential platform for the release of other therapeutics for anti-scarring, wound healing, anti-inflammation, among a host of other applications.

NDs have previously been functionalized with cytochrome *c*, DNA, and various protein antigens [23–25]. However, these agents were typically not released, and

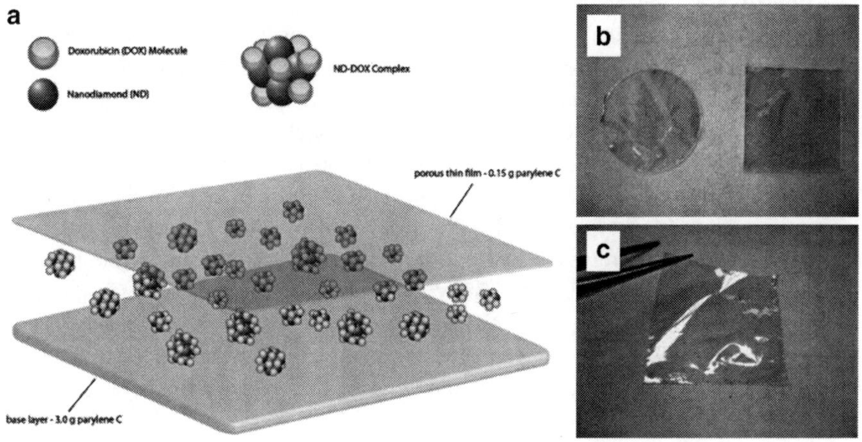

Fig. 8.1 (**a**) A schematic of the hybrid film patch. NDs and DOX molecules bound through potent physical interactions to allow for remarkable sustained release properties in various configurations are deposited atop a base layer of parylene which also serves as the device backbone. A final parylene film was then deposited for additional elution control. (**b**) Hybrid films with a 10 g base layer of varied size and shapes are shown here. (**c**) The patch exhibits inherent flexibility and a thin physical profile for reduced invasiveness during implantation

remained functionalized against the ND surface for non-therapeutic studies. We have previously demonstrated the ability to functionalize NDs with the apoptosis-inducing chemotherapeutic agent, DOX, and anti-inflammatory immunosuppressant glucocorticoid, dexamethasone, as well as the capability of releasing these agents with preserved drug efficacy, introducing the concept of NDs as drug delivery platforms [8, 9]. However, to enhance the applicability of the NDs as foundational components for device fabrication, we have generated a localized elution device to address a broad range of physiological disorders and treatment challenges including difficulties involving tumor heterogeneity, blood circulation and unsustainable controlled release over a prolonged timeframe [26]. Therefore, the ND–parylene hybrids possess particular significance and relevance toward oncological and anti-inflammatory translation, among of host of other fields that may include wound healing, tissue engineering, regenerative medicine, pain management, etc. Furthermore, the biocompatible properties, tailorability, and localization capacity of our flexible parylene C encapsulated DOX–ND hybrid film (Fig. 8.1b, c) address several of these challenges and offer a promising method and avenue for future drug delivery strategies.

For elution and bioassay studies, a conformal and impenetrable base layer of parylene C was deposited on top of pre-cut glass slides. Within the parylene deposition system, the parylene C dimer (di-para-xylylene) was pyrolized into its monomer form (*para*-xylylene) and subsequently deposited at room temperature under vacuum, where the monomers spontaneously polymerized [22, 27]. As this process took place under ambient conditions, the functionality and structure of the DOX–ND conjugates were not altered, indicating the potential of applying this fabrication protocol towards the packaging of several classes of therapeutics.

The base layer served as a flexible foundation upon which the microfilm device was constructed, and simultaneously provided an impermeable and pinhole-free platform for unidirectional drug elution. As newly deposited parylene is hydrophobic (Fig. 8.2a), making it challenging for solvent penetration into the device to effect drug release, additional surface processing was performed to enhance drug deposition uniformity and elution properties. As such, the parylene layers were oxidized via oxygen plasma treatment, which has been shown to increase surface roughness while adding CO_3^- and carbonyl (C = O) groups, effectively creating a hydrophilic surface (Fig. 8.2b) [28]. Oxidization of parylene C surfaces has been shown to remain in a preserved hydrophilic fashion after treatment, while increasing the level of cell adhesion which may also enhance drug delivery efficacy [29]. An appropriate amount of a DOX–ND solution composed of a 4:1 ratio of NDs and DOX of concentrations 330 and 66 μg/mL, respectively, was then added to the base layers via solvent evaporation at room temperature to produce a final concentration of 6.6 μg/mL in solution (Fig. 8.2c).

A semi-porous layer was then deposited as an elution limiting element (Fig. 8.2d) on top of the DOX-ND layer. To achieve the fabrication of this component, we utilized reduced parylene dimer amounts, which have been shown to impact film deposition via pinhole formation, resulting in the release of encapsulated compounds [30]. Therefore, with smaller dimer masses, the dimensions and amount of pinholes were increased, acting as an adjustable physical barrier for controlled drug release. Furthermore, this additional thin film provided a structural

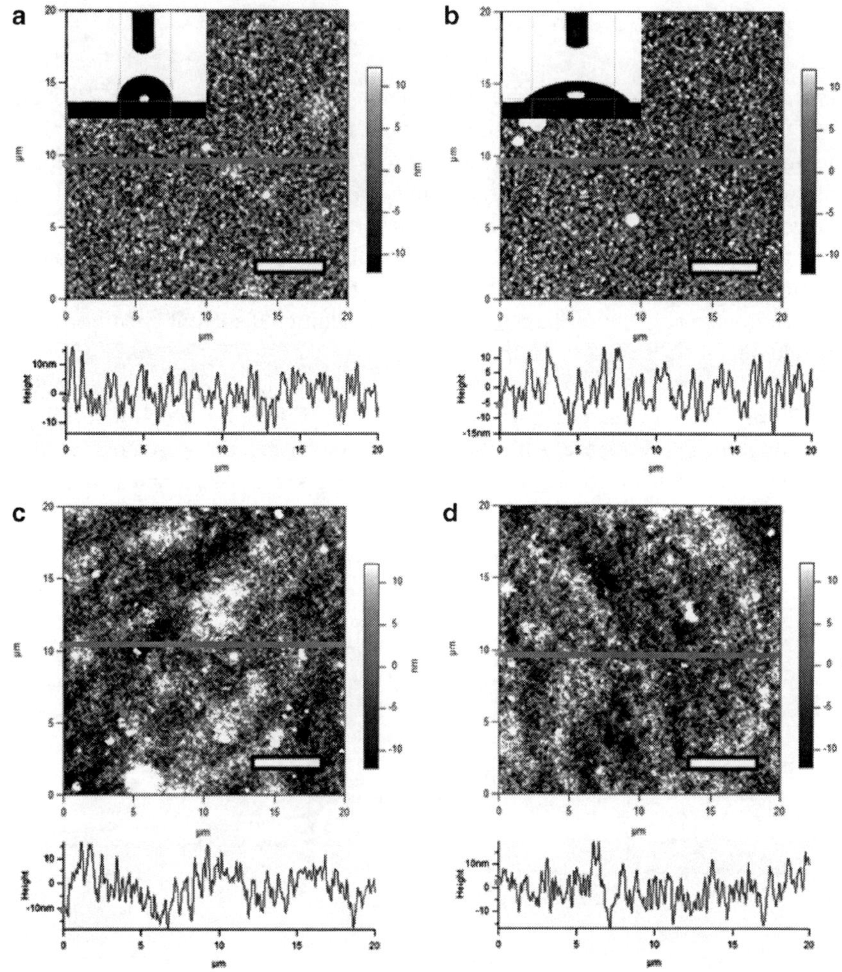

Fig. 8.2 (**a**, **b**) Plasma treatment served as a vital pre-cursor to enabling comprehensive device wetting and drug release. Atomic force microscope (AFM) images revealed the coarse grained polymer structure of native as well as plasma-treated parylene C, with an RMS roughness of 6.245 and 9.291 nm, respectively. RMS roughness values were obtained through Asylum's MFP-3D/Igor Pro software. *Insets*: static contact angle measurements revealed the hydrophilicity transformation associated with plasma treatment which was a necessary process step to allow drug release. (**c**) DOX–NDs were deposited on a plasma treated layer of parylene C which promoted surface wetting. The images portray uneven coating of the DOX–ND conjugates at the nanometer scale, possibly due to the general roughness of the underlying parylene or the inherent agglomerating properties of the NDs themselves. (**d**) DOX–NDs were covered with an additional thin and semiporous layer of parylene C. The structure and conformation of the underlying DOX–NDs were unaffected by the thin film of parylene C deposition; scale bars = 5 μm

platform that simultaneously protected the underlying DOX–ND layer and also acted as a foundation for additional device modifications. The final layer of parylene C was also treated with oxygen plasma to promote solvent penetration and drug release.

8.3.1 Slow-release Analysis

Microfilm samples were immersed in nanopure water and placed in a 37°C and 5% CO_2 incubator to mimic physiological conditions. Upon incubation, the pH of nanopure water was observed to be reduced to a value of pH 3–4, verified through litmus testing (EMD colorpHast). The decline in pH enabled a greater release of the packaged therapeutic than in standard room temperature (data not shown). This chapter will focus on the slow release potential of the parylene-ND-based patches with preserved drug activity while continued work is being done to determine the relationship between elution factors such as pH, additional solvent conditions, and drug release.

DOX solubilized in water generates an absorbance signal from approximately 375–575 nm, with a peak at approximately 480 nm (Fig. 8.3a) which was utilized as a readout for DOX release. Absorbance values under 350 nm were not recorded, as parylene C is known to absorb strongly at lower wavelengths and would have interfered with the signal.

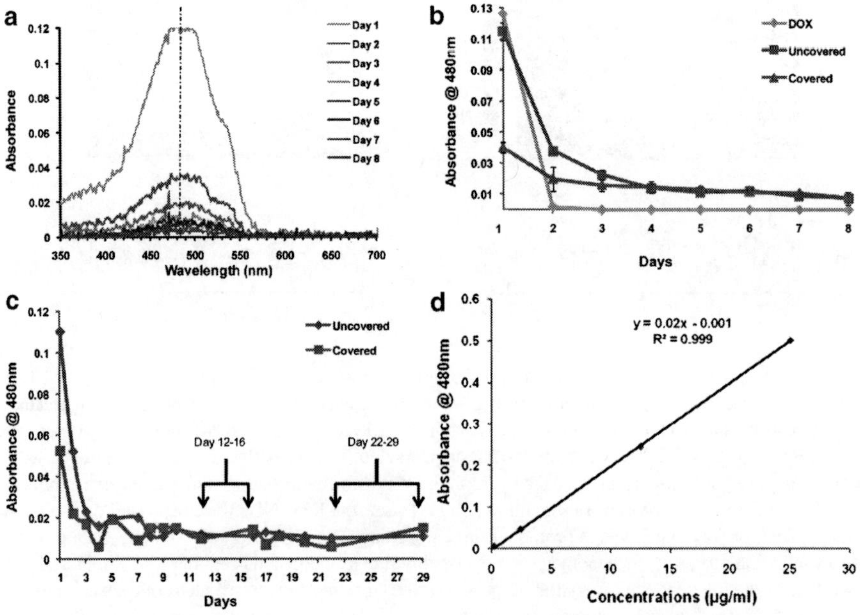

Fig. 8.3 (**a**) Eluate collected at every 24 h interval for samples uncovered with the top eluting parylene element was analyzed for drug release, with burst release often observed for incomplete devices. (**b**) Three initial 8-day delivery trials were performed for uncovered and covered microfilm devices and optical absorbance measured at 480 nm. (**c**) A long-term trial revealed a sustained release effect. (**d**) Spectroscopy data at 480 nm is plotted against respective concentrations of DOX, which demonstrates a linear correlation between absorbance measurements and eluted DOX concentrations.

Films that did not possess the deposited thin parylene layer released the majority of the deposited drug within the first day of testing, whereas films that contained the layer demonstrated a more controlled and constant release of therapeutic (Fig. 8.3b, c). The rapid release of drug within the first day served as an indicator of burst release which is a commonly observed issue in drug delivery. A large initial dosage of a cytotoxic drug such as DOX can result in severe complications such as cardiotoxicity as well as patient mortality in extreme cases. The ability to reduce burst release was an important characteristic of the microfilm device. Controlled and localized delivery offers several advantages over conventional systemic drug administration, including the ability of maintaining a desired concentration over prolonged periods of time with a single administration [3]. Moreover, DOX has poor penetration into tumor tissues, due to low diffusion rates caused by small interstitial spaces and strong intracellular binding [15]. Due to this effect, steep DOX concentration gradients are typically observed when injected in vivo, with the highest concentrations localized near the microvasculature [15]. A gentler concentration gradient can instead be developed via continuous treatment [15]. Drugs with poor penetration, such as DOX, also have low cell death thresholds, as they primarily impact cells located at the periphery, resulting in inefficient tumor killing. The maintenance of continuous treatment can alleviate this challenge by eliminating additional layers of cells via extended timeframes, without supplemental therapies [31]. As an additional benefit that addresses a major complication surrounding cancer treatment, sustained and localized elution can prevent oncogenic regrowth/recurrence between chemotherapy sessions [31].

Small dosages of drugs applied through continuous infusions can mediate a gradual increase in penetration while reducing additional toxicity issues which typically complicate the course of treatment [31]. Large doses of DOX can result in nausea, vomiting, alopecia, myelosuppression, and eventual congestive heart failure [16, 17] through cardiotoxic effects. With a continuous infusion of DOX, patients in prior clinical investigations experienced lower cardiotoxicity and reduced occurrence and severity of side effects than patients under other treatment courses, even at higher cumulative dosages with no noticeable effects upon drug efficacy [16, 17]. In fact, it has been suggested that cardiotoxicity might be related to peak plasma levels of the therapeutic instead of the cumulative drug dosage [16].

In pursuit of the benefits attributed to sustained drug release, the hybrid films were tested for their initial release profile over the first 8 days (Fig. 8.3b). Uncovered DOX–ND complexes eluted at least three times more DOX over the first day than films with the additional elution control layer, which released the drug molecules at a nearly constant rate after 24 h. In a clinical setting, the muted initial release to reduce burst elution via the covered films may aid in reducing symptoms associated with spiked levels of drugs that result from direct drug administration. After 8 days, a large portion of DOX–ND complexes on both uncovered and covered patches remained upon visual inspection, possibly due to large aggregations of NDs surrounding an inaccessible DOX core or physical entrapment onto the oxidized parylene surfaces [8].

In order to evaluate the long-term performance of the microfilm device, the release trials were repeated over a 1-month long period (Fig. 8.3c). In these studies, a similar initial trend comparable to the 8-day results was observed. Two important consequences of the initial burst release of DOX–ND from the uncovered film affected drug preservation and dosage. The decreased ability of the uncovered film in sequestering the DOX molecules was a direct cause for the elevated release rates observed over the first 3 days. Samples were allowed to elute for a period longer than 24 h at specific times in order to determine the extended dosage levels, specifically at days 12–16 and 22–29, which are highlighted in Fig. 8.3c. During the aforementioned analysis time-frames, covered films released a greater total amount of DOX than uncovered films, primarily due to the uncovered film's drug reservoir being depleted at an early stage from its large initial release. Since equal amounts of DOX–ND were coated onto both films, the increased elution from the last data point inferred that the covered hybrid films would be capable of continued steady release for a longer period of time compared to uncovered films due to a substantial amount of drug that remained in reserve.

In addition, the decreased drug preservation resulted in increased initial drug dosage, which is a particularly important issue that should be avoided during the initial phases following implantation. An immediate release of DOX following implantation has been shown to cause several negative side effects previously discussed. The ND–parylene microfilm device is envisioned to circumvent these effects, because the initial elution of DOX from the hybrid film is gradual and tapered, rather than abrupt coupled with rapid depletion, which is often referred to 'burst release.' The robustness and stability of the microfilm devices were confirmed visually throughout the experiment and ongoing trials are examining the continued extension of elution duration to maximize and/or tune the length of microfilm functionality. Sequential and combinatorial release are also potential strategies that can be implemented using this microfilm. In cases of the treatment of disorders in addition to cancer, such as anti-inflammation or wound healing, preventing burst release can play a role in reducing the cases of infection which can certainly introduce additional complications which interfere with timely recovery.

Based upon the linear relationship equating spectroscopy measurements with DOX concentrations in water displayed in Fig. 8.3d, the approximate dosage of the eluate was determined. Uncovered DOX–ND devices initially released almost 90% of the DOX–ND complexes into 2 mL of nanopure water over the first 24 h to a concentration of approximately 5.75 μg/mL, with a steep decline in elution to roughly 400 ng/mL after one week. Conversely, samples with a thin parylene coat maintained a more constant delivery rate, ranging from roughly 2 to 450 ng/mL in the same time frame. This continuous elution can potentially alleviate drug loss and treatment inefficiency through blood circulation, extravasation, or other methods of excretion [32].

8.3.2 *Biological Performance Assay*

In order to examine the biological efficacy and mechanisms of the microfilm, a DNA electrophoresis assay was performed in vitro. DNA laddering assays monitoring the presence of apoptotic responses were conducted with RAW 264.7 murine macrophages in various culture conditions, due to their natural recruitment to foreign bodies such as implants as well as well-established responses to DOX incubation.

The patterned degradation of DNA that is characteristic of apoptosis appeared as a result of exposure to DOX, and was observed in the gel for the positive control and all samples containing DOX–ND (Fig. 8.4). To show the progression of apoptosis over time, two sets of samples were harvested after 16 (lanes 1–4) and 20 h (lanes 5–8) of growth. Lanes 1:5, 2:6, 3:7, and 4:8 correlate to the negative and positive controls, the uncovered device, and the covered porous device, respectively. The capacity of DOX–ND complexes to naturally reduce DOX elution rates could be clearly observed when comparing lanes 3, 4, 7 and 8 to the 2.5 μg/mL positive controls. While positive controls prompted potent DNA fragmentation as a reference readout, the DOX–ND devices displayed a more gradual and delayed onset of apoptosis, envisioned to potentially reduce the severe side effects that result from a sudden spike in DOX dosage [16].

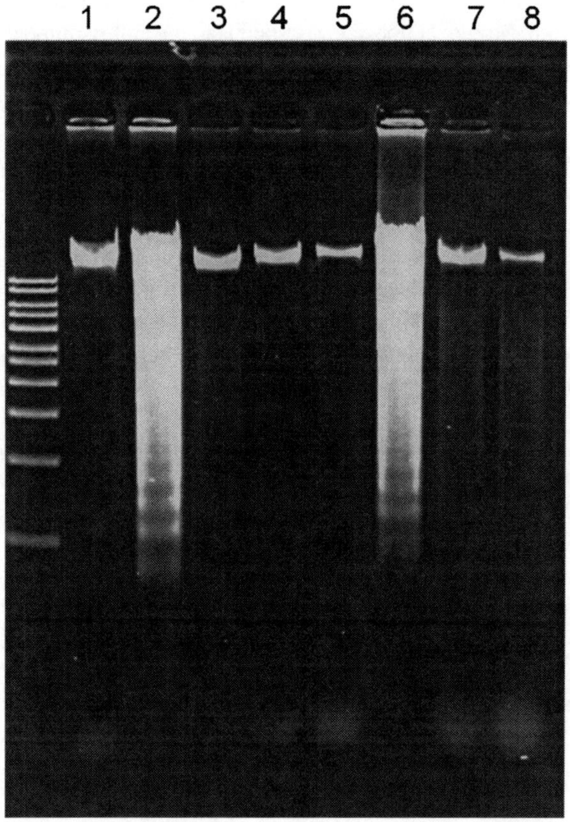

Fig. 8.4 Prolonged release and correlated delays in DNA laddering mediated by released Dox was shown. Gel electrophoresis DNA assay using RAW 264.7 murine macrophages incubated for 16 h (lanes 1–4) and 20 h (lanes 5–8) on glass (lanes 1, 5), parylene with 2.5 μg/mL of DOX (lanes 2, 6), 33 μg/mL of DOX–ND on parylene C (lanes 3, 7), and 33 μg/mL of DOX–ND sandwiched between a base and permeable layer of parylene C (lanes 4, 8). This study was able to demonstrate the sustained release capabilities of the ND microfilm patch, shown by the gradual increase in DNA laddering.

In addition, the DOX–ND devices were loaded with over 13 times the concentration of DOX compared to the positive control, further signifying the controlled release performance of the microfilm device. Therefore, the assay additionally reveals the slow-elution effects that are native to the DOX–ND complex, demonstrating the importance of integrating the NDs into the device. Furthermore, the relative degrees of banding in the gel suggested that the onset of apoptosis was dependent upon DOX dosage. This observation was most apparent when comparing uncovered and porous devices in lanes 3 and 4 (16 h) with lanes 7 and 8 (20 h). At 16 h initial fragmentation was observed in the uncovered device as a result of greater DOX–ND elution and a higher concentration of DOX–ND released into the solution due to the lack of a porous parylene layer, which further impeded upon the onset of apoptosis. At 20 h, fragmentation occurred in both uncovered and porous samples. The laddering assay showed a clear correlation with the spectroscopy data, revealing the different relative rates of elution from porous and uncovered substrates, further demonstrating the sequestration abilities of the film. Moreover, the data substantiated the ability of the device to deliver at least 13 times more DOX to a localized region. Most importantly, the combined effects of localized delivery and gradual therapeutic elution of a large reservoir of drug offered a safer, yet more sustained and potent drug delivery device.

Hybrid parylene-ND-based microfilms were constructed as a flexible, robust, and slow drug-release microfilm device with tunable dimensions that were intended to serve as implants or as stand-alone devices for specific therapies such as antitumor patches. This device configuration offers a platform upon which several therapeutic drug delivery devices can be developed to treat a spectrum of physiological disorders. Microfilm devices were capable of releasing a continuous amount of drug for at least 1 month. We believe that by altering DOX–ND deposition amounts and the thickness of the permeable parylene layer, dosage amounts and thus, total release timeframes can be calibrated and controlled. As drug release can be correlated to drug concentration gradients, the patches can then be optimized to reduce elution rates should the local therapeutic concentration reach a defined threshold.

Downstream iterations of the microfilm device will include utilizing layer-by-layer techniques to construct well-ordered ND and drug multilayers on a polymer surface [9]. Also, modified parylene polymers (e.g. amine terminated dix-A) can also be explored for further modification of the microfilm surfaces. The versatility of the key components in this device, namely the biocompatible structural material, parylene, and the ND-based drug sequestering element, provides exciting prospects involving adjustable and extended timed release in a highly biocompatible format with combinatorial therapy studies.

8.4 Conclusions

As the DOX–ND complex is highly efficient for controlled drug release, the microfilm device developed may serve as a transformative technology for pre and postoperative therapies to induce localized therapeutic release and tumor apoptosis prior to surgical intervention, or reduce the recurrence of cancer after surgery. Furthermore, the encapsulated NDs may be additionally functionalized with fluorescent molecules to serve as an imaging agent release platform as well for cancer detection or other types of diagnostic applications after ND elution. Moreover, the flexibility of this device and scalability of the fabrication process enables the patch to be uniformly coated on any surface including a spectrum of implants (e.g. drug delivery) to increase the biocompatibility and functional capabilities (e.g. therapy, inflammatory suppression, regenerative medicine, etc.) of the coated device to address continually expanding domains of therapeutic delivery. The authors gratefully acknowledge support from the National Science Foundation, a V Foundation for Cancer Research V Scholars Award and National Institutes of Health grant U54 A1065359

References

1. Peer D, Karp JM, Hong S, Farokhzad OC, Margalit R, Langer R (2007) Nat Nanotechnol 18:751
2. Volodkin D, Arntz Y, Schaaf P, Moehwald H, Voegel JC, Ball V (2008) Soft Matter 4:122
3. Langer R (1990) Science 249:1527
4. Wood KC, Chuang HF, Batten RD, Lynn DM, Hammond PT (2006) Proc Natl Acad Sci USA 103:10207
5. Kim B, Park SW, Hammond PT (2008) ACS Nano 2:386
6. Deming TJ (2002) Adv Drug Deliv Rev 54:1145
7. Lacerda L, Bianco A, Prato M, Kostarelos K (2006) Adv Drug Deliv Rev 58:1460
8. Huang H, Pierstorff E, Osawa E, Ho D (2007) Nano Lett 7:3305
9. Huang H, Pierstorff E, Osawa E, Ho D (2008) ACS Nano 2:203
10. Schrand AM, Huang H, Carlson C, Schlager JJ, Osawa E, Hussain SM, Dai L (2007) J Phys Chem Lett B 111:2
11. Liu K-K, Cheng C-L, Chang C-C, Chao J-I (2007) Nanotechnology 18:325102
12. Yu S-J, Kang M-W, Chang H-C, Chen K-M, Yu Y-C (2005) J Am Chem Soc 127:17604
13. Härtl A, Schmich E, Garrido JA, Hernando J, Catharino SCR, Walter S, Feulner P, Kromka A, Steinmüller D, Stutzmann M (2004) Nat Mater 3:736
14. Yang W, Auciello O, Butler JE, Butler JE, Cai W, Carlisle JA, Gerbi JE, Gruen DM, Knickerbocker T, Lasseter TL, Russell JN Jr, Smith LM, Hamers RJ (2002) Nat Mater 1:253
15. Lankelma J, Dekker H, Luque RF, Luykx S, Hoekman K, Valk P, Diest PJ, Pinedo HM (1999) Clin Cancer Res 5:1703
16. Legha SS, Benjamin RS, Mackay B, Ewer M, Wallace S, Valdivieso M, Rasmussen SL, Blumenschein GR, Freireich EJ (1982) Ann Intern Med 96:133
17. Legha SS, Benjamin RS, Mackay B, Yap HY, Wallace S, Ewer M, Blumenschein GR, Freireich EJ (1982) Cancer 49:1762
18. Hahn AW, York DH, Nichols MF, Amromin GC, Yasuda HK (1984) J Appl Polym Sci Symp 38:55
19. Yamagishi F (1991) Thin Solid Films 202:39
20. Schmidt EM, McIntosh JS, Bak MJ (1988) Med Biol Eng Comput 26:96
21. Burkel WE, Kahn RH (1977) Ann N Y Acad Sci 283:419
22. Fortin JB, Lu T-M (2002) Chem Mater 14:1945
23. Huang L-CL, Chang H-C (2004) Langmuir 20:5879
24. Ushizawa K, Sato Y, Mitsumori T, Machinami T, Ueda T, Ando T (2002) Chem Phys Lett 351:105

25. Kossovsky N, Gelman A, Hnatyszyn HJ, Rajguru S, Garrell RL, Torbati S, Freitas SSF, Chow G-M (1995) Bioconjug Chem 6:507
26. Jain RK (2001) Adv Drug Deliv Rev 46:149
27. Gorham WF (1966) J Polym Sci A1 4:3027
28. Lee JH, Hwang KS, Kim TS (2004) J Korean Phys Soc 44:1177
29. Chang TY, Yadav VG, Leo SD, Mohedas A, Rajalingam B, Chen C-L, Selvarasah S, Dokmeci MR, Khademhosseini A (2007) Langmuir 23:11718
30. Spellman GP, Carley JF, Lopez LA (1999) J Plast Film Sheet 15:308
31. Tannock IF, Lee CM, Tunggal JK, Cowan DSM, Egorin MJ (2002) Clin Cancer Res 8:878
32. Tannock IF (2001) Cancer Metastasis Rev 20:123

Chapter 9
Protein–Nanodiamond Complexes for Cellular Surgery

Nanodiamond and Its Bioapplications Using the Spectroscopic Properties as Probe for Biolabeling

J. I Chao, E. Perevedentseva, C. C. Chang, C. Y. Cheng, K. K. Liu, P. H. Chung, J. S. Tu, C. D. Chu, S. J. Cai, and C. L. Cheng

Abstract Nanodiamonds have attracted great attentions lately for their superb physical/chemical properties and promising bio/medical applications. This versatile material in its nanoscale exhibits little cytotoxicity in the cellular level and is considered to be most biocompatible among carbon derivatives. Its surface can be easily modified; so, various functional groups can be generated to allow conjugation of various biomolecules of interest for applications. Nanodiamonds thus provide a convenient platform for bio and medical applications. Among the various useful characteristics of nanodiamods, their spectroscopic properties, such as Raman and fluorescence, are ideal for use as a biocompatible marker to probe the biointeractions. In this chapter, the possibilities of using nanodiamond as a probe for biolabeling/cellular surgery are discussed. For this purpose, functionalization and characterization methods of nanodiamond surfaces are developed. The interaction of proteins with targeted bioobjects are investigated in various models to test the feasibility of ND to use as nano-bio-probe. Further applications using the developed nano-bio-probe as a surgical tool in the nano scale is also proposed. This developed methods of using nanodiamond as a nano-bio-probe will provide a biocompatible biolabel in bio and medical applications.

9.1 Introduction

Carbon exists in many forms that provide various applications in the fields of electronics and biotechnology. Carbon paste, graphite, carbon fibers, porous carbon, and glassy carbon are used for biosensors; other newly developed promising carbon derivatives,

J.I Chao, C.C. Chang, and K.K. Liu
Department of Biological Science and Technology, National Chiao Tung University, Hsin-Chu, 30050, Taiwan

E. Perevedentseva, C.Y. Cheng, P.H. Chung, J.S. Tu, C.D. Chu, S.J. Cai, and C.L. Cheng (✉)
Department of Physics, National Dong Hwa University, Hualien, 97401, Taiwan
e-mail: clcheng@mail.ndhu.edu.tw

D. Ho (eds.), *Nanodiamonds: Applications in Biology and Nanoscale Medicine*,
DOI 10.1007/978-1-4419-0531-4_9, © Springer Science+Business Media, LLC 2010

such as carbon nanotubes, fullerenes, and diamond nanofilms, have all found their ways to the applications in nanotechnology [1–4]. Among the various forms of carbon, diamond is by itself unique and usually recognized by its exceptional hardness, thermal properties, excellent optical properties, and electrochemical inertness. Nanodiamonds (NDs) have attracted emerging attentions, focusing on their bio and medical applications recently. From the point of view of bioapplications, available with variable sizes in the nano-scale, diamond surfaces also provide a convenient platform for bioconjugation [5, 6]. The hydrogen–carbon bonding characteristics of most diamond surfaces are well studied [7] and convenient for bioconjugation either chemically (covalently or non-covalently) or physically (adsorption). It is also easy to modify the diamond surfaces to create various functional groups for bioconjugations. Recently, the physical and chemical properties of nanometer-sized diamond particles have attracted great attention for their promising applications in the field of nanobiotechnology [8, 9]. The methods of functionalization and modification of different nanoparticles are developed [10, 11]. Some works exist concerning the surface functionalization and modification of diamond films surface with biomolecules, such as DNA, different proteins, e.g., antigen–antibody interaction, fluorescent dyes labeled proteins, etc., were immobilized on the diamond thin films [11–14]. In this respect, nanometer-sized diamond particles allow easier biomolecules immobilization, while at the same time they possess reproducible electrochemical behavior and useful physical characteristics [6, 15]. As a particle in the nanoscale, it is readily available commercially for bio and medical applications.

The seemly promising applications also pose great public concerns on these nanoparticles' safety. The development of nanotechnologies and their wide applications bring up the concern of nanomaterial safety. Recent various experimental studies indicated inhaling nanoparticles may produce significant lung toxicity, and the potential of toxicity increases with decreasing particle size and increasing surface area [16–19]. The problem of cytotoxicity is directly related to the common problem of health impact of nanomaterials and hence the feasibility of the nanomaterial's application in biology or medicine. Direct investigations are necessary to understand the mechanisms of nanotoxicity for the nanomaterial to have any sensible bio/medical applications. There have been some investigations on the cytotoxicity of carbon nanoparticles [20] and nanodiamonds recently. NDs were cocultured with various cells such as lung cancer cells [21], kidney cancer cells [22], etc. All these studies indicate ND is relatively nontoxic in the cellular level. However, long-term toxicity or safety of NDs in the animal models still remains to be investigated.

Nanoparticles have potential as nano-sized probes and sensors for tissues (e.g., intravascular probes, probes for tumor targeting, etc.) and directly for cells for both diagnostic (e.g., imaging) and therapeutic purposes (e.g., drug delivery) [8, 9]. Their nontoxicity and biocompatibility are necessary conditions of applications for such purpose. Therefore, for any sensible application in biosystem, first, for the use in biosystems, the cytotoxicity should be addressed. Second, easy surface functionalization and modification as well as detection are necessary for nanoparticles' biosensing. It is hoped that the designed nanoparticle–biomolecules complexes can interact specifically or non-specifically with components of investigated bioobject and hence many interactions can be studied in detail. For this purpose, ND appears to be an ideal candidate as it possesses the necessary properties.

The strategies for diamond's applications in medicine involve developing a biocompatible biolabels with superb signals (fluorescence or Raman spectroscopy) for detections. It should be biocompatible and easy functionalized, with reliable detection signals and techniques. It is hoped that ND can be used as an alternative for molecular dyes or quantum dots (QDs), as these molecules or nanoparticles suffer cytoxicity and photobleaching at use. The biocompatible nanoprobe calls for conjugation of biomolecules on ND surface and the characterization methodologies and techniques. Various techniques are available for characterization. The obtained ND–biomolecule complex serves as a nano-bio-probe that allows observation of biointeractions or in a medical treatment (such as drug delivery or nanomedicine).

The interaction of the ND based nano-bio-probes with bio-objects (such as cells or bacteria) can be studied using spectroscopic methods, in particular Raman spectroscopy and fluorescence imaging. Raman scattering can be excited at any wavelength that does not photobleach the investigated systems. It has high spectral and spatial resolution, meaning spectral overlapping can be largely avoided. However, Raman signal intensity is usually low and the spectra can be complex for many nanoparticles and biomolecules, so the only few works using Raman detection for biosensing were concentrated mainly on surface-enhanced Raman scattering (SERS) detection [23]. Different Raman-active dyes were used to label the investigated systems. To realize the benefits of high-sensitivity and high-selectivity detection coupled with multiple labeling capabilities, the Raman spectroscopic fingerprint can be designed through choice of Raman dye label. The goal of those works was to isolate the dye-coated surface from the sensing environment and to stabilize the particles against agglomeration and mechanical degradation. Unfortunately, dye labeling poses uncertain toxicity to the biomolecules or cells that are under investigation. To avoid this complication, nanodiamonds are promising candidates. Raman characteristic peak of diamond at 1332 cm^{-1} is quite intense and narrow; it is not affected by the surface functional group or the connected biomolecules of interest. More importantly, Raman investigation is a noninvasive method and can be performed in ambient condition. We have demonstrated [24] that nanodiamond particles can label the cells. Typical Raman spectrum of diamonds exhibits a sharp peak located at 1332 cm^{-1} for phonon mode of the sp^3 bonding carbons. This peak is usually accompanied by two graphite modes due to graphitic structures (the D- and G-bands of sp^2 carbons at 1355 cm^{-1} and 1575 cm^{-1}, respectively) on the surfaces. Nevertheless, the graphitic structure usually can be removed during the carboxylation stage, with the help of strong acid washing, when functional groups are generated on the diamond surface. Therefore, this diamond peak is isolated, the Raman absorption cross-section is large enough, and can be used as an indicator for the location of diamond when the laser scans spatially across the sample. As the diamond size decreases, significant surface atoms dominate the particle, so the 1332 cm^{-1} will decrease relatively to the graphitic signals (the D- and G-bands). In our size-dependent Raman analysis, diamond particle size of 50 nm and above could provide 1332 cm^{-1} peak, clear enough for Raman detection. For smaller size diamonds, one suffers the competition of D- and G-band intensities. Nevertheless, it is still possible to use the broader D- and G-bands for detection.

In this chapter, the possibility and methodologies of using nanometers-sized diamond particles as nano-bio-probe will be discussed. The biocompatibility is demonstrated by the cytotoxicity test of carboxylated nanometer-sized diamonds (cNDs, average diameters 5 and 100 nm) in the A549 human lung epithelial cells. To emphasize its uniqueness, we present the interaction of cNDs with cells; and the undamaging penetration of cNDs into cells and its accumulation in cells; the distribution of cND within the cells is visualized via detection of nanodiamond spectroscopic signals (Raman and fluorescence). The principles of using nanodiamond for cells labeling and the microscopic and spectroscopic methods of detection are discussed. The conjugation of diamonds and proteins is achieved via the functional groups created on the diamond surfaces through the carboxylation. Carboxylated nanodiamond conjugated with test protein, lysozyme, through physical adsorption is considered as a model nano-bio-probe and its interaction with bacteria *Escherichia coli* is analyzed. In another case, growth hormone receptors' interaction with growth hormone conjugated on cND surfaces were observed using Raman mapping technique. The developed nano-bio-probe allows visualization of the protein–bacteria or protein–cell interactions via the detection of diamond Raman or fluorescence signals can be achieved in the mean time preserving the original functions of the test proteins.

9.2 Nanodiamond's Cytotoxicity

9.2.1 The Cytotoxicity Test

To access the cytotoxicity of nanodiamond, nanodiamonds of various diameters were cultured with human lung cancer cells. The A549 cell line (ATCC, #CCL-185) was derived from the lung adenocarcinoma of a 58-year-old Caucasian male. These cells were cultured in RPMI-1640 medium (Invitrogen Co., Carlsbad, CA), which were supplemented with 10% fetal bovine serum (FBS), 100 units ml^{-1} penicillin, 100 μg ml^{-1} streptomycin, and L-glutamine (0.03%, w/v). The cells were maintained at 37°C and 5% CO_2 in a humidified incubator (310/Thermo, Forma Scientific, Inc., Marietta, OH).

The cells were plated in 96-well plates at a density of 1×10^4 cells/well for 16–20 h. Then the cells were treated with various concentrations of 5-cND or 100-cND for 4 h in serum-free RPMI-1640 medium. Here, cND denotes the carboxylated nanodiamond, and details are given in the next section. After the treatment, the cells were washed twice with PBS, and were recultured in complete RPMI-1640 medium (containing 10% PBS) for 2 days. Subsequently, the medium was replaced and the cells were incubated with 0.5 mg/ml of MTT in complete RPMI-1640 medium for 4 h. The surviving cells converted MTT to formazan that generates a blue-purple colour when dissolved in dimethyl sulfoxide. The intensity was measured at 565 nm using a plate reader (OPTImax; GE Healthcare, Little Chalfont, Buckinghamshire, UK) for enzyme-linked immunosorbent assays. The relative percentage of survived cells was calculated by dividing the absorbance of treated cells by that of the control in each experiment.

Figure 9.1 presents an optical view from conventional optical microscope of A549 cells untreated and treated with 1mg/ml of 100 nm cND. As seen from these images, the cells' morphologies were not altered when incubated with cNDs [24].

In cytotoxicity experiments, A549 human lung cancer cells and HFL-1 normal lung cells were tested with various concentrations of cND. We specifically used cNDs, as these would be the building blocks for further connection to biomolecules of interest. The results A549 tests are presented in Fig. 9.2. Both 5 and 100 nm

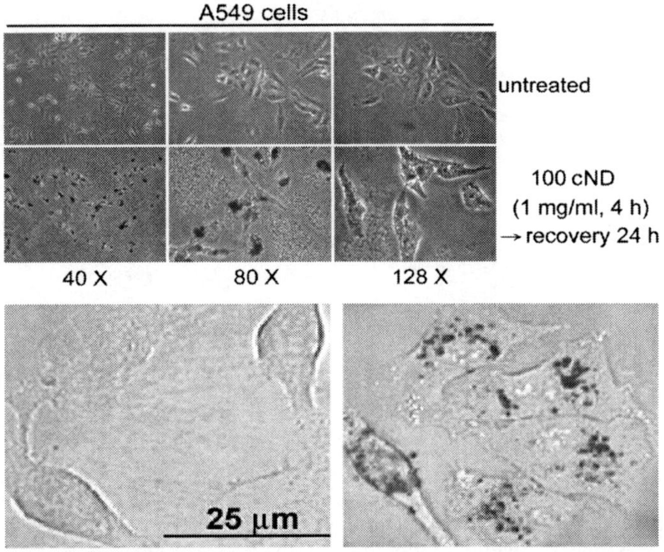

Fig. 9.1 A typical view from conventional optical microscope for A549 cells untreated and treated with 1 mg/ml of 100 nm cND (Top, reproduced with permission from [24]; copyright 2007, the Biophysical Society). An enlarged view of untreated and treated A549, Bottom left, Bottom right, respectively

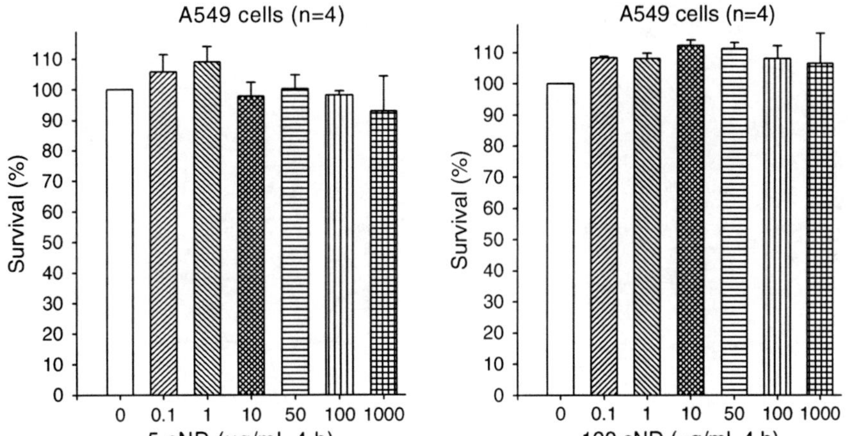

Fig. 9.2 Cytotoxicity test of cNDs on A549 Human lung epithelial cells. The cell survival rate as measured by MTT assay for both 5 nm cND and 100 nm cND treated A549 cells. Reproduced with permission from [24]; copyright 2007, the Biophysical Society

cNDs did not significantly induce the cell death in A549 even at cND concentrations went as high as 1 mg/ml.

The cytotoxicity test of multiwall carbon nanotubes (MWCNT) on A549 cell lines exhibits different results. At various concentration of CNT incubations, the cell death rate increases with increasing concentrations, as shown in Fig. 9.3. However, in our test, the CNT toxicity depends heavily on the sample qualities as the sample cleaning procedures varies, the toxicity effects may be more complex than a simple cytotoxicity test can be evaluated.

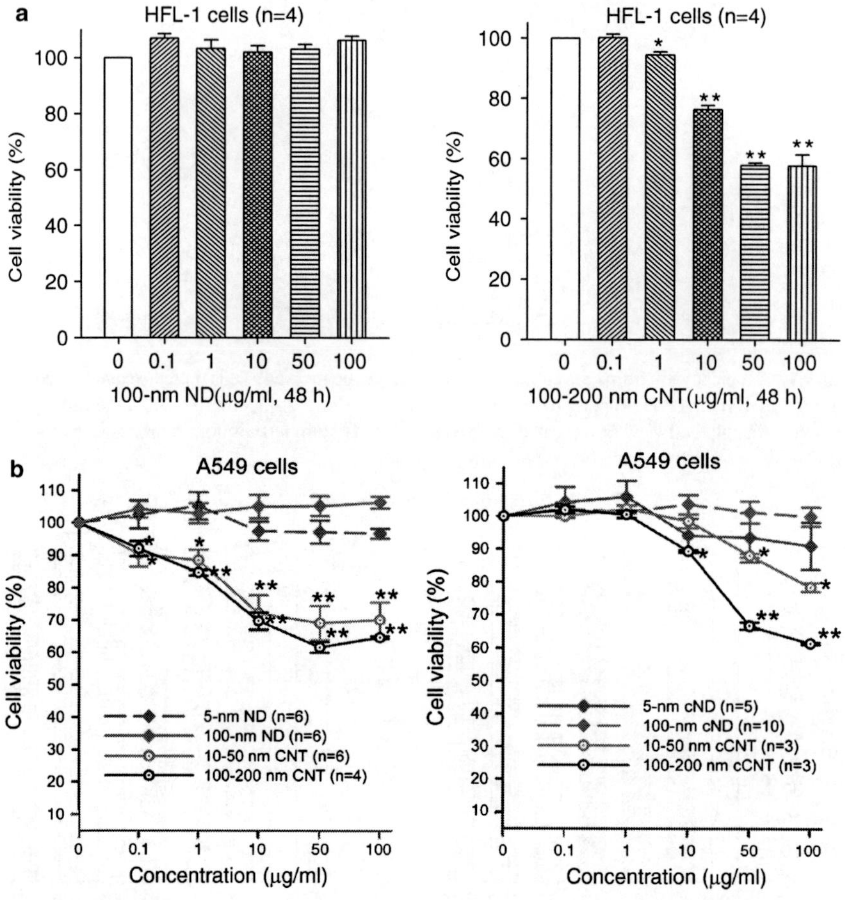

Fig. 9.3 (**a**), (**b**) Cytotoxicity test of NDs/cNDs and CNT/cCNT on normal Human lung cell (HFL-1). (**b**) Cytotoxicity test of ND, cNDs and CNT, cCNT on A549 Human lung epithelial cells. Reproduced with permission from [21]; copyright 2007, IOP publishing Ltd, UK

9.3 Using Nanodiamond as a Probe for BioDetections

The strategy of using nanodiamond for biodetection (labeling) is to create nanobiological probes (Nano-Bio-Probe) based on nanodiamonds (NDs). As depicted in Fig. 9.4, the ND–protein complex is formed using nanodiamonds' surface functional groups for conjugation with biomolecules of interest. The unique spectroscopic properties (such as Raman and fluorescence) of ND can be used as a probe for detection of biointeractions. The logic step is to first functionalize the surfaces of ND, then the conjugation of biomolecules onto the functionalized ND surface to create ND–biomolecule complex. As schematically illustrated in Fig. 9.4, the first step involves the functionalization of ND surface, followed by biomolecules conjugation to form a functional nano–bio probe. There have been various methods developed for the functionalization of ND surface [6, 11]. Different methods have developed to avoid agglomerations of ND to form large aggregates, that hinder the functionalization efficiency and hence the functionalities of the conjugated bio molecules [10, 25].

One of the important steps for ND's bioapplications is to functionalize the surface for further conjugation of biomolecules of interested. Surface functionalization provides functional groups needed for bioconjugation. The developed surface of diamond particles can contain a large number of surface inorganic groups such as ether (–C–O–C), peroxide (–C–O–O–), carbonyl (–C=O), and hydroxyl-type (–C–O–H) bonding; as well as hydrocarbon fragments. These surface molecular groups can be formed directly during nanodiamond's production; however, different methods of nanodiamond's controlled functionalization consisting of creation of surface functional molecular/ionic groups are developed. These methods include hydrogenation of nanodiamond surface with hydrogen-plasma and following oxidation, photochemical functionalization, and strong acids treatment, or combination of these methods [13, 26].

9.4 The Spectroscopic Properties of the Surface Functionalized Nanodiamond

9.4.1 Infrared Spectroscopy of the Surface Functionalized Nanodiamond

The surface properties of nanodiamond play decision role for the many nanodiamonds' applications, particularly, for bio- and medical applications. The surface functional groups substantially determine nanodiamond's interaction and conjugation with biomolecules. Spectroscopic method, such as infrared (IR) spectroscopy, is a useful noninvasive method to characterize the surface. However, identification and characterization of ionic and molecular groups on the surfaces of nanometer-sized crystals by IR spectroscopy is a challenging task. This is because the spectra of these groups are always complex owing to the variations of the crystallites in structure, shape, and size. The functional groups' electrochemical states determine the role of the surface; and these can be affected by the surface defects and curvature, and even in some structural peculiarities of the nanocrystal such as symmetry violation and lattice constant changes [27]. Nevertheless, the functionalization of ND creates surface functional molecular/ionic groups that can provide chemical/physical interaction of nanodiamond with organic molecules for bioprobing and biosensing applications. The surface modification with bioactive organic molecules for specific or nonspecific interaction with cell structures is a necessary step for the preparation of bioprobes/biosensors on the base of nanodiamond. The chemical properties of nanodiamond – functional groups composite, as well as its surface properties like crystal structure and defects, charge and polar group alignment, etc. determine the surface modification with biomolecules, which can be immobilized on the nanodiamond surface by physical adsorption [28, 29], by noncovalent or covalent chemical conjugation through linker molecules [5, 26, 30]. Carboxylated/oxidized diamond, due to the negative electron affinity of diamond and the functional groups on the particles surface, can electronically interact with protein molecules, in particular, through hydrophobic and ionic interactions as well as hydrogen bonding and van der Waals forces. The structural properties of conjugated molecules, in particular, adsorbed ones, also depend on the properties of adsorbing surface [31].

To gain more insight into this problem, diamond ultrafine particles with chemically functionalized (via strong acid treatment) surface are studied for diamond particles of various sizes, diameters varying from nanometer to near-micrometer scale. We concentrated on the well-developed method of functionalization using strong acids treatment, described initially by Ando et al. [32] and developed in many following works [5, 6, 26]. Strong acid treatment creates, first of all, carbonyl- (and hydroxyl-)

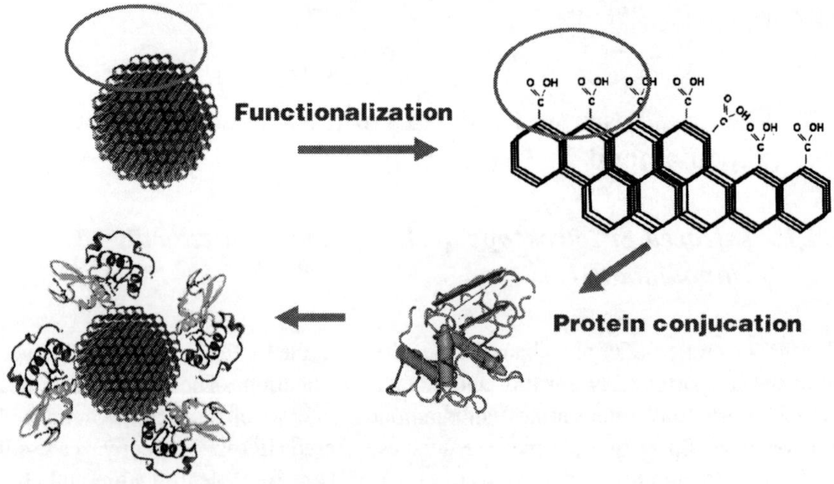

Fig. 9.4 Schematic representation of surface functionalized nanodiamond conjugation with protein to form ND–protein complex

groups on diamond surfaces. Such functionalized diamond also can be called carboxylated diamond (cND). The basic formation of the carboxyl and hydroxyl groups on the diamond can be formulated by the following equation [10]:

$$C \xrightarrow{+HNO_3} C-O-NO_2 \xrightarrow[-HNO_3]{+H_2O} C-OH \xrightarrow{\text{oxidation}} C=O$$

The surface properties of nanodiamond are also related to the problem of the nanodiamond aggregation, which is important for the preparation of well-dispersed nanodiamond suspensions. The nanodiamond aggregation is determined by surface functional groups as well as other physical–chemical properties of the surfaces, like ζ-potential and polar moment [27]. Nanodiamond size and surface structure also are important factors in the aggregation. For nanodiamond with size less then 10 nm, some additional mechanisms of tight aggregation have been discussed with the forming of crystal-like ordered structures [10, 27]. In general, the aggregates are formed through van der Waals interactions (which is stronger when the particles size is smaller, typically proportional to $1/r^6$); in addition, to the aggregation, functional groups can form interparticles hydrogen bonds (as well as other weak interactions), [10, 33] which add to the complexity of the system.

Recently, C=O stretching was used as a probe for protein configuration, because it is the most valuable band for protein diagnostic purposes [34, 35]; in other works, the C=O and COOH groups vibrational properties are studied associated with the polymer chains [36, 37]. Size-dependent properties of C–H vibration close connected to the crystal ordering of nanodiamond had been studied and found practical application for analysis of interstellar and circumstellar dust [38]. The spectroscopic properties of surface-functionalized nanodiamond have been investigated using Fourier transform infrared spectroscopy (FTIR). The surface C=O stretching frequency was observed on various nanometer-sized diamond particles and analyzed [15]. The main interest of that work is to analyze the size-dependent spectroscopic information on the adsorbed C=O on the nanodiamond surfaces as well as the pH-dependence on the C=O stretching frequencies. For this purpose, the size- and pH-dependent C=O stretching frequency (between 1680 and 1820 cm^{-1}) are studied for particles size from 5 nm to 500 nm ranges. The spectroscopic signal of the C=O on the surface of nanodiamond will provide a useful tool in the characterization of the functionalized diamond surface for biological or medical applications.

Figure 9.5 displays the observed C=O stretching frequency shift as a function of ND's sizes. We attributed the observed shifts as a result of hydrogen bond formation due to the interactions of the COOH groups among the carboxylated nanodiamond surfaces. In common case, both the peak position and shape of the C=O stretching are very sensitive to local environment. In the state free of any interaction, this band can exceed 1780 cm^{-1} and its shape is described by a Gaussian [39]. Formation of one hydrogen bond can cause this frequency red shifts by 20–25 cm^{-1}. In cyclic dimers in concentrated solutions, e.g. under extremely strong hydrogen bonding, this band is at 1710–1715 cm^{-1}, and has predominantly Lorentzian shape than Gaussian [39]. In the protein case, a protonated carboxyl group can form up to four hydrogen bonds with neighboring hydrogen bond donors and acceptors; the oxygen atom of COOH group may form up to two hydrogen bonds, and the hydroxyl oxygen and the hydroxyl hydrogen may each form one hydrogen bond. It has been shown that C=O stretching mode is sensitive to hydrogen bonding interaction [34–37, 39]. In our case, three different types of hydrogen bonds can occur as shown in schematic in Fig. 9.6: hydrogen bonds between carboxylic groups and water, hydrogen bonds between carboxylic groups of different nanoparticles, Fig. 9.6a and b and hydrogen bonds between carboxylic groups on the particle surface, Fig. 9.6c. The C=O stretching frequency shift due to hydrogen bonding is independent of the bonding types [37]. For deprotonated carboxyl group, COO^- symmetric, and asymmetric, stretching frequencies are characteristically located ~1410 cm^{-1} and ~1555 cm^{-1} [35].

The C=O stretching frequency was found to be sensitive to temperature [40]. Observed water molecules from temperature desorption confirmed the C=O stretching shifts was associated with the hydrogen bonding formation. Other molecules were also observed during the temperature treatments associated with desorption of loosely bonded surface atoms.

Figure 9.7 [40] presents the IR spectra of C=O stretching in the 1620–2200 cm^{-1} range measured in vacuum (10^{-6} torr) at different temperatures from room temperature up to 600°C, including heating and cooling processes. The two sets separated

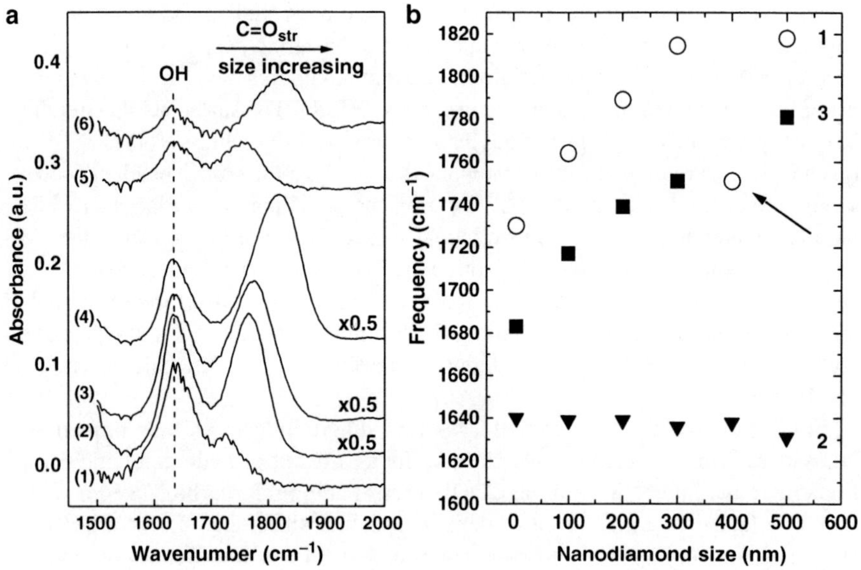

Fig. 9.5 (**a**) The surface C=O spectra of different size nanodiamonds: (1) 5 nm; (2) 100 nm; (3) 200 nm; (4) 300 nm; (5) 400 nm; (6) 500 nm; pH 4.5; (**b**) The dependence of C=O stretching frequency on the nanodiamond size: (1) C=O stretching frequency at pH 4.5; (2) O–H stretching at pH 4.5; (3) C=O stretching at pH 13. Reproduced with permission from [15]; copyright 2006, American Institute of Physics

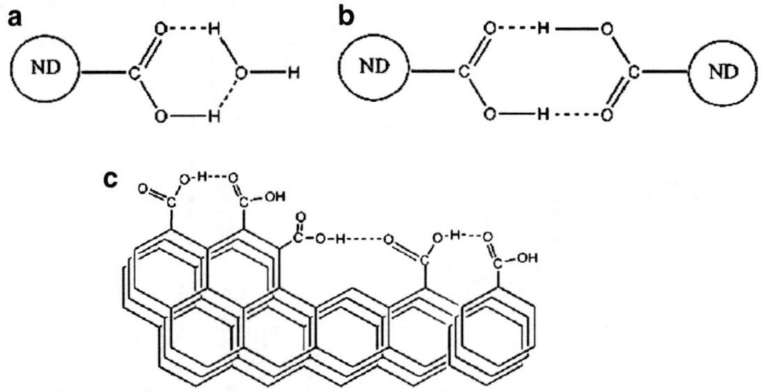

Fig. 9.6 Schematic representation of hydrogen bonds formation between carboxylated nanodiamond and medium molecules (**a**), between carboxylated nanodiamond's particles (**b**) and on nanodiamond surfaces (**c**). Reproduced with permission from [15]; copyright 2006, American Institute of Physics

measurements were performed for samples with or without preannealing in vacuum before the temperature-dependent measurements. For preannealing, samples were annealed in vacuum at 300°C in a 10^{-6} torr vacuum for longer than six hrs before the temperature-dependent measurements. Figure 9.7a and b shows the temperature shifts of the C=O peak for samples without preannealing and with preannealing, respectively. The spectral sets show clearly temperature dependence, and the behavior differs for samples, which have been and have not been subjected to preannealing. Some additional peaks emerged (or significantly increased) at 2150 and 2270 cm^{-1} after the temperature treatments (Fig. 9.7); these peaks' intensities increase with increasing temperature (Fig. 9.7a and b). The nature (attribution) of these peaks is not fully clear yet [41]. They could be attributed to sp-hybridization being characteristic to stretching vibration of medial –C≡C– and terminal –C≡C–H, and C≡C bonds, correspondingly [42]. These peaks usually belong to alkyne-containing molecular compounds, but are also detected from nanocarbon structures, such as carbon nanotubes (CNT) [43] when C≡C bonds arise on the CNT edges and break to stabilize the structure [44]. In our case, one could suppose, that temperature (up to 600°C) creates conditions for sp-hybridization form surface graphite with local defects similar to CNT structures. The surface carbon forming long chains of carbynes: polyenes (…–C≡C–C≡C–…) and polycumulenes (…=C=C=C=…) is less probable, as it requires special methods of elaboration/preparation and is very unstable [45, 46]. Existence of carbyne structures can be detectable with IR at 3300 cm^{-1} [47], but this peak is overlapped with OH vibrations in the 3000–3600 cm^{-1} range; with Raman in the 1900–2200 cm^{-1} range [48], but we did not observe corresponding peaks in the samples annealed at 300° and 600°C, that were subject to Raman investigation right after annealing.

The C=O frequency shifts for preannealed and nonannealed nanodiamonds are compared. In Fig. 9.8, the frequencies of C=O stretching as a function of temperature are presented. Figure 9.8a displays the temperature dependence for nonannealed

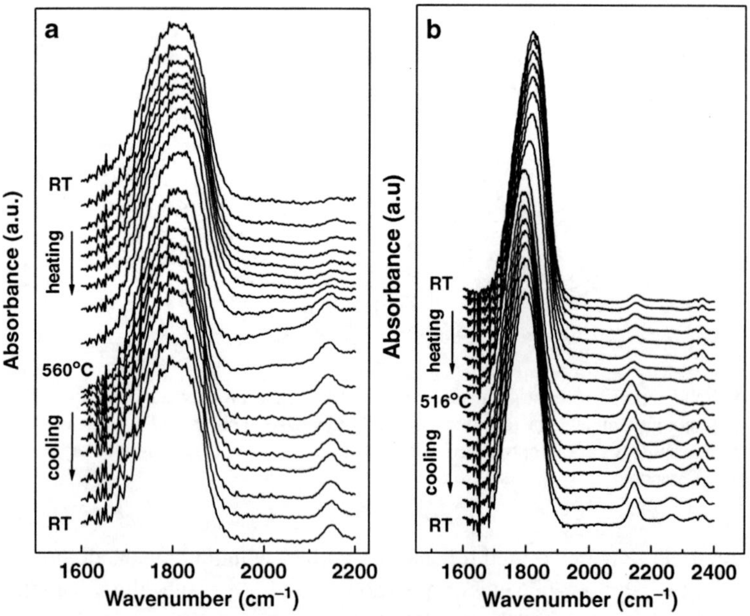

Fig. 9.7 The IR spectra of C=O, in the 1600–2320 cm^{-1} range, measured at different temperatures. Spectra for sample measured (**a**) with, (**b**) without pre-annealing at 300°C for 6 h. Reproduced with permission from [40]; copyright 2008, Elsevier B.V

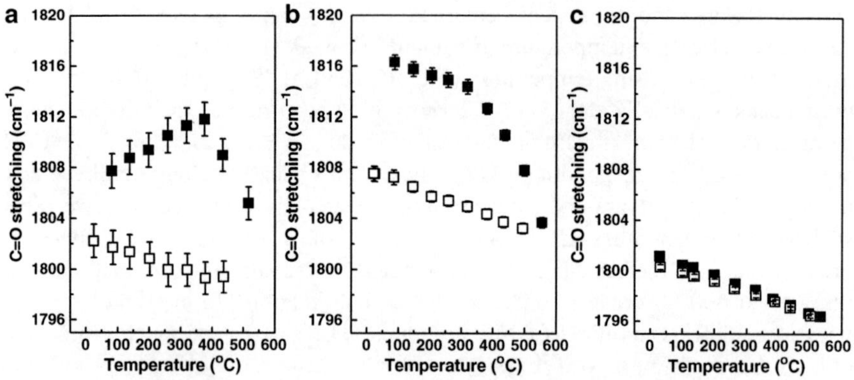

Fig. 9.8 Temperature dependence of C=O stretching frequencies of 100 cND. (**a**) Without and (**b**) with pre-annealing at 300°C, and (**c**) with pre-annealing at 600°C. (■) – Heating, (□) – Cooling. Reproduced with permission from [40]; copyright 2008, Elsevier B.V

carboxylated nanodiamond, demonstrating the blue shift in the temperature range from 25 to 350–400°C and following red shift at temperatures in the 400–600°C range. Figure 9.8b is for sample preannealed at 300°C in vacuum, and Fig. 9.8c, at 600°C preannealing, shows red shift of the C=O stretching frequency in the whole measured temperature range. In addition to the different temperature behavior of annealed and nonannealed samples, the C=O frequency shifts are not reversible and are different for heating and cooling processes, except the sample pre-annealed at 600°C. In this case, the C=O stretching becomes reversible as Fig. 9.8c illustrates.

In Fig. 9.8a, the IR spectra were taken in vacuum for sample without preannealing. The C=O frequency increases from room temperature up to 400 °C. When comparing with the preannealed sample's (Fig. 9.8b) and with the temperature desorption spectra in Fig. 9.9, the observed increasing C=O stretching is evidently caused by desorption of water from room temperature up to 400°C. The water desorption reaches a plateau at 400°C, this agrees well with the observed increasing C=O frequency also stops at 400°C. Desorption of water destroys hydrogen bonding network on the carboxylated nanodiamond surface which causes the increase of the C=O stretching. Therefore, an increase of the C=O frequency is observed. The observed shift from room temperature to 400°C is only ~5 cm^{-1}. However, both from size-dependent investigation [15] and from protein C=O studies [35], forming one hydrogen bond will lower the C=O frequency by ~20–25 cm^{-1}. In this preannealing stage, the observed blue shift of the C=O is only ~5 cm^{-1}, suggesting the effect can involve some other mechanisms. At temperature below 400°C H-bonds breaking increases C=O stretching frequency and some other interactions compete and decrease it; this effect may involve desorption of other molecules, as have observed in Fig. 9.9. Chemical desorption is less probable as the investigating temperature is relatively low for breaking the chemical bonding. In the temperature

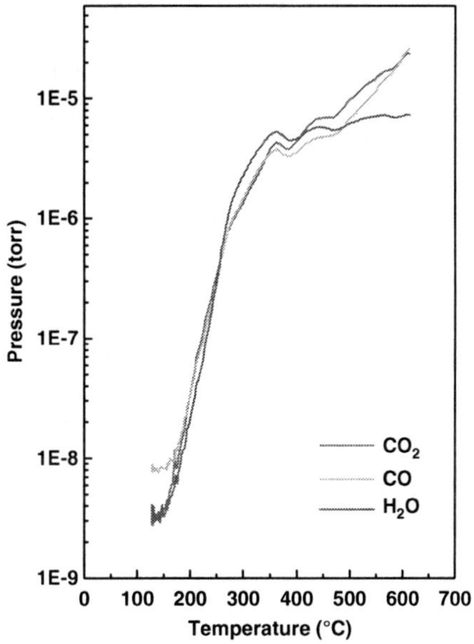

Fig. 9.9 Temperature desorption spectra of the thermally annealed cND. The recorded spectra show CO, CO_2, and water desorption from the cND surface from room temperature to 600 °C. Reproduced with permission from [40]; copyright 2008, Elsevier B.V

desorption spectra, Fig. 9.9, we observed both CO_2 and CO desorption from the surface in addition to water. So the surface desorption of water as well as CO_2 and CO is responsible for the observed blue shift of C=O by 5 cm^{-1} during the preannealing stage. For the sample subject to preannealing in vacuum, Fig. 9.8b, the C=O frequency exhibits red shift when nanodiamond is subjected to temperature annealing in vacuum. The other lateral interactions affecting C=O stretching are in competition with the hydrogen bonds, and can become more significant at higher temperatures, e.g., desorption of CO_2 or CO molecules. This desorption continues to raise at temperatures higher 450–400°C, Fig. 9.9, when water desorption has become more or less steady. In this case we can assume the cooling process observed in a nonreversible fashion, as compared to heating process, is also caused by further desorption of CO_2 and CO on the surface. To prove this, the sample was preannealed to 600°C and then subject to the temperature effect as illustrated in Fig. 9.8c. After preannealing to 600°C, nearly all physisorbed molecules have been desorbed, the results observed is then pure temperature effect measured and the results are as shown in the Fig. 9.8c. The temperature effect causes the C=O frequency to shift ~5 cm^{-1} in the temperature range from room temperature to 600°C and the process is reversible.

The lateral interactions affecting C=O stretching in addition to hydrogen bonds depend on these groups density as well as on neighbors' composition. In the temperature desorption experiment in the 100–600°C temperature range, CO, CO_2, H_2O, also, H_2 (not shown) are observed. In the temperature desorption, the same molecules can originate from decomposition of COOH surface groups, occurring in temperature range 220–600°C [49]. The thermal properties of detonation nanodiamond treated with strong acids are studied in the 100–700°C temperature range [49]. The stability of surface groups has been analyzed. The thermal shift of C=O stretching is observed together with the disappearing with temperature some other IR spectral lines in the 1000–1700 cm^{-1} range. From the temperature desorption spectroscopic measurements at the temperature above 220°C, oxygen-containing complexes were shown to begin decompose. The CO_2-yielded complexes, such as carboxylic acids, anhydrides, and lactones, were found exhibiting desorption maxima near 300°C; CO-yielded complexes for sp^3-bonded carbon corresponding to hydroxyl groups in tertiary alcohol and ketonic groups have desorption maximum near 600°C. The full decomposition of oxygen-containing groups occurring at 900°C resulted in noticeable surface graphitization of nanodiamond. Our measurements were performed in the temperature range up to 600°C, so already some decomposition can contribute. However, the assumption of decomposing surface molecular groups should only be applied to the weakly bonded ones on perhaps the defect/disordered carbon sites.

9.4.2 Raman Spectroscopy of Nanodiamond

In Fig. 9.10, Raman spectra and SEM images of nanodiamonds of different sizes are presented [50]. For the 5–50 nm nanodiamond, characteristic G-band at 1590 cm^{-1}, and D-band at 1350 cm^{-1} dominate; while for 100–500 nm nanodiamonds,

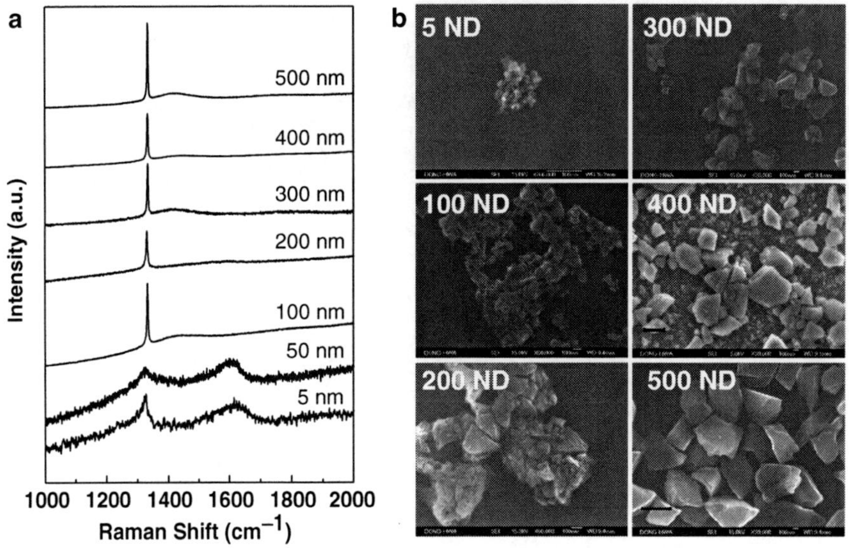

Fig. 9.10 (**a**) Raman spectra of nanodiamonds of different sizes at laser excitation 532 nm. Reproduced with permission from [50]; copyright 2007, Elsevier B.V.; (**b**) SEM images of corresponding nanodiamond powders

intense diamond peak at 1332 cm^{-1} with trace graphitic/amorphous signature exists. The nondiamond phase of ultrafine nanodiamond usually represents the shell with monocrystal diamond core inside [27]. However, for size larger than 50 nm, the structure is usually explained as diamond polycrystal structure with graphitic structure on surface. We observed that nanodiamonds with sizes 100 nm and larger, the Raman spectra have similar characters with sharp and intense diamond peak, while the spectra for 5–50 nm nanodiamonds appear differently. For the smaller diamonds, graphitic or amorphous structures dominate the spectra.

The Raman signal of diamond (1332 cm^{-1}) is an ideal marker for detection. The advantage of using nanodiamonds' unique Raman signal is it does not overlap with any other signal from the bio-objects. As demonstrated in Fig. 9.11, the formed ND–Lysozyme complex exhibits its own Raman signal, which is composed of diamond's signal and other signal from the proteins. Therefore, upon forming the ND–protein complex, the diamond signal is not affected. This unique signal serves as perfect marker for detection. With modern advancement in Raman mapping algorithm, the Raman mapping can be performed rapidly. This adds to the feasibility of using ND as a detection label in a biointeraction.

The Raman spectral mapping developed is also very sensitive due to high Raman crosses section of the diamond peak (1332 cm^{-1}). As seen in the Fig. 9.12, one single ND can be located in the A549 cancer cell using this mapping technique, and the ND's Raman signal does not interfere or overlap with any of the Raman signal for the observing cells.

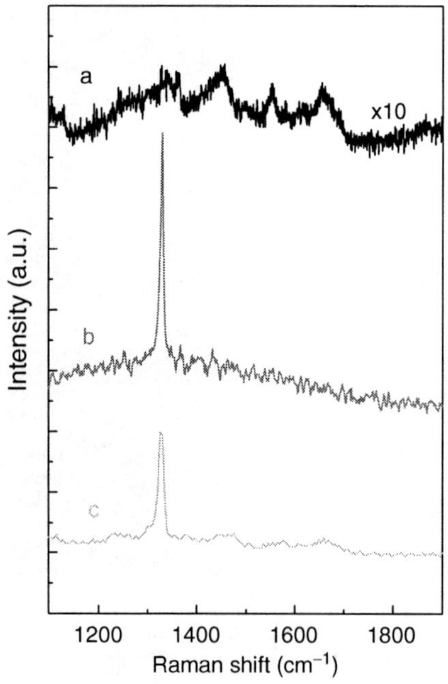

Fig. 9.11 Raman spectra of (**a**) lysozyme; (**b**) 100 cND; and (**c**) Lysozyme–cND complex. Reproduced with permission from [71]; copyright 2007, IOP publishing Ltd, UK

9.4.3 Fluorescence of Surface Functionalized Nanodiamond

The photoluminescence (PL) of diamond is studied with constant interest for many years and attracts much attention lately [51, 52]. The study of nanodiamond luminescence from defects generated by proton beam irradiation following thermal annealing [53, 54] stimulated many possible applications of implanted diamond in electronics and optoelectronics. Strong energy treatment creates nitrogen-vacancy defect centers, which emit broadband luminescence in the red region [55]. It is estimated that the strong energy treatment could increase the luminescence for 100 nm nanodiamond by 10^2-fold compared to untreated nanodiamond [56]. A few observations on the intrinsic fluorescence of nanodiamond exist for ultrafine diamond with size less than 10 nm [57, 58].

However, high energy treatment results in complication in the sample preparation and hence hinders its applications. We focused on the particle size-dependent luminescence properties and demonstrated that the luminescence without complicated high-energy/annealing treatment can also be used for fluorescence imagining (see Section below). In our work [50], photoluminescence (PL) from the natural defects/impurities, surface and bulk, using various excitation wavelengths was studied for

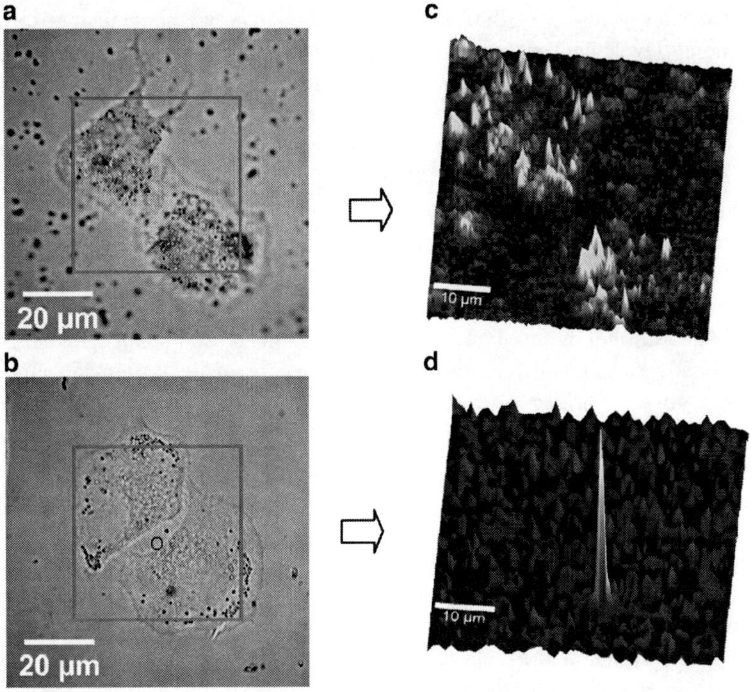

Fig. 9.12 Raman mapping on A549 Human lung epithelial cell. (**a**) A typical scanning electron microscope image of 100-nm cND with the diamonds directly deposited on a single crystal silicon wafer. (**b**) Raman spectrum of 100-nm diamond. (**c**) Optical image of a A549 cell interacted with 100-nm cNDs. The sharp Raman signature in (**c**) can be used as an indicator to locate the diamond position in the cells. (**d**) The diamond Raman peak intensity distribution versus distance across the line indicated in (**c**) scanning with a step 0.5 μm. Reproduced with permission from [24]; copyright 2007, the Biophysical Society

nanodiamond particles with size from 5 nm to near-micrometer ranges. The PL was found to be size and laser wavelength dependent. We attribute the luminescence of 5–50 nm nanodiamond to natural defects, predominantly generated on the surfaces. For nanodiamonds with size larger than 100 nm, the vacancy and divacancy centers in bulk diamond phase as well as structural inhomogenities of nondiamond phase are competitive origins for photoluminescence.

Figures 9.13a and b depict the PL spectra of different sizes nanodiamonds obtained using 488-nm and 532-nm excitation wavelengths, respectively. For 5 and 50 nm nanodiamonds, wide structureless bands in the range 550–700 nm are observed, and the maximum depends on the excitation wavelength. For 5 nm, the maximum of PL exited by 488-nm and 532-nm wavelengths is near 556 nm and 630 nm, correspondingly. The PL shifts for 50-nm nanodiamonds to 620 nm at 488 nm excitation wavelengths and 640 nm at 532 nm excitation wavelengths. Similar band was previously observed for ultrafine nanodiamond by several groups [58] and considered to be associated with radiative recombination via a system of continuously distributed energy levels in the band gap of the diamond nanoclusters [58, 59]. These energy levels can arise from structural disorder on the nanodiamond surface and the presence of carbon in the threefold-coordinated state and of dangling bonds on the surface of the nanoclusters [58]; or from the intrinsic defects of a single-crystal diamond, which contribute to the structureless PL in the same spectral range [58]. For nanodiamonds with sizes 100–500 nm the Raman spectra look similar and reveal high content of diamond phase as illustrated in Fig. 9.10 in the Raman spectra. Moreover, their PL spectra change significantly as compared to the 5–50 nm diamonds. Figure 9.13a shows the PL is emitted in the range of 580–720 nm with maximum at 660–680 nm for 532-nm excitation. The 488 nm also excites PL in red region, but additional intense band with maximum near 520 nm appears. Both wide spectral bands have local maxima; fitting the spectra using Lorenzian algorithm allows recognizing the peaks characteristic for different diamond structural peculiarities. The peaks in the range 500–740 nm are observed for 488 nm excitation and in the range 580–720 for 532-nm excitation.

For nanodiamond with size 5 nm, the part of surface carbons is extremely essential and the physical–chemical properties are accordingly determined by their surface state; therefore, the surface treatment can affect it [58]. In contrast to 5-nm nanodiamonds, for nanodiamonds larger than 100 nm only small fractions of the atoms reside on the surface and thus not affected by the acid treatment. The agreement between Raman and PL spectra of nanodimonds, size 5 and 50 nm, supports the same PL mechanisms for them; essentially surface structure defects are involved

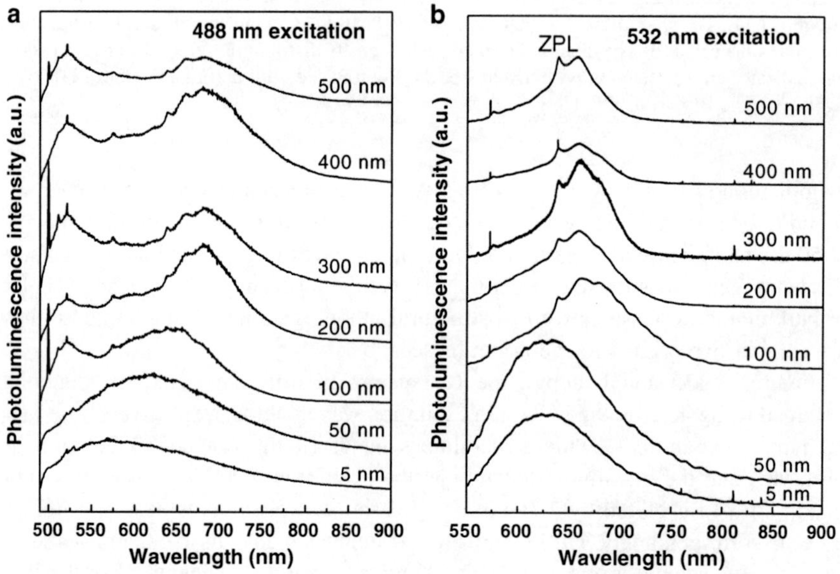

Fig. 9.13 The PL spectra of different sizes nanodiamonds obtained using (**a**) 488 nm and (**b**) 532 nm excitation wavelengths. Reproduced with permission from [50]; copyright 2007, Elsevier B.V

in PL emission for 5 nm and 50 nm alike. The positions of the maximum emission in range 600–700 nm for excitations 488 and 532 nm are presented in Fig. 9.13 as a function of sizes. As seen in this figure, the origins of luminescence nature vary for nanodiamonds of different sizes. The PL peak maximum positions, in the range 510–530 nm, excited at 488-nm excitation wavelength does not vary much with different sizes (Fig. 9.13a). However, at both 488- and 532-nm excitation wavelenghts the peak maximum positions, in range 600–700 nm, exhibits red shift with nanodiamond size increasing.

Diamond is a wide-band gap semiconductor with a variety of luminescent centers associated with donor-acceptor pairs or structural defects. Gildenblat et al. [60] considered the recombination of these pairs as responsible for luminescence. The donor–acceptor pairs and the location are associated with crystal defects. Therefore, difference in PL characteristics can be attributed to the variation of the average distance between donor–acceptor pairs, depending on crystal structure and size of nanodiamond; however, this consideration would hold only for the PL emission in the range 500–550 nm [52, 60]. Impurities, such as boron or nitrogen [52, 55, 61, 62] have been identified to be responsible for red diamond luminescence. The nitrogen–vacancy color centers have been observed in all types of natural and synthetic diamonds [51, 52], including nanodiamonds [56] and some diamond-like materials [52]. These color centers almost all are studied with samples exposed to high-energy ion or electron irradiation following high-temperature annealing to generate the color centers [53, 56]. In our experiment, we found these color centers can also be formed inside diamond during synthesis. The PL from different defect color centers, without special treatment or doping to create them, has been discussed by Jakoubovskii et al. [52]: the 1.945 eV (639 nm) and 2.156 eV (576 nm) peaks, attributed to the negative and neutral charge states of nitrogen–vacancy centers, $(N–V)^-$ and $(N–V)^o$, appear under 2.14 eV (515 nm) excitation. The peak near 639 nm is the zero-phonon line (ZPL) of $(N–V)^-$ centers with sideband in the range 600–740 nm. Together with sideband, additionally series of PL lines in the same range 1.81–2.2 eV (564–687 nm) can be resolved, and were assigned to presumable radiative transitions from different excited states to the same ground state arising due to defects, in particular, a divacancy-related centers. This mechanism can be applied to the predominant PL from 5 to 50 nm nanodiamonds; however, for 5–50- nm nanodiamond, we think surface defects are more important, whereas, for nanodiamonds bigger than 100 nm, corresponding defects can be located both on surface and in the bulk.

To explain some observed peculiarities of nanodiamond PL, the effect of quantum confinement discussed previously for carbon nanostructured materials, in particular, by Pócsik and Koós [64] should be considered. Owing to quantum confinement, the energy levels shift up by decreasing the width of the well (the size of the structural units: nanocrystals, nanoclasters, etc.). PL is related to light emission as a result of radiative recombination from molecule-like carbon units with different local bonding environments. The radiative recombination emits PL bands with different peak energies and lifetime; and constructs the widely distributed inhomogeneously broadened PL spectrum. As result of this inhomogeneity, the PL spectrum narrows with decreasing excitation photon energy. Therefore, the PL peak position does not depend on excitation energy until the excitation energy is far from the PL band, but when the excitation wavelength falls in the PL band, the band maximum shifts to longer wavelength. This result and the corresponding concept explain the difference in spectra observed at 488- and 532-nm excitations for all studied sizes of nanodiamonds. The PL in the range 500–550 nm, is emitting at excitation 488 nm and disappearing at excitation 532-nm; while at excitation with 633-nm, PL in the range 640–750 completely survives (not shown); narrowing of spectra in the range 580–740 nm at 532-nm excitation is observed comparing with 488-nm excitation.

The intrinsic defects- and impurities-originated photoluminescence from nanodiamond particles of various sizes (5–500 nm) were observed. Luminescence spanning from 500 to 800 nm can be excited with 488- and 532-nm wavelengths lasers without any high-energy treatment to create the defects/impurities. The luminescence is found to be particle-size and laser-wavelength dependent, suggesting different kinds of defects/impurities predominantly sustain the luminescence from nanodiamonds of different sizes.

9.4.4 Confocal Fluorescence Imagining Using of Nanodiamond in Cell

9.4.4.1 Confocal Fluorescence Microscopic Investigation

There have been many applications using commercial semiconductor nanocrystals (quantum dots, QDs) for their well-defined spectral characteristics [65]. Here, we provide an alternative to use the nontoxic, easy functionalized/conjugated cNDs for biolabeling. In a previous study [24], a comparison of the fluorescence (excited with 488-nm wavelength laser) of commercial quantum dots (QD-IgG) verses cND/protein-attached cNDs was investigated. A broad fluorescence with maximum intensity centered at around 525 nm was observed for 100 cND/cND–lysozyme. The ratio of fluorescence intensity of QD-IgG/cND–lysozyme was estimated to be about 30. Despite the weakness of the cND's fluorescence as compared to the QDs, it has been reported high-energy treatment of ND creates more defect centers, and the fluorescence can be enhanced by two orders of magnitude that make it closer to most QDs [56]. We mainly focus on the "untreated" cNDs and argue that the fluorescence is strong enough for confocal fluorescence measurements. In addition, in our separated experiment, we demonstrated the fluorescence is strong enough for the flow cytometry applications (results not shown).

During the interaction of nanodiamonds with A549 cells, it appears that the diamond particles aggregated in the area of the cell's cytoplasm. A second spectroscopic method was applied to show the exact location of the cND within the cell that can be identified using the natural fluorescence of nanodiamonds. We demonstrate the confocal fluorescent images in Figs. 9.14 and 9.15. In Fig. 9.14, the cytozol and

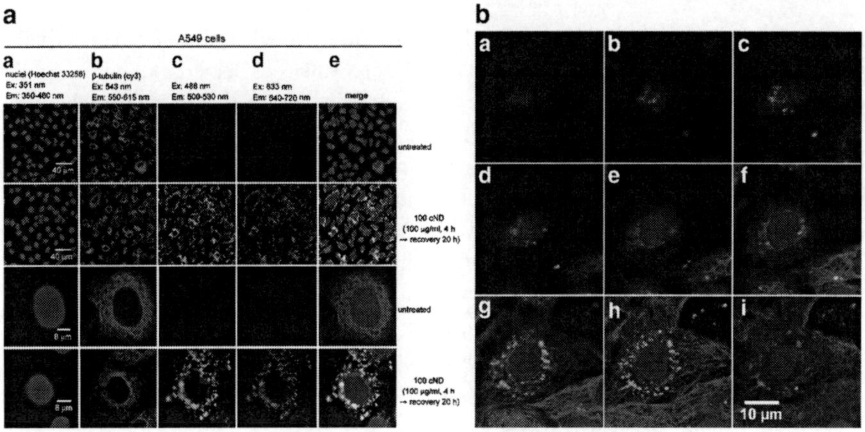

Fig. 9.14 The confocal fluorescence images of A549 cell and carboxylated 100- nm diamond. (**A**) (**a**), The cell nuclei were dyed with Hoechst 33258 to reveal the position of the nucleus. (**b**) The cell tissues were dyed with anti β-tubulin (cy3) to reflect to cytoskeleton of the cells. (**c**) The cells were interacted with 100 nm cNDs, excited with 488-nm wavelength, and the emission was collected in the range of 500–530 nm. (**d**), Same as in c, but exciting wavelength was 633 nm and emission was collected in the range 640–720 nm. (**e**) Merging the images of (**a–d**). (**B**) Cross-sectional scan on a single A549 cell. A series of confocal fluorescence images at changing position in *z*-direction with step 1 μm from the top (*upper left*) down (*bottom right*) in one of the A549 cells. (**A**), reproduced with permission from [24]; copyright 2007, the Biophysical Society

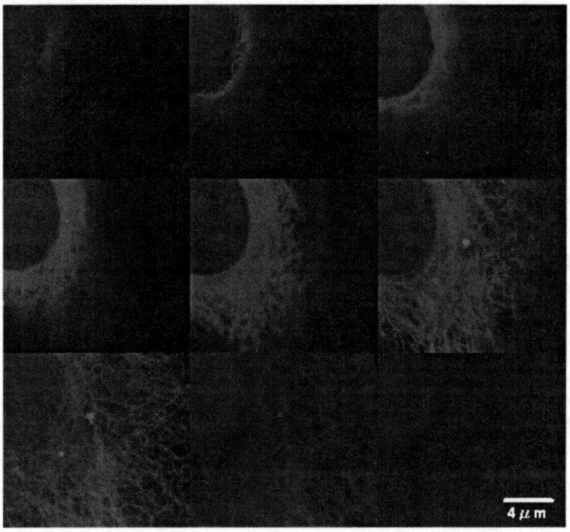

Fig. 9.15 The position of a single aggregated cND can be clearly visible (shown in *green*). At this concentration of diamond (1 μg/ml), the observed light signal from the 100 cNDs was excited with wavelength 488 nm and the emission was collected in the range of 500–530 nm. Reproduced with permission from [24]; copyright 2007, the Biophysical Society

nuclei were stained with the anti-β-tubulin and Hoechst 33258, respectively (Fig. 9.14a). Intensive signal exhibited by 100-cNDs was collected in the same cells (at excitation wavelength 488 nm the emitted light was collected at wavelengths 500–530 nm; when excitation at 633 nm, the exited signal was detected in the range 640–720 nm). When merge all images of Fig. 9.14A (a–d), the results in **e** demonstrate the dye can be replaced by 100 cNDs in our experiments. The different colors emission has been reported before at various defect centers . Through the confocal fluorescence image mapping, we confirm the coexisting of the nanodiamond particles with the A549 cells. These observed images of cND presumably could be also due to some scattering as the excitation wavelength is very close to detection range. In a separated experiments, we have obtained the cNDs fluorescence spectra using various wavelength lasers excitation, and at 488-nm excitation the emission is maximum in ranges 500–530 nm and 580–680 nm (Fig. 9.13), at 532-nm excitation the emission is observed in range 580–700 nm, and at 633 nm – in range 640–720 nm, results are not shown here.

To further trace the interaction of cNDs and the cell, a 3-D scan was performed with confocal fluorescence microscope. The excitation was at 488 nm, and the emission was collected in the range of 500–530 nm, a series image of anti-β-tubulin and cND in different positions on the *z*-axis are plotted in Fig. 9.14b. When scanning the confocal microscope in vertical direction with step 1 μm from top to bottom, the distribution of the anti-β-tubulin surrounding the nucleus can be clearly visible. When cNDs penetrate inside the cells, they indeed resided near the cytoplasm as revealed by different *z*-position fluorescent images. The fluorescent images completely traced the contour of the β-tubulin around the outside rim of the nucleus. Next, we further show that the natural fluorescent cND provide fluorescence signal strong enough for observation even at low concentration. In Fig. 9.15, the same experiments as in Fig. 9.14 was performed, except the cND concentration was 1 μg/ml. We located one or two cND and cND aggregates within the cell as revealed by a series of confocal fluorescence imaging in the *z*-direction.

9.5 Bio Molecules Conjugation to Nanodiamond

9.5.1 Physical Adsorption

We demonstrated protein physically adsorbed on cND in the model case of protein lysozyme adsorption on carboxylated nanodiamond. Protein lysozyme (AMRESCO, USA) in concentration 180–200 μM was dissolved in phosphate buffer saline (PBS, pH = 6.5–7). The protein concentration was checked with UV/Visible spectrometer (Jasco V-550) using the solution adsorption at 280 nm [63]. A molar absorbance of lysozyme at 280 nm ($3.7547 \times 10^4 M^{-1} cm^{-1}$) served to calibrate the protein concentration by the measured absorbance at the Soret band maximum. The initial concentration of lysozyme in solution was measured before adsorption, then carboxylated nanodiamond was added to the solution in concentration 4–10 mg/ml. To ensure equilibration of the adsorption, the protein solution and the diamond powder were thoroughly mixed together with a shaker for 2 h, after that the mixture was several times centrifuged and the sediment was washed with deionized water to ensure the collected Lysozyme–cND does not contain the mixture of separated lysozyme and cND, but the desired Lysozyme–cND complex. This process was further checked by measuring the residual concentration of protein in supernatant using UV-visible spectroscopy after the separation of Lysozyme–cND complex. The amount of lysozyme on nanodiamond surface was estimated by the difference between initial and residual protein concentrations in solution. The maximum amount of lysozyme that can be adsorbed by 1 mg of 100-nm carboxylated nanodiamond was calculated to be $80 \pm 10\,\mu g$, which corresponds to a few thousands (~2000) lysozyme molecules on surface of one nanodiamond particle, neglecting the aggregation of the nanodiamonds.

To obtain the infrared spectra of the nanodiamond with adsorbed lysozyme, the samples were dropped on silicon substrate and dried, then IR spectra were measured. Nanodiamond and lysozyme–nanodiamond complex particles sizes were characterized using dynamic light scattering method (DLS, Brookhaven BI-200SM, with BI-9000 digital correlator) and compared. The nanoparticles suspensions were previously filtered with a 200-nm filter to exclude larger particles.

9.5.2 Covalent Bonding

To demonstrate this, we studied the interaction of carboxylated nanodiamond–GH complex with A549 human lung epithelial cells. Fish growth hormone (recombinant fish growth hormone of the yellow grouper, rEaGH) is covalently bonded to the carboxylated nanodiamond (cND), named as cND–rEaGH. The dissociation constant K_d (the inverse of binding constant) between human GH/hGHR dimmers [66] are 16 and 0.25 nM. In heterogeneous (nonhomologous) GH/GHR binding could be lower by 5–500-folds [67]. The K_d between rEaGH/hGHR is about 8 μM in the worst case. In this case, the binding constant is still relatively higher than other cases. Meanwhile, qualitative cell proliferation assay has indicated that the rEaGH interacted with human GHR. The functions of fish growth hormone are similar to those of mammalian growth hormone in vitro [68]. To start the preparation, nanodiamond (with average diameter 100 nm, GE, USA) was carboxylated via standard procedure of strong acids treatment to create surface functional groups [15]. The rEaGH, the analog of human GH, was synthesized following the methods developed by us previously [68, 69]. The synthesized rEaGH was conjugated to cND surface using the reagents of 1-ethyl-3-[3-dimethylaminopropyl] carbodiimide hydrochloride (EDC) and *N*-Hydroxysuccinimide (Sulfo-NHS) (Pierce Led.) and these reagents linked carboxyl group of nanodiamond and amino group of rEaGH and formed a covalent amide bond [70]. The prepared sample was centrifuged at high speed (~12000 rpm); the nonconjugated GH in the supernatant was discarded.

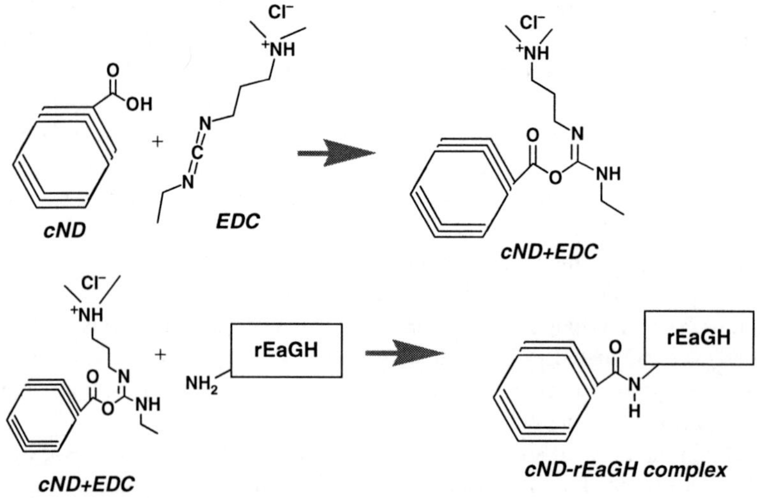

Fig. 9.16 The schematic of EDC linking of rEaGH on the carboxylated nanodiamond to form cND–eEaGH complex

The sediments was collected and washed several times with deionized water to wash away nonconjugated GH. This process was repeated several times to ensure we have cND–rEaGH complex, but not the mixture of cND and rEaGH (Fig. 9.16).

9.6 The Interaction of ND–Protein Complex with Biological Cells

9.6.1 Protein Lysozyme Interaction with Bacteria E. coli

Although the observation of SEM is sufficient to demonstrate the interaction of lysozyme and *E. coli*, the necessary vacuum condition may pose complication for sample preparation and may disturb the biosystems. Raman spectral mapping provides an alternative to observe the interaction of *E. coli* and lysozyme in situ. Observed interaction clearly demonstrates that nanodiamond can be used for biolabeling using the detection of Raman signal. The investigated samples' Raman spectra are displayed in Fig. 9.11. Figure 9.11a is the Raman spectrum of protein lysozyme. In the 1400–1700 cm^{-1} regions, some weak peaks were found due to the amides in the protein, amino acids, CH and CH_2 groups. Figures 9.11b and c shows the Raman spectra of cND and Lysozyme–cND, respectively. As seen in the Raman spectra, nanodiamond exhibits a sharp peak located around 1332 cm^{-1} is intense and clear, suitable for labeling. To further elaborate this, *E. coli* suspended with nanodiamond or Lysozyme–cND complex were stirred for 30–200 s to provide better conditions for bacteria and nanodiamond (or and Lysozyme–cND) interaction.

The suspensions were dropped on silicon substrate, rapidly dried, and optical microscope images were observed (Fig. 9.17a). In this figure, both *E. coli* with cND and with Lysozyme–cND complexes are shown. The optical images allow identification of only the relative large *E. coli* bacteria, but neither the lysozyme nor the Lysozyme–cND. The same areas of the samples were scanned with confocal Raman spectrometer, and the distribution of diamond Raman signal (1332 cm^{-1}) was mapped; the Raman mapping of the Lysozyme–cNd is presented in the bottom images of Fig. 9.17a. In this Raman mapping, the bright spots locate the positions of the Lysozyme–cND complex via the mapping of diamond Raman signature. The advantage of using nanodiamond for labeling is to neglect the complicate and weak Raman signal of lysozyme. When the microscopic and Raman images are overlapped, the overlapping of nanodiamond and bacteria images in Lysozyme–cND-treated sample indicates the interaction of lysozyme with bacteria, but visualized with nanodiamond's Raman signal. For *E. coli* suspension treated with nanodiamond without lysozyme, there is no correspondence between nanodiamond's and bacteria's images, so *E. coli* interaction with nanodiamond without lysozyme was not observed, whereas the interaction of *E. coli* with Lysozyme–cND complex is clearly observed in Fig. 9.17a. Note that *E. coli*, both in water and in lysozyme solution, as expected, did not display any signal in corresponding wavenumber range (1332 cm^{-1}) locked for the mapping. The images of *E. coli* interacted with

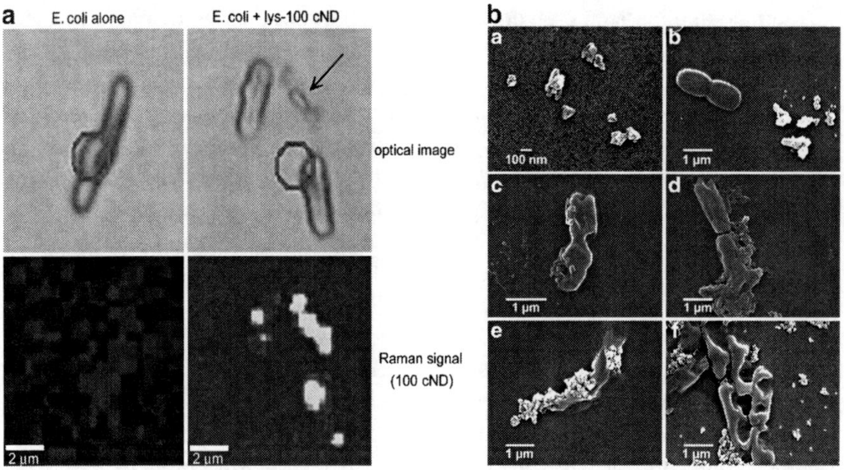

Fig. 9.17 (**A**) The interaction of *E. coli* with cND–lysozyme complex as viewed with conventional optical microscope (objective 100×) and confocal Raman spectrometer. In the optical image, *E. coli* can be seen. The Raman signal of the diamond peak (1332 cm^{-1}) was locked and scanned across a 10 µm × 10 µm area and the distribution of diamond signal intensity was plotted. The location of the nanodiamond is indicated in yellow. (**B**) SEM images of 100 cND (**a**); *E. coli* mixed with nanodiamond (**b**); *E. coli* treated with Lysozyme–cND, interaction for 30–90 s (**c**, **d**); *E. coli* treated with Lysozyme–cND interaction for 150–200 s (**e**, **f**). Nanodiamond concentration was 25 µg/ml. [(**A**) Reproduced with permission from [24]; copyright 2007, the Biophysical Society, (**B**) reproduced with permission from [71]; copyright 2007, IOP publishing Ltd, UK]

Lysozyme–cND complex are essentially different, as seen in Fig. 9.17b, the SEM images of the interacted cND and Lysozyme–cND with *E. coli*. The nanodiamond Raman signal is clearly associated with bacteria, and signal's intensity distribution draws the individual bacteria's shape indicating intensive interaction with the bacteria. In fact, as seen from SEM image of Fig. 9.17b, under this circumstance, *E. coli* would be badly destroyed. Raman imaging mapping is in good agreement with SEM observation as presented. That is, diamond Raman signal visualizes bacteria–lysozyme interaction, namely, the nanodiamond labels the protein lysozyme. The presented method of labeling and detection is simple, does not require complicate procedure of sample preparation for measurements, neither chemical treatment (fixing of bacteria, etc.) nor physical (vacuum condition). This method allows observing the interaction in living conditions, in vivo. Different bioactive molecules for specific and non-specific interaction with investigated object can be immobilized on nanodiamond via adsorption, as well as chemical linking method [71].

9.6.2 Observing the Growth Hormone Receptors (GHR) Interacting with Growth Hormone (GH) through ND Labeling

The interaction of surface growth hormone receptor of A549 human lung epithelial cells with growth hormone was observed using nanodiamond's unique spectroscopic signal via confocal Raman mapping. The growth hormone molecules were covalent-conjugated to 100-nm diameter carboxylated nanodiamonds, which can be recognized specifically by the growth hormone receptors of A549 cell. The Raman spectroscopic signal of diamond provides direct and in vitro observation of growth hormone receptors in physiology condition in a single cell level.

The gene of growth hormone (GH) is expressed not only in the pituitary gland, but also in benign and malignant tumors [72]. Meanwhile, the GH receptors have been detected in a variety of cancer cells [73]. It is though that the GH signal transduction is initiated by the GH-induced growth hormone receptor (GHR) dimerization [74]. Previous studies indicated that both the production GH from endocrine and autocrine stimulated cancer development via GH–GHR signal transduction pathway [75]. This implies that GH/GHR may be involved in the pathogenesis of human colorectal cancer [76]. Namely, the expression level of GHR may represent the cancer development stage of certain tumors. Therefore, it is important to identify the GHR level in the tumor cells. The detection of growth hormone receptor can be important and the detection technique may facilitate us to identify early phase carcinoma. In this study, we synthesized a highly sensitive nanodiamond–GH complex, which can be used to probe the number of GHR in the single molecular level. This may help us to monitor the status of cancer development in the cellular level.

Among various spectroscopic techniques, Raman spectroscopy is recently effectively used in many bio and medical studies [77]. It is considered to be a noninvasive and relatively nondestructive method. With confocal configuration, it can achieve high spectral (~1 cm^{-1}) and spatial (roughly half of the excitation wavelength) resolution, meaning spectral overlapping can be largely avoided. The advantages of Raman spectroscopy permit the developing of biolabeling techniques for bioapplications using biocompatible nanoparticles with intense and simple Raman signal. To these requests, nanodiamonds respond satisfactory for its simple Raman signal; they are biocompatible [13], chemically stable, and convenient for bioconjugation, either chemically (covalently or non-covalently) [5, 26] or physically (adsorption) [6, 26]. Typical Raman spectrum of diamonds exhibits a sharp peak at 1332 cm^{-1} for phonon mode of the sp^3 bonding carbons. The diamond peak is isolated, and the Raman absorption cross-section is large [78]. Therefore, this peak can be used as an indicator for the location of diamond. When a nanodiamond and biomolecule are conjugated, the largely complicated Raman spectra of this complex system can be represented by the simple diamond signal, i.e., the 1332 cm^{-1} diamond line can be used as a marker to locate/label the biomolecule. In this work, characteristic Raman "fingerprint" of ND allows locating the physical positions of the created ND–GH complex upon interaction with the GHR.

To demonstrate this, we studied the interaction of carboxylated nanodiamond–GH complex with A549 human lung epithelial cells. Fish growth hormone (recombinant fish growth hormone of the yellow grouper, rEaGH) is covalently bonded to the carboxylated nanodiamond (cND), named as cND–rEaGH. The method was described in Sect. 9.5.2. The prepared cND–rEaGH complexes were fixed on silicon wafer surface and analyzed with confocal Raman spectroscopy (Witec Alpha 300, Germany, using a 488-nm wavelength laser excitation).

The interaction of A549 cells with cND–rEaGH complex, or with cND, was observed via confocal Raman mapping. Some isolated Raman signals of the cND, rEaGH, and A549 were selected for the mapping of the GH and GHR interaction. Shown in Fig. 9.18, the Raman shifts near 1440–1460 and 1660–1670 cm^{-1}, correspondingly, are essentially the same for both the GH from the CH_2 groups and α-helix structure of amide I [79], and the A549 cell primarily from the CH of lipids and amide I [80]. Figure 9.18a depicts an intense and sharp 1332 cm^{-1} Raman shift for the bulk diamond from the cND. A broad shoulder at 1437 cm^{-1} is from the graphitic structure (sp^2 carbons) existed on the cND surface. Owing to large Raman cross-section of the sp^2 carbon, the broad peak is visible even only small fraction of the graphitic structure exists. Figure 9.18b is the Raman spectrum of the GH. Figure 9.18c displays the spectrum for the combination of the rEaGH and cND where both diamond and amide signals exist. Figure 9.18d is the Raman spectrum from a single A549 cell. Both A549 and GH exhibit the same peaks at near 1441 and 1667 cm^{-1}. The A549 lung cancer cell line was derived from the lung adenocarcinoma of a 58-year-old Caucasian male. The cells were cultured on Si substrate kept in Petry dish with RPMI-1640 cell growth medium (Invitrogen Co., Carlsbad, USA). The cND–rEaGH (or cND) was then added in necessary quantity to create the cND–rEaGH (or cND only) concentration ~10 μg/ml. The sample was incubated at 37°C and 5% CO_2 in a humidified incubator for 4 h. The substrate with adhered cells was twice washed with PBS (phosphate-buffered saline, pH7.4) to

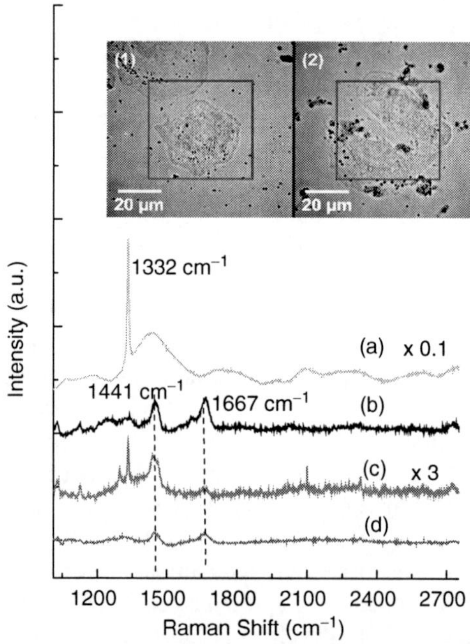

Fig. 9.18 Raman spectra of (**a**) 100 cND, (**b**) rEaGH, (**c**) 100 cND–rEaGH, and (**d**) A549. Inset: (1) Optical image (×60) of an A549 cell incubated with cND, (2) Optical images (×60) of two A549 cells incubated with cND–rEaGH. Reproduced with permission from [81]; copyright 2007, American Institute of Physics

remove the cND–rEaGH not reacted with GHR; or cND not penetrated into cells. Experimentally, this time is enough for penetration of cND into cells as well as for cND–rEaGH interaction with cells. We have proved that the GHR proteins were expressed in A549 cells (data not shown). The optical images of the cells are shown in the inset of Fig. 9.18. In the inset (1), a single cell with cND applied and incubated; in inset (2) the cells were applied with cND–rEaGH and incubated. Both optical images do not reveal any difference. However, Raman spectra reveal the diamond Raman signal can be easily observed in biosystem in ambient condition without further sample preparation is required. In addition, the conjugation of biomolecule does not alter the diamond signal, nor does it change the amide signals from the proteins in the biomolecules. When the cNDs were applied to incubate with the A549 cells, they are observed penetrating inside the cell. The exact mechanism of the cNDs cell endocytosis may be similar to the uptake and pathway mechanism of carbon nanotube [81]

This endocytosis can be observed with Raman mapping. The mapping was performed in a xy-plane of $50 \times 50\,\mu m^2$ area, 1-µm step, and at various z-axis positions. The series of z-axis scans were performed from $z = 10\,\mu m$ to $z = -10\,\mu m$, 1-µm step, with the $z = 0$ arbitrarily set at the Si substrate surface. Figure 9.19a and b illustrate this observation. The optical images of the cell(s) are shown in the insets of Fig. 9.18.

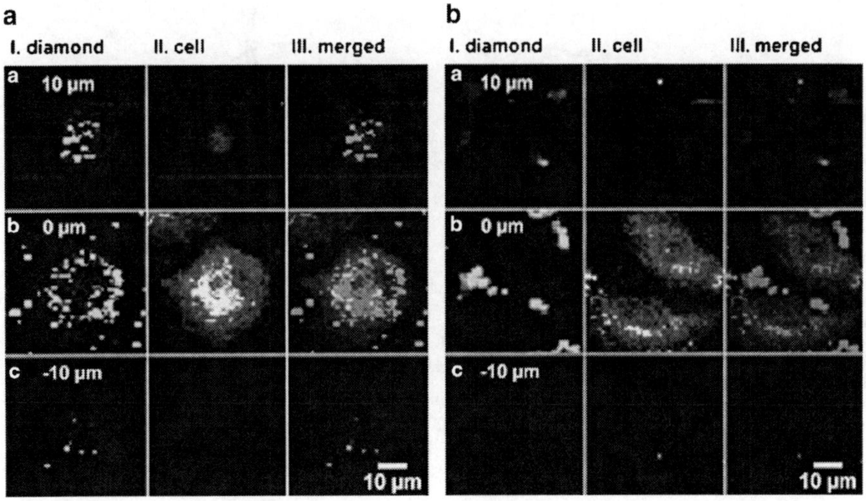

Fig. 9.19 (**A**) Confocal Raman mapping image of A549 cell and cND (**B**) A549 and 100 cND–rEaGH carboxylated nanodiamond complexes. The images are obtained by mapping the Raman signal using 488 nm laser excitation. The images shown are at different z-position scans; (**a**) at $z = 10\,\mu m$ with diamond collected in 1320–1340 cm^{-1} and cell collected in 1432–1472 cm^{-1}; (**b**) at $z = 0\,\mu m$ position; (**c**) at $z = -10\,\mu m$ position. Reproduced with permission from [81]; copyright 2007, American Institute of Physics

Figure 9.19, shows only three positions for $z = 10$, 0, and $-10\,\mu m$, respectively (full mapping of all z-positions are available but not shown). The images are constructed from a selected spectral line intensity on the Raman spectra that is unique for the samples (ND, GH, GHR or cell). Therefore, in Fig. 9.19, the intensity of each point come from the selected Raman spectrum intensity collected at this point. Column I of Fig. 9.19 maps the Raman signal of cND, whereas column II is the Raman signal locked on the A549 (the amide peak at 1432–1472 cm^{-1}). As seen in Fig. 9.19a, when the two columns were merged (column III), the cND signals overlap with the cell, indicating the cND can penetrate inside the cell. The z-positions mapping reveals that the cND reside near the nucleus of the cell. However, when the cND–rEaGH complexes were applied the effects are different, shown in Fig. 9.19b. With the GH conjugated on the cND, the whole complex resides only on the surfaces of the cell. Hormone-binding domain of growth hormone receptor is extracellular domain [82]. Therefore, the GH/GHR complex forming is on the extracellular part of membrane. This is consistent with our observation. Figure 9.20 presents a schematic representation of the interaction mechanism of the cND and cND–REaGH with the A549 cells. This observation provides strong evidence that growth hormone receptors exist on the surface, and the GH/GHR interaction can be labeled with the diamond Raman signal. This investigation suggests Raman mapping is a useful technique to observe biomolecule interaction with cells. When this method combines with surface enhanced Raman spectroscopy (SERS) or surface enhanced resonant Raman

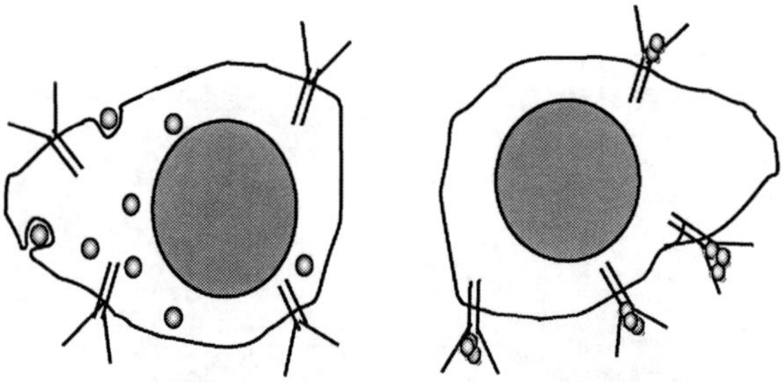

Fig. 9.20 Schematic representation of the interaction of cND and cND–rEaGH with A549 lung cancer cells. The cNDs can penetrate into the cell cytoplasm, while the cND–rEaGH interact with the growth hormone receptor on the cell membranes and reside only on the outer sites of the cells, as observed in the Raman mapping shown in Fig. 9.19

spectroscopy (SERRS), signal can be enhanced by orders of magnitude; it is possible to observe one single molecule interacts with a single cell [83, 84].

9.7 Nano Surgery Using Nanodiamond–Protein

As shown in the above sections, ND can be a very efficient biocompatible nanoparticle with unique spectroscopic properties for biolabeling. Nanodiamond (ND) exhibits intrinsic fluorescence and its unique Raman signal is both convenient for biolabeling. Meanwhile, the impurity, nitroso (C–N=O) inside the ND can be photolyzed by two-photon absorption, releasing NO to facilitate the formation of a sp^3 diamond structure in the core of ND and transforming it into a sp^2 graphite structure [85]. Such a conformational transition enlarges the size of ND from 8 nm into 90 nm, resulting in a popcorn-like structure. This transition reaction may be useful as nano-knives in nanomedicine application.

In this section, ND (average diameter 4–6 nm) was synthesized from the detonation of the mixture of trinitrotoluene (TNT) and hexogen (DNX). As shown in Fig. 9.21, both AFM and TEM images indicate that the size of the ND particles is uniform (Fig. 9.21a). A magnified view inside red circle is shown as an inset in the upper right corner, which indicates that the ND contains a 6-nm diamond core and around 1-nm thick graphite shell. The lattice spacing is about 2 Å which may correspond to {111} plane of diamond structure. The inset in the lower right corner is a diffraction pattern from the nanodiamonds in the view. It shows a typical electron diffraction ring for diamond. The inset in the low left corner is an AFM image that shows the surface morphology of the nanodiamond. Figure 9.21b is an SEM image of nanodiamond after laser irradiation. The average size of laser radiated nanodiamond is about 90 nm. Inset is a TEM image shows the magnified view of nanodiamond after laser irradiation [86].

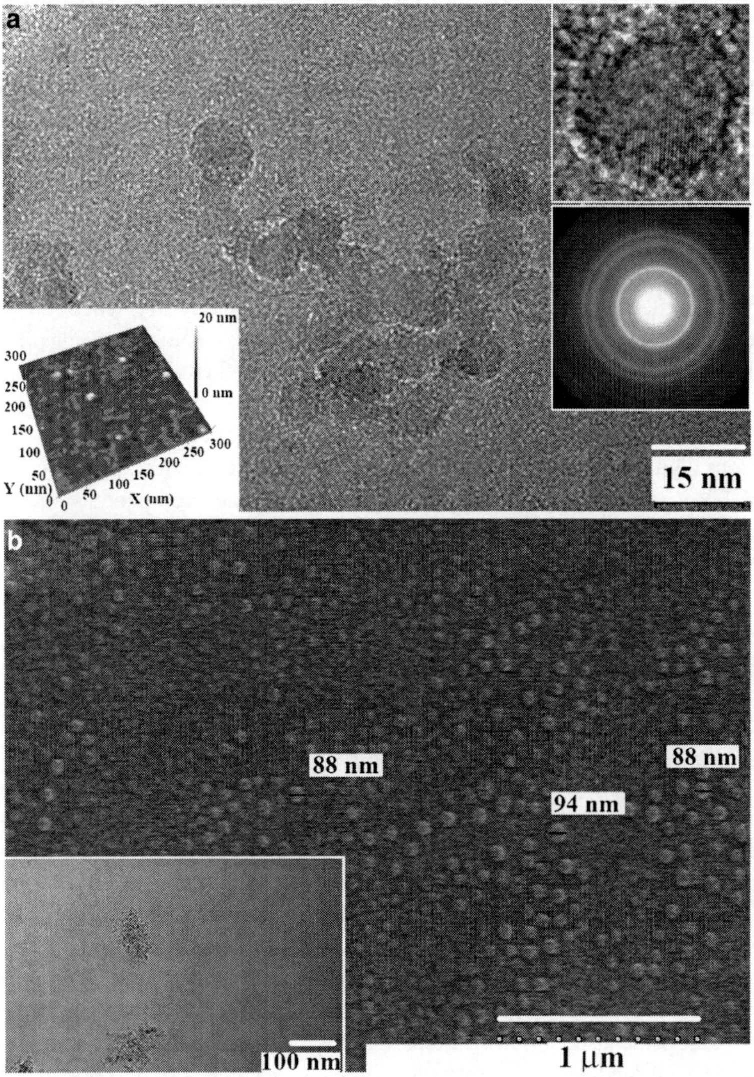

Fig. 9.21 (**a**) TEM image of nanodiamond. A magnified view inside red circle is shown as an inset in upper right corner. The lattice spacing is about 2 Å which may correspond to {111} plane of diamond structure. The inset in the low right corner is a diffraction pattern from the nanodiamonds in the view. It shows a typical electron diffraction ring for diamond. The inset in the low left corner is an AFM image that shows the surface morphology of the ND. (**b**) SEM image of nanodiamond after laser radiated. The average size of laser-radiated nanodiamond is about 90 nm. Inset is a TEM image shows the magnified view of nanodiamond after laser radiated. Reproduced with permission from [86]; copyright 2008, American Institute of Physics

The transformation of ND into popcorn (or onion-like) structures was achieved by laser irradiation of the ND solution for 30 s with 532 nm; 20 ns, full-width half maximum (FWHM), Nd:YAG, pulse laser (LS2137U/2, Lotis TII Ltd., Minsk, Belarus); with 140-mW average power; 10-Hz repetition rate; 2-mm beam size. A popcorn-like conformational change of ND was observed after the laser irradiation. The size of ND changed from 8 nm, Fig. 9.21a, to approximate 90 nm (Fig. 9.21b). This observation is consistent with our previously observation [85].

The feasibility of ND as a nano-knife in a nanosurgery can be depicted as the following. To do this, ND is conjugated with growth hormone (GH), method described in the previous section (Sect. 9.5.2), to form ND–GH, where GH is one of the typical growth factors for certain normal tissues and carcinoma. It exerts regulatory functions in controlling metabolism, balanced growth, and differentiated cell expression by acting on specific receptors, growth hormone receptor (GHR), in liver, or on cartilage cell surface, triggering a phosphorylation cascade. Thus, numerous signaling pathways are modulated and specific gene expression dictated [87]. GHR is one of the general targets of cancer drug development, as blocking or inhibiting the function of GHR may be a basis for cancer therapy.

The lung cancer cell A549 was used as a model system to be treated with ND-linked GH. The graphite surface of ND was first carboxylated by mixing ND with nitrate/sulfate (9:1) at 70°C and stirred for 24 h. The excess acid was neutralized with 0.1 N NaOH and washed with ddH_2O. The GH was prepared in our laboratory using recombinant techniques as described previously [68, 69]. The carboxylated ND (cND) molecules were linked by peptide bonding with GH via the zero-length cross-linkers 1-ethyl-3- [3-dimethylaminopropyl] carbodiimide hydrochloride (EDC) and *N*-Hydroxysuccinimide (Sulfo-NHS) (Pierce Chemical Comp., USA). The reaction was monitored by changes in the fluorescence spectra of the unique fluorescence of ND and auto-fluorescence of GH. This was followed by examination of MALDI-TOF mass spectra. Each ND particle was bound with two molecules of GH, forming an NDGH complex (data not shown). The NDGH complex (19 μM) was incubated with the A549 lung cancer cells in culture. After 8 hours of incubation, the cells were washed with phosphate buffered saline to remove the nonspecific binding complex. The NDGH complex bound to A549 cell membrane avidly. The cells were then irradiated with the same laser power mentioned above. The results show that approximately 60% of the cells died within 24 h after the irradiation (Fig. 9.22a). In contrast, less than 10% of the controls cells were found dead under comparable treatment. As shown in Fig. 9.18b, the cell death can be attributed to the explosion of ND on the cell surface.

The exact mechanism causing the cell death is not clear. It could have been due to the shock wave generated in the phase transformation, or the heat absorbed when laser irradiation from the trapped gases, or the laser heat treatments created some atomic carbon radicals that are toxic the cells. Although the mechanism is not clear, nevertheless, this opens a new possible application in nanomedicine. Especially, when the developed Raman mapping using ND as a nano-bio probe, combined with the laser nano surgery techniques in the residual cancer cells removal during surgical operations, this presents a promising method of detection the cancer, and destroy it in the nanoscale, or in the single cell level.

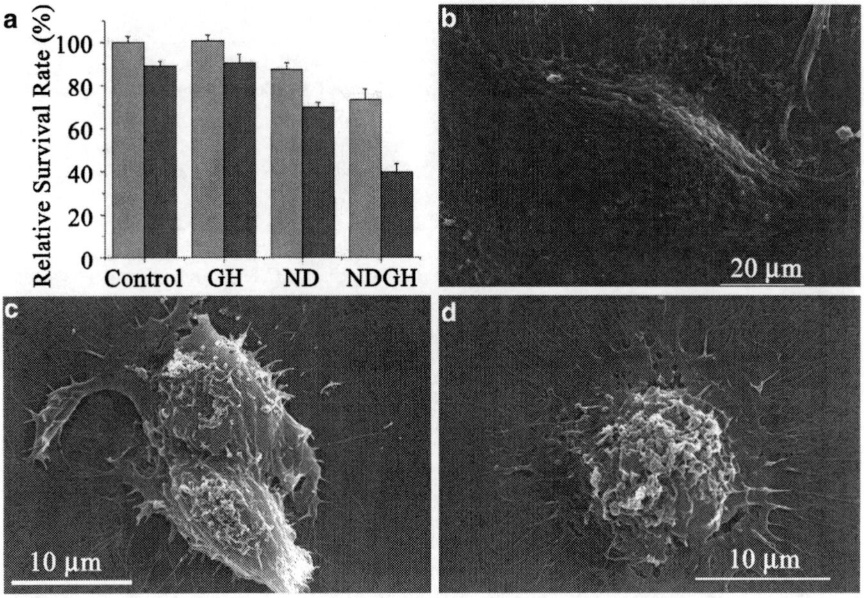

Fig. 9.22 (**a**) Cell viability assay, (**b**) SEM images of the A549 cell lines, which were not treated with ND, (**c**) treated with ND and (**d**) irradiated with laser following ND treatment. The *green* bar denotes the cells without laser irradiation and the *red* bar denotes the cell after laser irradiation and incubation for 24 h. Reproduced with permission from [86]; copyright 2008, American Institute of Physics

9.8 Conclusion

In this chapter, we have demonstrated the feasibility of using nanodiamond as a nano-bio probe. The basic cytotoxicity and spectroscopic properties were examined in detail. Nanodiamonds and protein complexes were developed and characterization methods were established. The spectroscopic, Raman, fluorescence properties of the nanodiamonds are ideal for biolabeling. We realized the applications to detect the interactions of protein lysozyme with bacterial *E. coli*; and the interaction of growth hormone with growth hormone receptors on human cancer cells A549. The developed ND–GH complexes were applied to bind with cancer cells A549, followed by laser irradiation to destroy the cancer cells. This method shows promising applications for protein–nanodiamond complexes for cellular surgery in nanomedicine.

Acknowledgments The authors thank Pei-Hsin Chen for Cell culture; Ching-Chung Chou for laser treatment, ND–GH complex synthesis; Hsueh-Liang Chu for growth hormone preparation; Tzu-Cheng Lee for some of ND functionalization works. The authors also appreciate the financial support of this research by National Science Council of Taiwan, ROC in a National Nano Science and Technology program under Grant No. NSC-(95-97)-2120-M-259-003.

References

1. Poh WC, Loh KP, Zhang WD, Triparthy S, Ye JS, Sheu FS (2004) Langmuir 20:5484–5492
2. Carlise JA (2004) Nat Mater 3:668–669
3. Wang J (2005) Electroanalysis 17:7–14
4. Sotiropoulou SG, Gavalas V, Vamvakaki V, Chaniotakis NA (2003) Biosens Bioelectron 18:211–215
5. Ushizawa KS, Mitsumori Y, Machinami T, Ueda T, Ando T (2002) Chem Phys Lett 351:105–108
6. Chung P-H, Perevedentseva E, Tu J-S, Chang CC, Cheng C-L (2006) Diam Relat Mater 15:622–625
7. Cheng C-L, Chen C-F, Shaio W-C, Tsai D-S, Chen K-H (2005) Diam Relat Mater 14:1455–1462
8. Huang H, Pierstorff E, Osawa E, Ho D (2007) Nano Lett 7:3305–3314
9. Huang H, Pierstorff E, Osawa E, Ho D (2008) ACS Nano 2:203–212
10. Krüger A, Kataoka F, Ozawa M, Fujino T, Suzuki Y, Aleksenskii AE, Vul' AYa, sawa E (2005) Carbon 43:1722–1730
11. Zhong YL, Loh KP, Midya A, Chen ZK (2008) Chem Mater 20:3137–3144
12. Clare TL, Clare BH, Nichols BM, Abbott NL, Hamers RJ (2005) Langmuir 21:6344–6355
13. Yang W, Auciello O, Butler JE, Cai W, Carlisle JA, Gerbi JE, Gruen DM, Knickerbocker T, Lasseter TL, Russell JN, Smith LM, Hamers RJ (2002) Nat Mater 1:253–257
14. Knickerbocker T, Strother T, Schwartz MP, Russell JN, Butler J Jr, Smith LM, Hamers RJ (2003) Langmuir 19:1938–1942
15. Tu J-S, Perevedentseva E, Chung P-H, Cheng C-L (2006) J Chem Phys 125:174713–174717
16. Warheit DB (2004) Mater Today 7:32–35
17. Colvin VL (2003) Nat Biotechnol 21:1166–1170
18. Nel A, Xia T, Mädler L, Li N (2006) Science 311:622–627
19. Jia G, Wang H, Yan L, Wang X, Pei R, Yan T, Zhao Y, Guo X (2005) Environ Sci Technol 39:1378–1383
20. Schrand AM, Huang H, Carlson C, Schlager JJ, sawa E, Hussain SM, Dai L (2007) J Phys Chem B 111:2–7
21. Liu K-K, Cheng C-L, Chang CC, Chao J-I (2007) Nanotechnology 18:325102
22. Yu SJ, Kang MW, Chang HC, Chen KM, Yu YC (2005) J Am Chem Soc 127:17604–17605
23. Mulvaney SP, Musick MD, Keating CD, Natan MJ (2003) Langmuir 19:4784–4790
24. Chao J-I, Perevedentseva E, Chung P-H, Liu K-K, Cheng C-Y, Chang CC, Cheng C-L (2007) Biophys J 93:2199–2208
25. Krueger A, Ozawa M, Jarre G, Liang Y, Stegk J, Lu L (2007) Phys Stat Sol A 204:2881–2887
26. Huang LCL, Chang H-C (2004) Langmuir 20:5879–5884
27. Kulakova II (2004) Phys Solid State 46:636–643
28. Huang TS, Tzeng Y, Liu YK, Chen YC, Walker KR, Guntupalli R, Liu C (2004) Diam Relat Mater 13:1098–1102
29. Puzyr' AP, Pozdnyakova IO, Bondar' VS (2004) Phys Solid State 46:761–763
30. Kossovsky N, Gelman A, Hnatyszyn HJ, Rajguru S, Garrell RL, Torbati S, Freitas SSF, Chow G-M (1995) Bioconjug Chem 6:507–511
31. Roach P, Farrar D, Perry CC (2005) J Am Chem Soc 127:8168–8173
32. Ando T, Inoue S, Ishii M, Kamo M, Sato Y, Yamada O, Nakano T (1993) J Chem Soc Faraday Trans 89:749–751
33. Xu X, Yu Z, Zhu Y, Wang B (2005) J Solid State Chem 178:688–693
34. Diomaev AK, Braiman MS (1995) J Am Chem Soc 117:10572–10574
35. Nie B, Stutzman J, Xie A (2005) Biophys J 88:2833–2847
36. ElMiloudy K, Benygzer M, Djadoun S, Sbirrazzuoli N, Geribaldi S (2005) Macromol Symp 230:39–50

37. Dong J, Ozaki Y, Nakashima K (1997) Macromolecules 30:1111 Sukhishvili S, Granik S (2002) Macromolecules 35: 301; Kim B, Peppas N (2002) Macromolecules 35: 9545
38. Chen Y-R, Chang H-C, Cheng C-L, Wang C-C, Jiang JC (2003) J Chem Phys 119:10626–10632
39. Diomaev AK (2001) Biochemistry (Moscow) 66:1269–1276
40. Chu C-D, Perevedentseva E, Yeh V, Cai S-J, Tu J-S, Cheng C-L (2009) Diam Relat Mater 18:76–81
41. Ivanov-Omskii VI, Andreev AA, Frolova GS (1999) Phys Solid State 33:569–573 in Russian
42. Coates J (2000) In: Meyers RA (ed) Encyclopedia of analytical chemistry. Wiley, Chichester, pp 10815–10873
43. Tran NE, Lambrakos SG (2005) Nanotechnology 16:639–646
44. Chen CW, Lee MH (2004) Nanotechnology 15:480–484
45. Ravagnan L, Piseri P, Bruzzi M, Miglio S, Bongiorno G, Baserga A, Casari CS, Bassi AL, Lenardi C, Yamaguchi Y, Wakabayashi T, Bottani CE, Milani P (2007) Phys Rev Lett 98:216103
46. Casari CS, Bassi AL, Ravagnan L, Siviero F, Lenardi C, Piseri P, Bongiorno G, Bottani CE, Milani P (2004) Phys Rev B 69:075422
47. Dischler B, Bubenzer A, Koidl P (1983) Solid State Commun 48:105–108
48. Ravagnan L, Siviero F, Lenardi C, Piseri P, Barborini E, Milani P, Casari CS, Bassi AL, Bottani CE (2002) Phys Rev Lett 89:285506
49. Butenko YV, Kuznetsov VL, Paukshtis EA, Stadnichenko AI, Mazov IN, Moseenkov SI, Boronin AI, Kosheev SV (2006) Fuller Nanotub Carbon Nanostruct 14:557–564
50. Chung P-H, Perevedentseva E, Cheng C-L (2007) Surf Sci 601:3866–3870
51. Davies G, Lawson SC, Collins AT, Mainwood A, Sharp S (1992) Phys Rev B 46:13157–13170
52. Iakoubovskii K, Adriaenssens GJ (2000) Phys Rev B 61(15):10174–10182
53. Prawer S, Devir AD, Balfour LS, Kalish R (1995) Appl Opt 34:636–640
54. Mita Y (1996) Phys Rev B 53:11360
55. Jelezko F, Tietz C, Gruber A, Popa I, Nizovtsev A, Kilin S, Wrachtrup J (2001) Single Mol 2:255–260
56. Yu S-J, Kang M-W, Chang H-C, Chen K-M, Yu Y-C (2005) J Am Chem Soc 127:17605
57. Zhao FL, Gong Z, Liang SD, Xu NS, Deng SZ, Chen J, Wang HZ (2004) Appl Phys Lett 85:914
58. Kompan ME, Terukov EI, Gordeev SK (1997) Fiz Tverd Tela (Phys Solid State) 39:2156
59. Aleksenski AE, Osipov VY, Vul' AY, Ber BY, Smirnov AB, Melekhin VG, Adriaenssens GJ, Iakoubovskii K (2001) Phys Solid State 43:145–150
60. Gildenblat GS, Grot SA, Badzian A (1991) Proc IEEE 79:647–668
61. Gruber A, Dräbenstedt A, Tietz C, Fleury L, Wrachtrup J, von Borczyskowski C (1997) Science 276:2012–2014
62. Chang HC, Chen KC, Kwok S (2006) Astrophysics J. 639:L63–L66
63. Pace CN, Vajdos F, Fee L, Grimsley G, Gray T (1995) Protein Sci 4:2411–2423
64. Pócsik I, Koós M (2001) Diam Relat Mater 10:161–167
65. Gao X, Cui Y, Levenson RM, Chung LWK, Nie S (2004) Nat Biotechnol 22:969–976
66. Uchida H, Banba S, Wada M, Matsumoto K, Ikeda M, Naito N, Tanaka E, Honjo M (1999) J Mol Endocrinol 23:347–353
67. Beattie J, Phillips K, Shand JH, Brocklehurst S, Flint DJ, Allan GJ (2002) Mol Cell Biochem 238:137–143
68. Chang C-C, Tsai C-T, Chang C-Y (2002) Protein Eng 15:437–441
69. Chang CC, Cheng MS, Su YC, Kan LS (2003) J Biomol Struct Dyn 21:247–255
70. Grabarek Z, Gergely J (1990) Anal Biochem 185:131–135
71. Perevedentseva E, Cheng C-Y, Chung P-H, Tu J-S, Hsieh Y-H, Cheng C-L (2007) Nanotechnology 18:315102 7pp
72. Mol JA, Garderen EV, Selman PJ, Wolfswinkel J, Rijinberk A, Rutteman GR (1995) J Clin Invest 95:2028–2034
73. Ilkbahar YN, Wu K, Thordarson G, Talamantes F (1995) Endocrinology 136:386–392

74. Lobie PE (1999) In: Bengtsson BA (ed) Signal transduction through the growth hormone receptor, growth hormone. Kluwer Academic Publishers, Boston, pp 17–35
75. Feldman M, Ruan W, Cunningham BC, Wells JA, Kleinberg DL (1993) Endocrinology 133:1602–1608
76. Wu X, Wan M, Li G, Xu Z, Chen C, Liu F, Li J (2006) Eur J Cancer 42:888–894
77. Banerjee HN, Zhang L (2007) Mol Cell Biochem 295:237–240
78. Knight DS, White WB (1989) J Mater Res 4:385–393
79. Havel HA, Chao RS, Haskell RJ, Thamann TJ (1989) Anal Chem 61:642–650
80. Notinger I, Verrier S, Haque S, Polak JM, Hench LL (2003) Biopolymers 72:230–240
81. Cheng C-Y, Perevedentsevs E, Tu J-S, Chung P-H, Cheng C-L, Liu K-K, Chao J-I, Chen P-H, Chang C-C (2007) Appl Phys Lett 90:163903
82. Colosi P, Wong K, Leong SR, Wood WI (1993) J Biochem Mol Biol 268:12617–12623
83. Zhou Z, Wang G, Xu Z (2006) Appl Phys Lett 88:034104
84. Kneipp K, Kneipp H, Itzkan I, Dasari RR, Feld MS (2002) J Phys Condens Matter 14:R597–R624
85. Lin K-W, Cheng C-L, Chang H-C (1998) Chem Mater 10:1735–1737
86. Chang C-C, Chen P-H, Chu H-L, Lee T-C, Chou C-C, Chao J-I, Su C-Y, Chen J-S, Tsai J-S, Tsai C-M, Ho Y-P, Sun K-W, Cheng C-L, Chen F-R (2008) Appl Phys Lett 93:033905
87. Garderen EV, Schalken JA (2002) Mol Cell Endocrinol 197:153–165

Chapter 10
Microfluidic Platforms for Nanoparticle Delivery and Nanomanufacturing in Biology and Medicine

Owen Loh, Robert Lam, Mark Chen, Dean Ho, and Horacio Espinosa

Abstract Nanoparticles are rapidly emerging as promising vehicles for next-generation therapeutic delivery. These highly mobile nanomaterials exhibit large carrier capacity and excellent stability which, when combined with innate biocompatibility, have captured the focus of numerous research efforts. As such, the ability to deliver well-controlled subcellular doses of these functional nanoparticles, both for fundamental research at the single cell level and in related device manufacturing, remains a challenge. Patterning these nanomaterials on biologically compatible substrates enables both novel biological studies and nanomanufacturing avenues through precise spatial control of dosing. Delivering them directly to live cells enables further studies where transfection remains a challenge. This chapter describes a unique tool for functional nanoparticle delivery, called the Nanofountain Probe. The Nanofountain Probe is capable of both direct-write nanopatterning of these materials with sub-100-nm resolution and targeted in vitro injection to individual cells. To motivate the discussion, a brief overview of microfluidic tools developed to deliver nanoparticles is presented. We then focus on the function of the Nanofountain Probe and its application to functional nanodiamond-based biological studies and nanomanufacturing. Development and application of the Nanofountain Probe and other nanomaterial delivery systems will be critical in developing future nanoscale devices and arrays that harness these nanoparticles.

O. Loh, R. Lam, M. Chen, D. Ho (✉), and H. Espinosa
Departments of Biomedical and Mechanical Engineering, Northwestern University,
2145 Sheridan Road, Evanston, IL, 60208-3107, USA
e-mail: d-ho@northwestern.edu

H. Espinosa (✉)
Department of Mechanical Engineering, Northwestern University,
2145 Sheridan Road, Evanston, IL 60208-3111, USA
email: espinosa@northwestern.edu

D. Ho (ed.), *Nanodiamonds: Applications in Biology and Nanoscale Medicine*,
DOI 10.1007/978-1-4419-0531-4_10,

10.1 Introduction

Nanoparticles have proven to be highly effective vehicles for delivery of biological and diagnostic agents. Their extreme surface-area-to-volume ratios and reactivity facilitate conjugation of large payloads [1]. Additionally, their subcellular size makes them highly intriguing for cell-uptake-mediated drug delivery. Combined with innate biocompatibility (e.g., diamond [2–6] and gold [7, 8]), these nanomaterials merit the tremendous attention they receive in the research community. Examples of applications under investigation include the use of nanoparticle–agent conjugates for gene regulation [9, 10] and delivery of a variety of therapeutics [5, 11–14]. Their potential impact on drug delivery is emergent in early experimental results showing moderated drug activity through controlled release from the nanoparticles [5, 12, 15].

Diamond nanoparticles, also known, previously, as nanodiamonds (NDs) or detonation diamonds after their method of production, have recently gained notoriety for their unique properties. These include proven biocompatibility [2–6], uniform particle distributions [16], large surface-area-to-volume ratio [17], near-spherical aspect ratio, and thoroughly explored carbon surface for various bio-agent attachments [18–20]. NDs have been functionalized with a range of assorted biological agents, such as therapeutics, proteins, antibodies, and DNA [5, 18, 21–25]. Like NDs, gold nanoparticles have proven biocompatible [7, 8] and are finding more and more applications in biodetection and chemical sensing, therapeutic delivery, labeling, and imaging [26]. In a recent example, gold nanoparticles were heterogeneously functionalized with antisense oligonucleotides and synthetic peptides [9]. The oligonucleotide functionalization allows the particles to pass freely into cells, without transfection agents. The peptides subsequently serve to improve cellular uptake and direct intracellular localization.

While a number of nanoparticle-based therapeutics have been approved for clinical use or are undergoing clinical trials, a far greater number are in the preclinical discovery and developmental stages [1, 27]. Beyond agent delivery, applications under investigation include biological labeling, biodetection, tissue engineering, targeted tumor destruction, and imaging contrast enhancement [27].

Owing to the numerous established and potential biological applications for nanoparticles, there is a need for tools which can precisely and consistently deposit controlled doses of these materials. Within the developmental phase, the ability to precisely deliver subcellular volumes of functionalized nanoparticles enables novel single cell studies in dosing, adhesion, and phenotype. For example, patterned arrays of gold nanoparticles were used to selectively control cell adhesion [28]. Also, virus-conjugated nanoparticle arrays were used to study cell infectivity on a single cell level [29]. Manufacturing of functional nanoparticle-based devices requires similar capabilities. As an added challenge, these devices often require massively parallel arrays of nanopatterned structures to be fully functional.

This chapter begins with a brief overview of tools developed for submicron patterning and direct in vitro delivery of functional nanoparticles. We then focus on a novel microfluidic device, called the Nanofountain Probe (NFP, [30–32]), capable of direct-write nanopatterning of liquid molecular and colloidal inks with sub-100-nm resolution, as well as in vitro injection of functional nanoparticles. Past demonstrations of direct-write nanopatterning using the NFP include proteins [33] and DNA [34] in buffer solution, nanoparticle suspensions [35, 36], and thiols [30, 32]. As a case study with particular relevance to biology and medicine, we focus on the use of the NFP as a tool to deliver precise doses of functionalized diamond nanoparticles in two modes: (1) direct-write nanopatterning of drug-coated nanodiamonds on glass substrates for subsequent cell culture and dosing studies; and, (2) targeted single cell in vitro injection of fluorescently tagged nanodiamonds.

10.2 Tools for High Resolution Functional Nanoparticle Delivery

10.2.1 Nanopatterning on Solid Substrates

A number of direct transfer and directed self-assembly techniques have been designed to deliver nanoparticles to a substrate in a controlled manner [37]. A sample of these techniques is reviewed in the following.

10.2.1.1 Directed Self-assembly

In directed self-assembly, templates consisting of regions of greater and lesser affinity for the nanoparticles are created on a substrate. The nanoparticles then selectively adhere to regions of greater affinity to create the desired pattern. Techniques used to create the templates include electron beam and photolithography [38], dip-pen nanolithography [39–42], and microcontact printing [43, 44]. For example, electron beam irradiation was used to selectively reduce regions of self-assembled monolayers of NO_2-terminated silane on an SiO_2 substrate to NH_2-termination [38]. Citrate-passivated gold nanoparticles then selectively assembled onto the protonated NH_2-terminated regions. While sub-100-nm features are possible, the strong nanoparticle–substrate binding required for well-defined self-assembled features may not be desirable in cases where the drug or other agent must be released. Furthermore, while a variety of substrate–nanoparticle conjugations have been developed for directed assembly, the possibilities may be somewhat more limited for agent-coated nanoparticles.

Langmuir–Blodgett is a technique typically used to assemble continuous uniform thin films [45]. Amphiphilic molecules are first spread over a water surface, forming a monolayer at the liquid–air interface. The substrate is dipped in water, penetrating the surface layer of molecules, then withdrawn in a controlled manner. As the substrate is withdrawn, a monolayer from the surface of the water is extracted with it, leaving a uniform coating. However, by controlling the dewetting of dilute monolayers of gold and silver nanoparticles, well-ordered parallel line and spoke patterns could be formed with line widths down to the micrometer range [46].

10.2.1.2 Direct-write Nanopatterning

In direct-write techniques, the nanoparticles are transferred directly from the tool to the substrate, Some examples include dip-pen nanolithography [47–49], NFPs [35], microcontact printing [50–52], electrohydrodynamic [53] and ink [54] jet printing, nanopipettes [55, 56], microarrayers [28], surface patterning tools [57], and various forms of lithography [58, 59]. In microcontact printing, a polymer stamp containing the desired pattern in relief is coated with the desired ink. The stamp is then brought into contact with the substrate, allowing ink transfer. Depending on the feature sizes on the stamp, even single particle resolution has been achieved [52]. Ink jet printing techniques use similar technology as found in commercial printers. Here, the ink consists of a solution of nanoparticles. Using this technique nanodiamonds in solution, which are the focus of the case study below, were printed with minimum feature sizes on the order of 100 μm [60, 61]. Use of submicron nozzles and electrohydrodynamic jet printing allows for deposition with higher resolution [53].

10.2.2 In vitro Nanoparticle Injection

In addition to rigorous control over delivery factors (i.e., dosing), continuous in vitro nanoscale transfections without supplemental chemical procedures would be desirable in single cell studies. Previous methods of transfection include, but are not limited to: carrier-mediated transfer [62]; biological [63, 64], chemical [65, 66], or electrical plasma membrane permeabilization [67]; and direct injection [68–70]. Each method's advantages and disadvantages have been reviewed elsewhere [69, 71]. Within single cell studies, direct injection methods remain attractive due to their precise targeting and loading capability with virtually any biologically relevant agent. However, there are still several drawbacks associated with traditional cellular injection methods, such as the need for specialized equipment that require large pressure differences and cellular damage during injection due to the relatively large micropipette needle, typically on the micrometer scale [69, 71]. In order to address these issues, several methods employing nanoscale tools have been developed. These include nanoneedles [72–76], microcantilevers [77–80], optical nanoinjection [81], and electrochemical attosyringes [82]. In the unique cases of nanoneedles and microcantilevers, the intended injection material must first be immobilized on the exterior surface of the probe. Therefore, it must remain bound during the injection process, and then subsequently released within the cell. This generally requires specialized chemical modification of the probe surface [74, 76], a process which can be complicated for agent-coated nanoparticles. Another newly fabricated method, capable of injection in a parallel manner, utilized arrays of tightly packed upright nanosyringes [83]. The syringes are loaded with the cargo and act as a bed of needles upon which the cells are cultured. This method was used to deliver plasmids and quantum dots.

10.3 Case Study: Nanofountain Probes for Functional Nanoparticle Delivery

As an alternative to the aforementioned delivery methods, the Nanofountain Probe (NFP [30–32]) makes use of a unique design to achieve both sub-100-nm patterning and in vitro single cell injection of functionalized nanoparticles and biomolecules in liquid solution. The NFP is an atomic force microscope (AFM)-based delivery probe (Fig. 10.1). Liquid molecular "inks" stored in an on-chip reservoir are fed through integrated microchannels to apertured dispensing tips (Fig. 10.1a, inset) by capillary action. This allows continuous delivery either to a substrate for direct-write nanopatterning (Fig. 10.1b), or to a cell for in vitro injection (Fig. 10.1c). For direct-write nanopatterning, the sharp apertured tip geometry allows for a unique combination of resolution and generality in its ability to pattern a broad range of organic and inorganic molecules and nanoparticle solutions. Past demonstrations of direct-write nanopatterning include proteins [33] and DNA [34] in buffer solution, gold nanoparticles in aqueous suspension [35], thiols [30, 32], and drug-coated nanodiamonds [36]. Furthermore, under control of the AFM [75, 79], a ubiquitous research tool, accurate NFP tip placement and real-time force measurements are achievable with respective resolutions of nanometers and nanonewtons.

10.3.1 Direct-write Patterning of Functionalized Nanodiamonds

The NFP was used to precisely place repeatable doses of drug-coated NDs (drug–NDs) with sub-100-nm resolution [36]. NDs were coated with DOX, a commonly used chemotherapeutic which acts through intercalation with a cell's DNA, causing

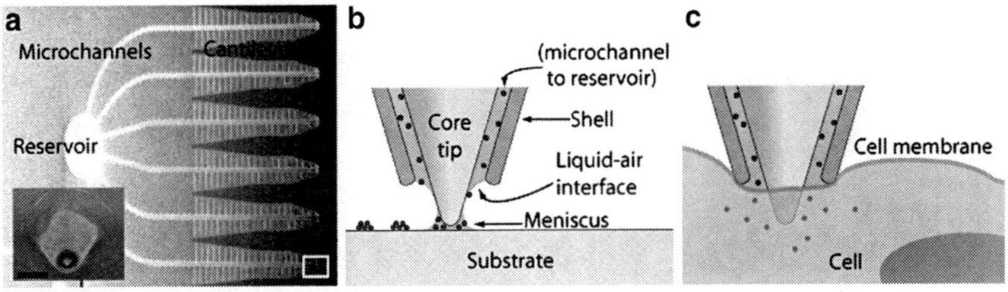

Fig. 10.1 Overview of nanoparticle "ink" delivery using the NFP. (**a**) SEM images showing a quarter section of the NFP chip (scale bar is 2.5 μm). Liquid ink is stored in an on-chip reservoir and fed through enclosed microchannels to apertured writing tips (*inset*) by capillarity. (**b**) For direct-write nanopatterning, the tip is brought into contact with the substrate where an ink meniscus forms. (**c**) For in vitro cellular injection, the tip is introduced to the cell membrane with a prescribed insertion force

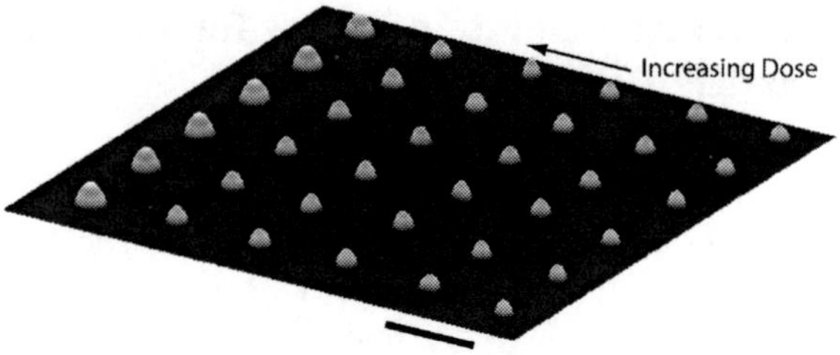

Fig. 10.2 Dot array of drug-coated nanodiamonds patterned on a glass substrate using the NFP [36]. The array is patterned with incrementally increasing feature size for spatial control of dosing. Scale bar is 4 μm

fragmentation and eventual apoptosis. Dot arrays of drug–NDs were patterned directly on glass substrates using the NFP (Fig. 10.2). To create each dot, the NFP was brought into contact with the substrate for a prescribed dwell time, then lifted and translated to the next point in the array. The feature size (both dot diameter and height) depend strongly on the square-root of the dwell time. Thus, the dose is readily controlled by the dwell time. Preservation of drug activity through the patterning process was confirmed through a TUNEL assay, with cells cultured on substrates patterned with dot arrays of drug–NDs showing significantly increased apoptosis [36]. While DOX was demonstrated as a case study, the broader utility of the technique is clear, as NDs have proven capability of carrying a variety of drugs and bioagents [5, 18, 21–25].

Nanopatterning drug–ND conjugates affords extremely precise quantitative and positional control of dosing. It was recently shown that DOX–NDs embedded in parylene are capable of controlled drug release over a period of months [15]. These embedded microfilms are currently being pursued as implants for targeted drug delivery. An attractive enhancement would be to replace the continuous drug–ND films currently used in these devices with patterned arrays using multiple drugs. This allows high fidelity spatial tuning of dosing in intelligent devices for comprehensive treatment. Furthermore, explicit patterning control of NDs would offer avenues for the construction of novel biological-nanoparticle assays [28] and in the immediate future nanomanufactured materials via seeding and nucleation of diamond thin film growth [84, 85].

10.3.2 Direct In vitro Injection

Beyond nanopatterning for biological studies and nanomanufacturing, the ability to directly inject doses of drug bound nanoparticles (e.g., NDs) into cells allows further study of the response of a single cell to a given dose. The ability of the NFP to facilitate these studies was demonstrated through injection of fluorescently tagged NDs.

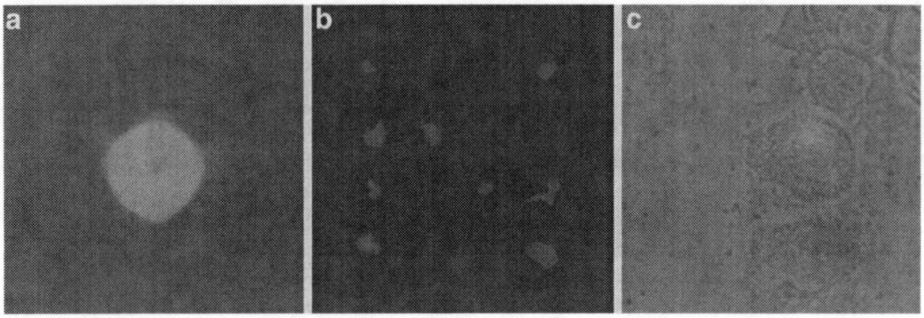

Fig. 10.3 Examples of in vitro single cell injection using the NFP. (**a**) Fluorescence image of an individual macrophage injected with fluorescein. (**b**) Fluorescence image of multiple individual macrophages injected with fluorescein in an "N"-shaped array. (**c**) Overlayed bright field and fluorescence images of an RKO colorectal carcinoma cell injected with fluorescently tagged nanodiamonds

Here, the positional accuracy and force sensitivity of the AFM, a pervasive tool in research, are leveraged to guide the NFP during targeted cell injection. As an initial feasibility study, live macrophages were injected with an aqueous fluorescein solution (Fig. 10.3a, b). Imaging of the cells after injection showed strong confinement of the fluorescent dose to individual cells, suggesting minimal damage to the cell membrane. Fluorescently tagged NDs were then prepared and similarly injected into a variety of cancerous and wild-type cell lines (Fig. 10.3c) [36]. Again, strong confinement of the injected dose was observed, and the ability to target specific regions of the cell demonstrated.

10.4 Discussion and Concluding Remarks

Despite their myriad applications in biology and medicine, nanoparticles are inherently difficult to deliver in precise subcellular doses. The apertured core-shell tip geometry of the NFP allows for a unique combination of patterning resolution and generality in its ability to pattern and deliver a range of liquid solutions. Combined with the positional and force sensitivity of the AFM, this geometry further enables minimally invasive delivery of functional nanoparticles directly to live cells. The combination of multiprobe arrays [30–32] and on-chip reservoirs for continuous ink delivery enables parallel nanomanufacturing for extended periods. The enabling capabilities of the NFP technology for future nanomanufacturing of drug delivery devices and single cell nanomaterial-mediated drug delivery studies are emergent in early examples of drug-coated nanodiamond delivery [36]. Lastly, future studies involving the patterned injection of DNA plasmids, siRNA, therapeutics and conjugated NDs also have exciting implications in determining and directing cellular response.

References

1. Zhang L, Gu F, Chan J, Wang A, Langer R, Farokhzad O (2008) Clin Pharmacol Ther 83:761–769
2. Schrand AM, Huang H, Carlson C, Schlager JJ, Osawa E, Hussain SM, Dai L (2007) J Phys Chem B 111:2–7
3. Bakowicz K, Mitura S (2002) J Wide Bandgap Mater 9:12
4. Liu KK, Cheng CL, Chang CC, Chao JI (2007) Nanotechnology 18:10
5. Huang H, Pierstorff E, Osawa E, Ho D (2007) Nano Lett 7:3305–3314
6. Schrand AM, Dai L, Schlager JJ, Hussain SM, Osawa E (2007) Diam Relat Mater 16:2118–2123
7. Shukla R, Bansal V, Chaudhary M, Basu A, Bhonde R, Sastry M (2005) Langmuir 21:10644–10654
8. Tshikhudo T, Wang Z, Brust M (2004) Mater Sci Technol 20:980–984
9. Patel P, Giljohann D, Seferos D, Mirkin C (2008) Proc Natl Acad Sci USA 105:17222–17226
10. Rosi N, Giljohann D, Thaxton C, Lytton-Jean A, Han M, Mirkin C (2006) Science 312:1027–1030
11. Gelperina S, Kisich K, Iseman M, Heifets L (2005) Am J Respir Crit Care Med 172:1487–1490
12. Morgan T, Muddana H, Altinoglu E, Rouse S, Tabakovic A, Tabouillot T, Russin T, Shanmugavelandy S, Butler P, Eklund P, Yun J, Kester M, Adair J (2008) Nano Lett 8:4108–4115
13. Farokhzad O, Cheng J, Teply B, Sherifi I, Jon S, Kantoff P, Richie J, Langer R (2006) Proc Natl Acad Sci U S A 103:6315–6320
14. Murphy E, Majeti B, Barnes L, Makale M, Weis S, Lutu-Fuga K, Wrasidlo W, Cheresh D (2008) Proc Natl Acad Sci USA 105:9343–9348
15. Lam R, Chen M, Pierstorff E, Huang H, Osawa E, Ho D (2008) ACS Nano 2:2095–2102
16. Ozawa M, Inaguma M, Takahashi M, Kataoka F, Kruger A, Osawa E (2007) Adv Mater 19:1201
17. Dolmatov VY (2001) Usp Khim 70:687–708
18. Kruger A (2006) Angew Chem Int Ed Engl 45:6426–6427
19. Yeap WS, Tan YY, Loh KP (2008) Anal Chem 80:4659–4665
20. Kruger A, Liang YJ, Jarre G, Stegk J (2006) J Mater Chem 16:2322–2328
21. Huang H, Pierstorff E, Osawa E, Ho D (2008) ACS Nano 2:203–212
22. Huang LC, Chang HC (2004) Langmuir 20:5879–5884
23. Ushizawa K, Sato Y, Mitsumori T, Machinami T, Ueda T, Ando T (2002) Chem Phys Lett 351:105–108
24. Kossovsky N, Gelman A, Hnatyszyn HJ, Rajguru S, Garrell RL, Torbati S, Freitas SSF, Chow GM (1995) Bioconjug Chem 6:507–511
25. Krueger A, Stegk J, Liang Y, Lu L, Jarre G (2008) Langmuir 24:4200–4204
26. Murphy C, Gole A, Stone J, Sisco P, Alkilany A, Goldsmith E, Baxter S (2008) Acc Chem Res 41:1721–1730
27. Salata O (2004) J Nanobiotechnology 2:3
28. Jang K-J, Nam J-M (2008) Small 4:1930–1935
29. Vega R, Shen C, Maspoch D, Robach J, Lamb R, Mirkin R (2007) Small 3:1482–1485
30. Kim KH, Moldovan N, Espinosa HD (2005) Small 1:632–635
31. Moldovan N, Kim KH, Espinosa HD (2006) J Microelectromech Syst 15:204–213
32. Moldovan N, Kim K-H, Espinosa HD (2006) J Micromech Microeng 16:1935–1942
33. Loh O, Ho A, Rim J, Kohli P, Patankar N, Espinosa H (2008) Proc Natl Acad Sci USA 105:16438–16443
34. Kim K-H, Sanedrin RG, Ho AM, Lee SW, Moldovan N, Mirkin CA, Espinosa HD (2008) Adv Mater 20:330–334
35. Wu B, Ho A, Moldovan N, Espinosa HD (2007) Langmuir 23:9120–9123
36. Loh O, Lam R, Chen M, Moldovan N, Huang H, Ho D, Espinosa H (2009) Small 5:1667–1674
37. Ho A, Espinosa H (2008) Scanning probes for the life sciences. In: Bhushan B, Fuchs H, Masahiko T (eds) Applied scanning probe methods 8: scanning probe microscopy techniques. Springer-Verlag, Heidelberg

38. Mendes P, Jacke S, Critchley K, Plaza J, Chen Y, Nikitin K, Palmer R, Preece J, Evans S, Fitzmaurice D (2004) Langmuir 20:3766–3768
39. Liu X, Fu L, Hong S, Dravid V, Mirkin C (2002) Adv Mater 14:231–234
40. Demers L, Park S-J, Taton T, Li Z, Mirkin C (2001) Angew Chem Int Ed Engl 40:3071–3073
41. Piner RD, Zhu J, Xu F, Hong SH, Mirkin CA (1999) Science 283:661–663
42. Salaita K, Wang Y, Mirkin C (2007) Nat Nanotechnol 2:145–155
43. Bae S-S, Lim D, Park J-I, Lee W-R, Cheon J, Kim S (2004) J Phys Chem B 108:2575–2579
44. Zin M, Ma H, Sarikaya M, Jen A (2005) Small 1:698–702
45. Langmuir I, Blodgett K (1935) Kolloid-Zeitschrift 73:258–263
46. Huang J, Kim F, Tao A, Connor S, Yang P (2005) Nat Mater 4:896–900
47. Prime D, Paul S, Pearson C, Green M, Petty M (2004) Mater Sci Eng Biomim Mater Sens Syst 25:33–38
48. Roy D, Munz M, Colombi P, Bhattacharyya S, Salvetat J-P, Cumpson P, Saboungi M-L (2007) Appl Surf Sci 254:1394–1398
49. Wang W, Stoltenberg R, Liu S, Bao Z (2008) ACS Nano 2:2135–2142
50. Santhanam V, Andres R (2004) Nano Lett 4:41–44
51. Wu X, Chi L, Fuchs H (2005) Eur J Inorg Chem 18:3729–3733
52. Kraus T, Malaquin L, Schmid H, Riess W, Spencer N, Wolf H (2007) Nat Nanotechnol 2:570–576
53. Park J-U, Hardy M, Kang S, Barton K, Adair K, Mukhopadhyay D, Lee C, Strano M, Alleyne A, Georgiadis J, Ferreira P, Rogers J (2007) Nature 6:782–789
54. Murata K, Matsumoto J, Tezuka A, Matsuba Y, Yokoyama H (2005) Microsyst Technol 12:2–7
55. Iwata F, Nagami S, Sumiya Y, Sasaki A (2007) Nanotechnology 18:105301
56. Duoss E, Twardowski M, Lewis J (2007) Adv Mater 19:3485–3489
57. Vengasandra S, Lynch M, Xu J, Henderson E (2005) Nanotechnology 16:2052–2055
58. Harris D, Hu H, Conrad J, Lewis J (2007) Phys Rev Lett 98:148301
59. Shepherd R, Panda P, Bao Z, Sandhage K, Hatton T, Lewis J, Doyle P (2008) Adv Mater 20:1–6
60. Chen Y-C, Tzeng Y, Davray A, Cheng A-J, Ramadoss R, Park M (2008) Diam Relat Mater 17:722–727
61. Chen Y-C, Tzeng Y, Cheng A-J, Dean R, Park M, Wilamowski B (2009) Diam Relat Mater 18:146–150
62. Gregoriadis G, Buckland RA (1973) Nature 244:170–172
63. Schwarze SR, Hruska KA, Dowdy SF (2000) Trends Cell Biol 10:290–295
64. Uchida T, Yamaizumi M, Okada Y (1977) Nature 266:839–840
65. Graham FL, van der Eb AJ (1973) Virology 52:456–467
66. Chu G, Sharp PA (1981) Gene 13:197–202
67. Knight DE, Scrutton MC (1986) Biochem J 234:497–506
68. Eul J, Graessmann M, Graessmann A (1996) FEBS Lett 394:227–232
69. Celis JE (1984) Biochem J 223:281–291
70. Pepperkok R, Scheel J, Horstmann H, Hauri HP, Griffiths G, Kreis TE (1993) Cell 74:71–82
71. Stephens DJ, Pepperkok R (2001) Proc Natl Acad Sci U S A 98:4295–4298
72. Yum K, Na S, Xiang Y, Wang N, Yu MF (2009) Nano Lett 9:2193–2198
73. Han S, Nakamura C, Obataya I, Nakamura N, Miyake J (2005) Biochem Biophy Res Commun 332:633–639
74. Han S-W, Nakamura C, Kotobuki N, Obataya I, Ohgushi H, Nagamune T, Miyake J (2008) Nanomedicine 4:215–225
75. Obataya I, Nakamura C, Han SW, Nakamura N, Miyake J (2005) Nano Lett 5:27–30
76. Chen X, Kis A, Zettl A, Bertozzi CR (2007) Proc Natl Acad Sci U S A 104:8218–8222
77. Knoblauch M, Hibberd JM, Gray JC, van Bel AJE (1999) Nat Biotechnol 17:906–909
78. Tsulaia T, Prokopishyn N, Yao A, Victor Carsrud N, Clara Carou M, Brown D, Davis B, Yannariello-Brown J (2003) J Biomed Sci 10:328–336
79. Cuerrier CM, Lebel R, Grandbois M (2007) Biochem Biophys Res Commun 355:632–636
80. Belaubre P, Guirardel M, Garcia G, Pourciel JB, Leberre V, Dagkessamanskaia A, Trevisiol E, Francois JM, Bergaud C (2003) Appl Phys Lett 82:3122–3124

81. Stracke F, Rieman I, König K (2005) J Photochem Photobiol B 81:136–142
82. Laforge FO, Carpino J, Rotenberg SA, Mirkin MV (2007) Proc Natl Acad Sci USA 104:11895–11900
83. Park S, Kim Y-S, Kim W-S, Jon S (2009) Nano Lett 9:1325–1329
84. Guerin D, Ismat Shah S (1997) J Mater Sci Lett 16:476–478
85. Lifshitz Y, Lee CH, Wu Y, Zhang WJ, Bello I, Lee ST (2006) Appl Phys Lett 88:243114-1 –243114-3

Chapter 11
Biomechanics of Single Cells and Cell Populations

Michael A. Teitell, Sheraz Kalim, Joanna Schmit, and Jason Reed

Abstract Cells form the basic unit of life. Their health and activities can be quantified by a multitude of biochemical and biophysical techniques that measure responses to external or internal stimuli. Many experimental approaches attempt to integrate molecular mechanisms with changes in the mechanical properties of cells, such as visocoelasticity and compliance, to link cell function with structure. An emerging view of cellular heterogeneity is that even within homogenous cell populations, individual cells may exhibit unique behavioral characteristics that deviate significantly from the population average. Here, several approaches for quantifying biophysical cellular responses are briefly reviewed and linked to specific underlying molecular mechanisms. We succinctly describe each approach and then elaborate on a new interferometer-based method for higher-throughput biophysical analysis of single cells within populations.

Cells adapt to changes in their microenvironment through coordinated molecular and mechanical activities. A variety of methods have been developed to interrogate cellular mechanical properties, such as viscoelasticity and deformability, mainly at the single cell level of analysis. These approaches include atomic force microscopy (AFM), [1–4], bio-microrheology (BMR), [5–7], magnetic twisting cytometry (MTC), [8], magnetic pulling cytometry (MPC), [9], micropipette aspiration,

M.A. Teitell (✉) and S. Kalim
Department of Pathology and Laboratory Medicine, UCLA, Los Angeles, CA, USA
e-mail: mteitell@ucla.edu

M.A. Teitell and J. Reed
California NanoSystems Institute, UCLA, Los Angeles, CA, USA

M.A. Teitell
Jonsson Comprehensive Cancer Center, Broad Center of Regenerative Medicine and Stem Cell Research, and Molecular Biology Institute, UCLA, Los Angeles, CA, USA

J. Schmit
Veeco Instruments, Inc., Tucson, AZ, 85706, USA

J. Reed
Department of Chemistry and Biochemistry, UCLA, Los Angeles, CA, USA

D. Ho (ed.), *Nanodiamonds: Applications in Biology and Nanoscale Medicine*,
DOI 10.1007/978-1-4419-0531-4_11, © Springer Science+Business Media, LLC 2010

[10, 11], microplate stretching rheometry, [12], optical stretching rheometry, [13], optical tweezers [14, 15], and many derivative methods

11.1 Approaches in Single Cell Biomechanics

AFM can measure dynamic changes in cell membrane rigidity, which may be used to infer underlying changes in signal transduction pathways or cytoskeletal components that can then be validated by standard molecular biology techniques, [2, 3, 16]. An AFM microcantilever is usually fabricated from silicon-based materials (Si, Si_3N_4) with a sharp tip at the terminal end, only hundreds of angstroms in size. When brought in contact with the cell surface, microcantilever deflections over time are detected by the motion of a laser reflected off the top surface of the cantilever. There are two predominant modes of AFM analysis for live cells, contact mode and tapping mode. In contact mode, the AFM tip is drawn to the sample surface until contact is made, followed by a sweep over the sample surface under constant deflection, controlled by a feedback piezoactuator. Because there is contact with the surface, the microcantilever spring constant is typically less than the elastic forces maintaining the cell's shape, and this value is usually close to 0.1 nN/mm [17, 18]. The feedback signal, recording the change in force needed to maintain a constant deflection, is processed as an image to reconstruct the appearance of the cell scanned at nanometer resolution. In tapping mode, the microcantilever oscillates above the sample surface, controlled by a piezoactuator. The deflection forces are processed into an image of the cell surface. The elastic or Young's modulus can be calculated from the deflection measurement and known spring constant of the microcantilever. The elastic modulus is a measure of the compliance of a cell. The lower the elastic modulus, the more prone the cell is to deformation.

Bio-microrheology (BMR) typically utilizes 100 nm-range fluorescent microbeads placed inside a cell as probes, rather than probing at the plasma membrane. In force-induced or active, BMR probe motion is controlled by an applied force, which can be from an external source, such as a magnet or laser tweezers, or by an intracellular ATP-dependent driver, such as the molecular motor proteins kinesin, dynein, and myosin [6]. In contrast, thermal or passive BMR is based on an extension of concepts from the Brownian motion of particles in simple liquids [5–7]. Microbead probe introduction is either passive, by endocytosis, or active, by microinjection or ballistic injection. The effect, if appreciable, on probe motion or location based upon the manner of probe introduction has not yet been systematically analyzed. In passive BMR, rapid, real-time video microscopy records small probe tracking displacements over time in the horizontal plane to evaluate the cell interior. Probe trajectories are used to calculate the time-dependent mean-squared displacement (MSD) of each probe to determine the rheological properties of the complex intracellular fluid microenvironment, such as the viscoelasticity. With step increases in stress, such as from step increases in an applied force, the creep compliance of each cell can be determined. MSD calculations can also be used to determine individual probe motions, including trapped, subdiffusive, diffusive, combined diffusive and convective, and ballistic or purely convective particle trajectories, which can identify unique rheological subregions within single cells. Because the creep compliance is a measure of deformation with stress, this property can be used to calculate the interior elastic modulus and Young's modulus of a cell.

MTC and microplate stretching rheometry have been used in conjunction with chemical inhibitors, stimulants, or gene transfection to assess the role of cytoskeletal components, signal transduction pathways, and malignant cell transformation on cell rheologic properties. In MTC, ferromagnetic beads are attached to a cell membrane through a linked ligand that targets a specific membrane receptor. Using an applied magnetic field, the cell is stretched biaxially by the twisting of attached magnetic beads [19]. The bead rotation is measured by a magnetometer or optically [20] to calculate the elastic modulus of a cell. In microplate stretching rheometry, [21, 22], a single cell is grown between two borosilicate plates, one being rigid and the other flexible. As a cell deforms, due to changes in plate distance controlled by piezoelectric transduction, the deflection on the flexible plate, precalibrated for stiffness, is measured to calculate the viscoelastic properties of the cell.

Optical tweezers measure force displacement by focused laser beam trapping of beads attached to opposite ends of a single cell, which causes uniaxial stretching. Typically, the beads are a few microns in size and are attracted to a region with the highest photon density in a photon density gradient, created by a focused laser beam. [23]. Similarly, optical stretching uses counter-propagating divergent laser beams to trap and stretch individual cells [24]. However, optical stretching does not require pendant dielectric beads on the cell and uses two laser beams, which need not be focused. Cell stretching is achieved by asymmetric trapping of a cell, which imparts a resulting force, on the order of tens of pNs. Based on this stretching, stress profiles and deformability can be calculated for a cell.

In micropipette aspiration, [10], a suction pipette is used to aspirate a single cell and hold it firm at the pipette tip. The positive (suction) pressure required to do this is on the order of 100 Pa. By tracking the deflection of a bead attached to the cell opposite the aspirated end, the cortical tension, or the physical forces acting on the cell cortex that contribute to a defined cell shape, can be calculated.

11.2 Molecular Linkage to Cell Biomechanical Properties

Insights into global cell responses to specific stimuli can be gained by linking cell biomechanical properties with signal transduction, biochemical pathways, and the intracellular matrix, cytoskeleton, and cell membrane. For example, AFM has been used to investigate the relationship between Rho-A activation and cellular rigidity in human bronchial epithelial (HBE) cells. HBE cells transfected with a Rho-A inhibiting adenovirus that blocked normal Rho-A activity were significantly more deformable than control adenovirus infected cells [25]. These data seem logical since Rho-GTPases form a group of signaling proteins implicated in the regulation of cell motility, cell morphology, cytoskeletal reorganization, and tumor progression [26]. Perhaps less intuitively, the biomechanical effects of steroids have also been explored at the single-cell level using AFM. Endothelial cells were treated separately with progesterone, estradiol, testosterone, aldosterone, and prednisolone, but only estradiol and aldosterone altered cell rigidity and shape, with aldosterone enhancing and estradiol lessening cell rigidity [27, 28]. These selective biomechanical activities would be harder to predict than the activities of the Rho-GTPases, but they do parallel the selective biochemical, physiological, and clinical effects of different steroid compounds [29]. The relationship between environmental factors and cell responses, such as plasma sodium and aldosterone exposure for endothelial cells, is of great interest in vascular inflammation and has also been studied with AFM [30, 31]. Endothelial cells treated with increasing levels of sodium alone showed no changes in cell rigidity. However, when treated with aldosterone, cell stiffness increased in constant sodium conditions. These findings extend the ongoing work [32] that may implicate endothelial cell dysfunction and increased cell rigidity in hypertension for patients with cardiovascular disease.

Cells in our bodies are constantly subjected to interstitial (non-vascular) fluid flow between cells, which is a common environmental feature of multicellular organism physiology. Studies of mouse fibroblasts using conditions that model interstitial fluid flow revealed increased cellular rigidity. Mouse fibroblasts subjected to shear stress also stiffened through increased actin reorganization and Rho-kinase activation [33]. Paralleling this result, BMR measurements of mouse fibroblasts [34] showed increased cellular viscosity and rigidity as a consequence of Rho-kinase activation. Mouse fibroblasts treated with LPA, a Rho-kinase activator, doubled in cell rigidity, further supporting the influence of Rho-kinase activation on cell rigidity and motility, and validating these links between cell physiology and biomechanics [35].

Biophysical methods have also helped to dissect the complexities of cellular structural organization. Recently, molecular complexes between nesprins and Sun proteins, known as LINC complexes, have been suggested as key structural elements in human cells, linking the nuclear lamina or membrane to the cytoskeleton. BMR studies in mouse fibroblasts [36] using ballistically introduced microparticles were tracked with and without transfection of known LINC complex inhibitors. The transfected cells were less rigid than control cells maintaining LINC complexes, suggesting a potential biomechanical component for, at least, some diseases associated with disruption of nuclear and cytoskeletal architecture. In contrast, micropipette aspiration studies [37] of neutrophil responses to specific microtubule inhibitors that disrupt the cytoskeleton revealed a minimal dependence on the microtubule network for cell structural integrity. Neutrophils treated with colchicine and paclitaxel, microtubule targeting drugs, increased cellular viscosity by 30%, and cortical tension by 18%. These increases in viscosity and tension were attributed to the presence of a compensatory F-actin polymerization response to disrupted microtubulin. This response to maintain cell structural integrity highlights the global interconnectedness of the cytoskeleton network and the compensatory processes that cells utilize to avoid mechanical collapse and death.

Density differences in nuclear and cytoplasmic regions of stem cells have been detected using micropipette aspiration, and linked to cytokine signaling [38] and nuclear import of various biomolecules [39, 40]. Nuclei in human embryonic stem cells were six times more rigid than the surrounding cytoplasm, yet deformed more readily than did fibroblasts derived from embryonic stem cell differentiation, implicating a changing biomechanical nuclear environment with cell differentiation [41]. Determinations of density differences between cells may be of great practical importance as well, because treatment efficacy has been linked to nuclear or cytoplasm density in some forms of leukemia [42].

Our own recent studies [6, 7, 43] measuring changes in viscoelasticity and membrane rigidity in single cells treated with a microtubule-dissociating drug have led to a further understanding of the compensatory actions of the cytoskeletal network. AFM studies showed that upon treatment with nocodazole, a microtubule depolymerizing agent, mouse fibroblasts exhibited local rigidity, rather than collapse and loss of structural integrity. The source for compensatory stabilization was likely a delicate and shifting balance between unstable, rapidly replaced tyrosinated microtubules (Tyr-MTs), stable detyrosinated long-term MTs (Glu-MTs), and the intermediate filaments of the cytoskeletal network. Our data suggested the existence of a pliable, so-called "transitional" mechanical state of a cell that can adjust to defects in the cytoskeletal network to avoid mechanical catastrophe and death. Detailed knowledge of this network action would likely be important for the design of drugs to rationally target specific components of the cytoskeleton, such as in therapy for cancer [44]. Consistent with these AFM results, BMR studies also showed a metastable mechanical transition phase that cells rely upon after treatment with nocodazole to avoid structural collapse. Overall, these two examples utilized different biophysical approaches to arrive at a complementary, mutually reinforcing model of a dynamic cytoskeletal compensatory response to maintain mechanical integrity.

11.3 Biomechanics for Numerous Cells in a Population- An Emerging Need?

Beyond determining single cell biomechanical responses to chemical and physical stimuli, or characterizing the structural properties of a cell, there is an increasing interest in utilizing viscoelastic and cell deformability changes to distinguish between cell types and possibly diseased cell states for potential clinical utility [45]. Recent studies of single cancer cells indicate a number of characteristic and reproducible changes in viscoelasticity and cell rigidity correlated with malignant transformation [13, 46, 47], the epithelial to mesenchymal transition (EMT), [48, 49], chemotherapy exposure, and tumor cell responses [50]. Measurements with optical tweezers of several malignant cell types reproducibly showed increased deformability compared with non-malignant cells, distinguishing a diseased state from a healthy state, based on differences in this cell mechanical property alone.

Biomechanical analysis of cancer cells may also provide useful information beyond traditional pathology and genetic-based characterizations and tumor grading. For example, increased cancer cell deformability may portend metastatic behavior and support similar conclusions from genetic or biochemical tests, increasing the confidence of adopting specific therapeutic strategies. Increased cell deformability from malignant transformation has been shown for pancreas epithelial, bladder epithelial, and mouse fibroblast cells. Panc-1, a pancreatic carcinoma cell line, showed enhanced deformability when treated with sphingosylphosphorylcholine (SPC), a lipid known to promote cancer metastasis, using microplate stretching. AFM studies comparing malignant and non-malignant human bladder epithelial cells showed that the Young's modulus of the non-cancerous cells was an order of magnitude higher than that of the cancerous cells [51]. A reduced viscoelastic modulus was also shown by AFM in simian virus 40-transformed, and H-Ras-transformed mouse fibroblasts, in comparison to non-transformed fibroblasts [52]. AFM studies of metastatic lung, breast, and pancreatic cancer cells from patient's pleural effusions showed that these cells were 80% more deformable than benign mesothelial cells ex vivo [53]. A biomechanical comparison of benign versus cancerous breast epithelial cells using AFM revealed increased elasticity for the cancer cells, with a Young's modulus almost double that of the benign cells [18]. More recently, disease progression in prostate cancer was studied by AFM. Three prostate cancer cell lines and a benign prostate epithelial cell line from a patient with prostate enlargement showed distinct Young's moduli, with the three cancer lines exhibiting reduced rigidity compared to the benign line [54]. Interestingly, the varied metastatic potential between these three cancer lines was reflected in the differences obtained for Young's moduli of these cells, and might be used to assess the likelihood for disease progression. Overall, mechanical differences between normal and cancer cells, or even between different cancer cells, suggests a detectable, quantifiable means for distinguishing between cell types, [47], diseased states, [13, 55], and possibly aggressive behavior based on cell viscoelasticity, rigidity, and deformation measurements. Data thus far, from a variety of different cancer types, both in vitro and ex vivo, using several biophysical methods suggests a general increase in deformability for cancer cells of many tissue types compared to the non-malignant cells from which they are derived.

Biomechanical studies of cancer and paired noncancerous cells has increased our understanding of cancer cell migration and motility, responses to stress, and signaling pathway activities [56]. However, recent advances in the molecular and cell biology of cancer suggest that not all cancer cells function equally, resulting in current uncertainty over the malignant potential of individual cancer cells within a population, and an uneasiness about the best clinical approach to eradicate certain types of malignancy. This current uncertainty can be overly simplified into two models of cancer. One model is a clonal evolution model [57], which asserts that normal cells acquire a cancerous phenotype through the acquisition of multiple, often characteristic genetic or epigenetic alterations. A central feature of this traditional model is that all of the cells forming a cancer have a similar potential for reinitiating the tumor within a rather narrow range of variability. Eradicating a cancer using this model requires removing or killing all of the cancer cells from a patient. A competing model for cancer proposes the existence of relatively rare cells within a tumor, that are solely responsible for maintaining and reinitiating a tumor, rather than the bulk of the tumor cells themselves. These tumor-forming cells have been variously termed "cancer initiating cells" or "cancer stem cells" and a main focus of current research into cancer origins and therapy is focused on validating this model and identifying, characterizing, and eradicating these cells. This cancer stem cell hypothesis asserts that these cancer stem cells are solely capable of self-renewal and tumor propagation [58, 59]. In order to determine which of the competing models is most accurate for a given subtype of cancer, a large number of cells in a tumor would have to be prospectively analyzed one-by-one for their ability to self-renew and reinitiate a cancer. Bulk tumor cells and cancer stem cells could harbor unique biophysical properties for the same reasons that tumor versus non-malignant cells and cells of differing developmental origin can be identified using biomechanical profiling. Therefore, to potentially, prospectively identify rare cells in a population, such as cancer stem cells, or even to determine which cells in a population respond to certain treatments or altered environmental conditions, higher throughput methods for biomechanical analysis of live cells than currently exists are required. We and others are attempting to develop such new approaches.

11.4 Optical Profilometry for Higher Throughput Single Cell Biomechanical Profiling

We, along with the Gimzewski group at UCLA, devised a system that utilizes optical profilometry for nondestructive, near real-time quantification of biomechanical responses to force application or drugs [60, 61]. Optical profilometers, which are based on specialized microscopes with interferometric objectives, have been widely employed in the semiconductor, data storage, and MEMS fields for more than two decades [62]. These systems produce rapid, accurate, and reproducible nanoscale characterizations of surface roughness, form, film thickness, and more recently, dynamic behavior of MEMS and other moving devices. We have shown that with a proper configuration, optical profilers can sensitively and rapidly measure the biophysical properties of living biological systems.

Typically, optical profilers utilize light from a single source that is split, directed and reflected from a test object and a high quality reference surface to generate fringes from the interference of these two beams. Fringes created, as the objective is scanned along the optical axis that are detected by a camera, and analyzed to determine the shape of an object. In order to evaluate living cells, the system must be able to measure objects in fluid, which requires additional, specialized optical components to correct for the dispersive nature of the liquid medium. Also, the timescales of changing biomechanical properties can vary greatly, requiring a variety of measurement methodologies to maximize data, including strobed interferometry and video-rate deformation calculations. As discussed in the previous section, the biophysical responses of individual cells can differ significantly from one another in a particular environment, making measurements of large numbers of cells in parallel an important requirement. This presents challenges, both in how to optimize the field-of-view for lateral resolution and how each image is analyzed, as the software intelligently tracks and logs information from many different cells simultaneously.

The optical profiler we used for live cell studies is, in principle, an optical microscope with a 20× 0.28NA Michelson interference objective that allows for the recording of not only lateral features with typical optical resolution (1.16 μm for the 20× objective) but also height dimensions of reflective objects below the scale of one nanometer. Live cells are maintained in an environmentally controlled chamber with a glass observation window. The Michelson interferometer is composed of a beam splitter, reference mirror and compensating fluid cell to adjust for optical path differences induced by fluid in the observation chamber. During each measurement, the objective head is scanned vertically from the reflective surface below the cells to a height of 40 μ above the surface, such that each point in the volume passes through focus. The interferometer is aligned so that the interference intensity distribution along the vertical scanning direction has its peak (best fringe contrast) at approximately the best focus position (Fig. 11.1).

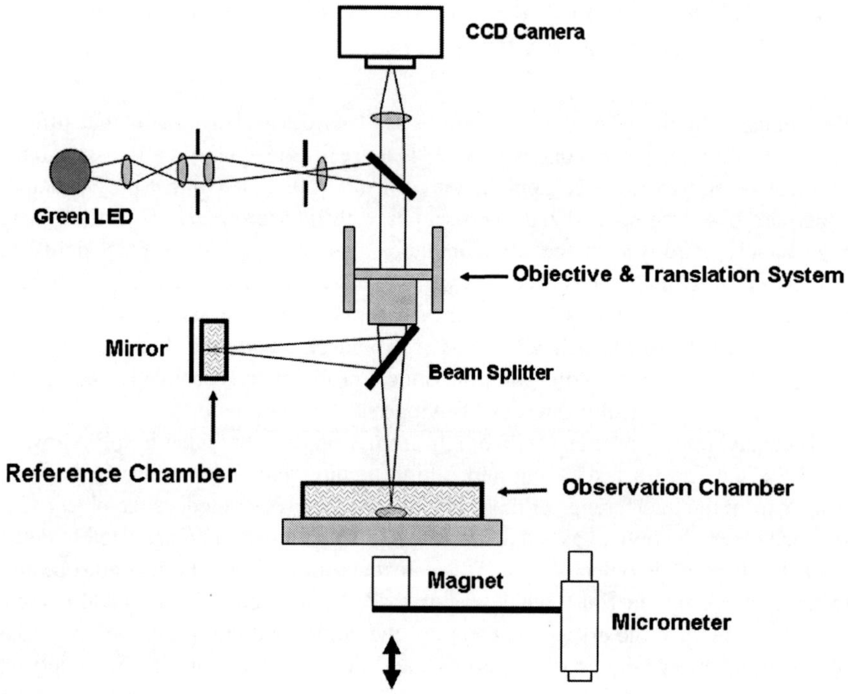

Fig. 11.1 Schematic of the interferometric profiler system used for live cell mechanical measurements

As shown in Fig. 11.2, three types of images can be obtained: a bright field microscopic image of the cell (*left*); a three-dimensional profile of microreflectors resting on the cell surface (*middle*); and an optical thickness image of the cell itself, which corresponds to the material density at every pixel (*right*). The interferometric height profile includes only the microreflector, since the cell-liquid interface is minimally reflective. Optical thickness, as discussed below, is calculated from the relative phase shift in the incident light as it propagates through the transparent film.

With this approach, the biomechanical properties of cells can be assessed in situ and in parallel using microindentation techniques. Samples are prepared for imaging by placing engineered magnetic microreflectors on top of cells and using these microreflectors as indentation probes for measuring nanometer displacements in cell height and position. Biomechanical properties such as the elasticity, viscocity and deformability of cells can be ascertained by imparting a magnetic field on the microreflectors (20pN – 20nN) to measure force displacement on the surface of the cell. Typically, contact indentation models, such as the Hertz model, are used to derive quantitative material properties of the cell from the force vs. indentation data. Other models can be used which account for adhesive properties of the cell surface, such as the Johnson-Kendall-Roberts (JKR) contact model [63]. A critical consideration for cell biomechanical studies is the dynamic range of the measurement technique. Mammalian cells exhibit a wide range of Young's moduli, from as soft as 10 Pa to as stiff as 100 kPa [64]. We estimate that the optical profiler can effectively measure samples with elastic moduli that vary from several Pa up to ~200 kPa, as currently configured. Live cells are known to exhibit complex frequency-dependent viscoelastic properties [21, 65, 66]. It is possible to use viscoelastic contact models to characterize the time-dependent elastic behavior of a cell, much like that routinely done for the study of soft polymers. In addition to the derived mechanical constants of the cell, interferometric imaging also provides direct biophysical data, such as cell thickness and position, for approximately ~1,000 cells, at any given time.

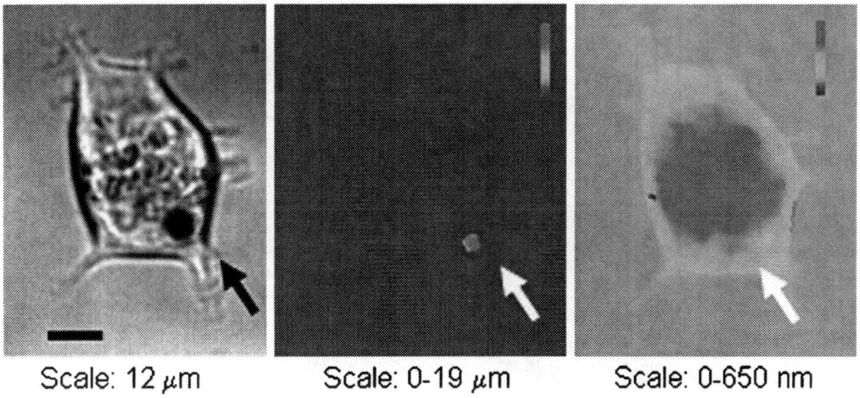

Fig. 11.2 Bright field (*left*) interferometric (*middle*) and optical thickness (right) images of an 8 micron microreflector (*arrows*) on a live NIH3T3 fibroblast in cell culture media

Optical thickness determined by interferometry is a well established technique to quantitatively and non-invasively measure local material density in transparent samples [67]. Figure 11.3 shows an optical thickness image of two live NIH3T3 mouse fibroblasts cells. The image shows details of each cell's internal structure, including the location of intracellular organelles. Cells are almost completely transparent in the visible spectrum, and the index of refraction of the fluid is close to the index of refraction of the interior, which does not allow much light to reflect off of the cell surface. Thus, almost all the light travels through the cell, and the signature of the material distribution within the cell is left on the measured phase. This signature is linear with the optical thickness of the cell because the whole wavefront traveling through the cell is interfered with the independent reference wavefront, and thus, this type of measurement is called quantitative phase imaging. Other quantitative phase imaging techniques are also being used for cell analyses, such as digital holography [68] and several interferometer variations [69]. The quantitative measurements of phase imaging avoids many issues of more traditional phase imaging techniques, such as Zernike phase contrast microscopy or differential interference contrast microscopy, that deliver only a qualitative measurement of cells.

Our initial studies [60, 61] evaluated the efficacy of using interferometry to profile live cells. We determined that the vertical motion in hundreds of NIH3T3 and HEK293T fibroblasts can be monitored simultaneously at a spatial resolution of <20 nm. Viscoelastic constants were determined by fitting the indentation curves to a time-dependent version of the three-factor model for a spherical indenter. For both cell types, the population viscoelastic constants were log-normally distributed. This result is in agreement with recent reports [64]. In contrast, most AFM/probe-based indentation studies have used small sample sizes ($n < 30$), and unlike optical profiling, do not have the sampling breadth to resolve the extended tail of the log-normal distribution. Failure to properly characterize this distribution can result

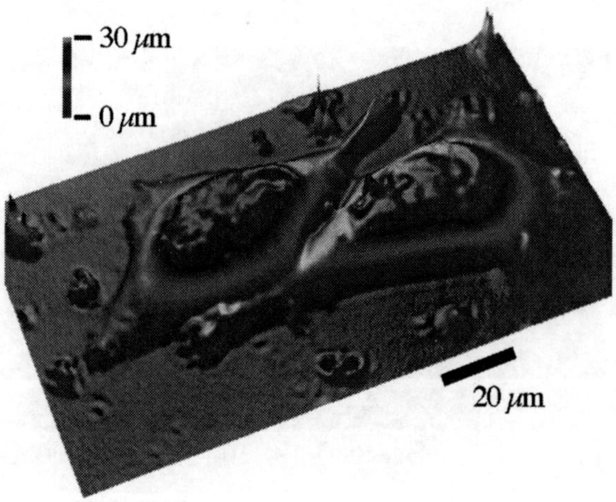

Fig. 11.3 Optical thickness image of live NIH 3 T3 cells in cell culture media

in erroneous conclusions when comparing experimental treatments or different cell types, especially when attempting to identify "outlier" cells that display distinct biophysical properties in a population. In addition, we achieved high throughput measurement of changes in viscoelastic behavior of NIH3T3 cells, when treated with cytochalasin-B, an actin depolymerizing drug. At low doses (0.1–1 μM), cytochalasin B does not produce large changes in the morphology of fibroblasts, although it does inhibit cell migration, [70, 71], and AFM indentation studies have reported minimal, if any, measurable change in Young's modulus [72]. In contrast, we determined that treated cells were more elastic, although their viscosity showed little or no decrease.

In a more recent study, the local redistribution of cell content was monitored as small indentions were made by highly magnetic probes, on the cell surface, using a rare earth magnet [61]. There was almost instantaneous redistribution of cell material, as a result of indentation on the surface of the cell, which was undetected with conventional optical microscopy alone. We analyzed the time-dependence of the content shift between specific regions of the cell by measuring the change in average optical thickness within four sub-regions of the cell body. The undriven regions responded at the same frequency as the driven regions, but with a temporal delay, as would be expected from a viscoelastic material. The amplitudes of motion of both the driven and undriven regions increased with time. Changes in local compliance were observed within 200s when force was applied cyclically to regions of the cell. It would be extremely difficult to measure these types of immediate viscoelastic parameters, at such a large scale, using conventional biophysical measurement techniques.

Our results show that optical profilometry achieves high throughput when measuring biomechanical properties using indentation normal to the cell surface. This represents a significant throughput advance over AFM, and other optical approaches, such as confocal microscopy or microfluidic optical stretchers, which cannot accurately measure mechanical properties of large arrays (hundreds to thousands) of cells simultaneously, with a single-cell specificity [72, 73]. The mechanical dynamic range and effective magnification of interferometric optical profiling equals or exceeds existing wide-field optical particle tracking techniques, [74], which implies that the two could be used in combination to conduct rapid, fully-3D mechanical probing of large arrays of live cells.

Acknowledgments Work on optical profilometry in the Teitell lab is generously supported by the UC Discovery/Abraxis Biosciences Biotechnology Award Bio07-10663. The authors thank Kayvan Niazi and Shahrooz Rabizadeh (Abraxis Biosciences) for encouragement and insightful discussions, and Daphne Weihs (Technion University), for critical reading and comments on the manuscript.

References

1. Benmouna F, Johannsmann D (2004) Langmuir 20:188–193
2. Dufrene YF (2008) Nat Rev Micro 6:674–680
3. Mahaffy RE, Park S, Gerde E, Kas J, Shih CK (2004) Biophys J 86:1777–1793

4. Mahaffy RE, Shih CK, MacKintosh FC, Käs J (2000) Phys Rev Lett 85:880
5. Crocker J, Hoffman B (2007) Meth Cell Biol 83:141–178
6. Weihs D, Mason TG, Teitell MA (2006) Biophys J 91:4296–4305
7. Weihs D, Teitell M, Mason T (2007) Microfluid Nanofluid 3:227–237
8. Gardel ML, Shin JH, MacKintosh FC, Mahadevan L, Matsudaira P, Weitz DA (2004) Science 304:1301–1305
9. Overby DR, Matthews BD, Alsberg E, Ingber DE (2005) Acta Biomaterialia 1:295–303
10. Hochmuth RM (2000) J Biomech 33:15–22
11. Hoffman BD, Massiera G, Van C, Kathleen M, Crocker JC (2006) Proc Natl Acad Sci USA 103:10259–10264
12. Suresh S, Spatz J, Mills JP, Micoulet A, Dao M, Lim CT, Beil M, Seufferlein T (2005) Acta Biomaterialia 1:15–30
13. Guck J, Schinkinger S, Lincoln B, Wottawah F, Ebert S, Romeyke M, Lenz D, Erickson HM, Ananthakrishnan R, Mitchell D, Kas J, Ulvick S, Bilby C (2005) Biophys J 88:3689–3698
14. Dao M, Lim CT, Suresh S (2003) J Mech Phys Solids 51:2259–2280
15. Svoboda K, Schmidt CF, Schnapp BJ, Block SM (1993) Nature 365:721–727
16. Muller DJ, Dufrene YF (2008) Nat Nano 3:261–269
17. Afrin R, Yamada T, Ikai A (2004) Ultramicroscopy 100:187–195
18. Li QS, Lee GYH, Ong CN, Lim CT (2008) Biochem Biophys Res Comm 374:609–613
19. Puig-De-Morales M, Grabulosa M, Alcaraz J, Mullol J, Maksym GN, Fredberg JJ, Navajas D (2001) J Appl Physiol 91:1152–1159
20. Bausch AR, Moller W, Sackmann E (1999) Biophys J 76:573–579
21. Desprat N, Richert A, Simeon J, Asnacios A (2005) Biophys J 88:2224–2233
22. Micoulet A, Spatz Joachim P, Ott Albrecht (2005) Chem Phys Chem. 6, 663–670.
23. Reece PJ (2008) Nat Photon 2:333–334
24. Guck J, Ananthakrishnan R, Mahmood H, Moon TJ, Cunningham CC, Kas J (2001) Biophys J 81:767–784
25. Wagh AA, Roan E, Chapman KE, Desai LP, Rendon DA, Eckstein EC, Waters CM (2008) AJP - Am J Physiol Lung cell Mol Physiol 295:L54–60
26. Sahai E, Marshall CJ (2002) Nat Rev Can 2:133–142
27. D'Ascenzo S, Millimaggi D, Di M Caterina, Saccani-Jotti G, BotrË F, Carta G, Tozzi-Ciancarelli MG, Pavan A, Dolo V (2007) Tox Lett. 169, 129-136.
28. Hillebrand U, Hausberg M, Lang D, Stock C, Riethmüller C, Callies C, Büssemaker E (2008) Pflugers Archiv 456:51–60
29. Miller VM, Mulvagh SL (2007) Trends Pharm Sci 28:263–270
30. Brown NJ (2008) Hypertension 51:161–167
31. Oberleithner H, Riethm ller C, Schillers H, MacGregor GA, de W Hugh E, Hausberg M (2007) Proc Natl Acad Sci USA 104, 16281–16286
32. Lee HW, Karam J, Hussain B, Winer B (2008) Curr Diab Rep 8:208–213
33. Lee JSH, Panorchan P, Hale CM, Khatau SB, Kole TP, Tseng Y, Wirtz D (2006) J Cell Sci 119:1760–1768
34. Kole TP, Tseng Y, Huang L, Katz JL, Wirtz D (2004) Mol Biol Cell 15:3475–3484
35. Sander EE, ten Klooster JP, van D Sanne, van d K Rob A, Collard JG (1999) J Cell Biol 147, 1009–1022
36. Stewart-Hutchinson PJ, Hale CM, Wirtz D, Hodzic D (2008) Exp Cell Res 314:1892–1905
37. Tsai MA, Waugh RE, Keng PC (1998) Biophys J 74:3282–3291
38. Kreis S, Munz GA, Haan S, Heinrich PC, Behrmann I (2007) Mol Can Res 5:1331–1341
39. Kutty RK, Chen S, Samuel W, Vijayasarathy C, Duncan T, Tsai J-Y, Fariss RN, Carper D, Jaworski C, Wiggert B (2006) Biochem Biophys Res Comm 345:1333–1341
40. Trompeter H, Schiermeyer A, Blankenburg G, Hennig E, Soling H (1999) J Cell Sci 112:4113–4122
41. Pajerowski JD, Dahl KN, Zhong FL, Sammak PJ, Discher DE (2007) Proc Natl Acad Sci USA 104:15619–15624
42. Valkov NI, Gump JL, Engel R, Sullivan DM (2000) Br J Haematol 108:331–345

43. Pelling AE, Dawson DW, Carreon DM, Christiansen JJ, Shen RR, Teitell MA, Gimzewski JK (2007) Nanomed: Nanotech. Biol & Med 3:43–52
44. Kopf-Maier P, Muhlhausen SK (1992) Chem-Biol Interact 82:295–316
45. Suresh S (2007) Nat Nano 2:748–749
46. Beil M, Micoulet A, von W Gotz, Paschke S, Walther P, Omary MB, Van V Paul P, Gern U, Wolff-Hieber E, Eggermann J, Waltenberger J, Adler G, Spatz J, Seufferlein T (2003) Nat Cell Biol 5, 803–811.
47. Elson E (1988) l. Ann Rev Biophys and Biophys Chem 17:397–430
48. Gupta GP, MassaguÈ J (2006) Cell 127:679–695
49. Thiery JP (2003) Curr Op Cell Biol 15:740–746
50. Lam WA, Rosenbluth MJ, Fletcher DA (2007) Blood 109:3505–3508
51. Lekka M, Laidler P, Gil D, Lekki J, Stachura Z, Hrynkiewicz AZ (1999) Eur Biophys J 28:312–316
52. Park S, Koch D, Cardenas R, Kas J, Shih CK (2005) Biophys J 89:4330–4342
53. Cross SE, Jin Y-S, Rao J, Gimzewski JK (2007) Nat Nano 2:780–783
54. Faria EC, Ma N, Gazi E, Gardner P, Brown M, Clarke NW, Snook RD (2008) Analyst 133:1498–1500
55. Liu J, Ferrari M (2002) Dis Markers 18:175–184
56. Makale M (2007) Birth Defects Res C Embryo Today 81(4) 329–343
57. Nowell P (1976) Science 194:23–28
58. Lapidot T, Sirard C, Vormoor J, Murdoch B, Hoang T, Caceres-Cortes J, Minden M, Paterson B, Caligiuri MA, Dick JE (1994) Nature 367:645–648
59. Park CH, Bergsagel DE, McCulloch EA (1971) J Natl Can Inst 46:411–422
60. Reed J, Frank M, Troke JJ, Schmit J, Han S, Teitell MA, Gimzewski JK (2008) Nanotechnology 19, 235101
61. Reed J, Troke JJ, Schmit J, Han S, Teitell MA, Gimzewski JK (2008) ACS Nano 2:841–846
62. Optical Inspection of Microsystems (2006) Osten W (Ed); CRC Press; Vol. 109
63. Johnson KL, Kendall K, Roberts AD (1971) Proc Royal Soc London a-Math and Phys Sci. 324:301–313
64. Balland M, Desprat N, Icard D, Fereol S, Asnacios A, Browaeys J, Henon S, Gallet F (2006) Phys Rev E 74
65. Rico F, Buscemi L, Trepat X, Garbulosa M, Rotger M, Farre R, Navajas D (2003) Eur Biophys J 32:306
66. Lenormand G, Millet E, Fabry B, Butler JP, Fredberg JJ (2004) J R Soc Interface 1:91–97
67. Davies HG, Wilkins MHF, Chayen J, Lacour LF (1954) Quart J Micro Sci 95:271–278
68. Marquet P, Rappaz B, Magistretti PJ, Cuche E, Emery Y, Colomb T, Depeursinge C (2005) Opt Lett 30:468–470
69. Popescu G, Ikeda T, Dasari RR, Feld MS (2006) Opt Lett 31:775–777
70. Rotsch C, Radmacher M (2000) Biophys J 78:520–535
71. Yahara I, Harada F, Sekita S, Yoshihira K, Natori S (1982) J Cell Biol 92:69–78
72. Cheezum MK, Walker WF, Guilford WH (2001) Biophys J 81:2378–2388
73. Carter BC, Shubeita GT, Gross SP (2005) Phys Biol 2:60–72
74. Fabry B, Maksym GN, Shore SA, Moore PE, Panettieri RA, Butler JP, Fredberg JJ (2001) J Appl Physiol 91:986–994

Chapter 12
Design of Nanodiamond Based Drug Delivery Patch for Cancer Therapeutics and Imaging Applications

Wing Kam Liu, Ashfaq Adnan, Adrian M. Kopacz, Michelle Hallikainen, Dean Ho, Robert Lam, Jessica Lee, Ted Belytschko, George Schatz, Yonhua Tzeng, Young-Jin Kim, Seunghyun Baik, Moon Ki Kim, Taesung Kim, Junghoon Lee, Eung-Soo Hwang, Seyoung Im, Eiji Ōsawa, Amanda Barnard, Huan-Cheng Chang, Chia-Ching Chang, and Eugenio Oñate

12.1 Introduction

The onset and recurrence of cancer is one of the major biomedical quandaries of our time. Currently, surgically removed tumors often leave behind a residual cancer cell population. As not all cancer cells can be detected to ensure complete tumor

W. Kam Liu (✉), A. Adnan, A. Kopacz, M. Hallikainen, D. Ho, R. Lam, J. Lee, T. Belytschko, and G. Schatz
Northwestern University, Evanston, IL, USA

Y. Tzeng
National Cheng Kung University (NCKU), Tainan, Taiwan

Y.-J. Kim, S. Baik, M. K. Kim, and T. Kim
Sungkyunkwan University (SKKU), Seoul, South Korea

J. Lee and E.-S. Hwang
Seoul National University (SNU), Seoul, South Korea

S. Im
Korea Advanced Institute of Science and Technology (KAIST), Daedeok Science Town, Daejeon, South Korea

E. Ōsawa
Nanocarbon Research Institute (NCRI), Shinshu University, Nagano Prefecture, Japan

A. Barnard
Commonwealth Scientific and Industrial Research Organisation (CSIRO), Clayton, VIC, Australia

H.-C. Chang
Institute for Advanced and Molecular Studies (IAMS), Academia Sinica, Taipei, Taiwan

C.-C. Chang
National Chiao Tung University (NCTU), Hsinchu, Taiwan

E. Oñate
International Center for Numerical Methods in Engineering (CIMNE), Barcelona, Spain

D. Ho (ed.), *Nanodiamonds: Applications in Biology and Nanoscale Medicine*,
DOI 10.1007/978-1-4419-0531-4_12, © Springer Science+Business Media, LLC 2010

removal, systemic and widespread chemotherapy is usually injected into the bloodstream to attempt to target the remaining cancer cells. This can result in devastating side effects because the cancer drugs flow freely throughout the bloodstream with a reduced ability to target-specific regions. This treatment kills both healthy and unhealthy cells, and thus the quality of life of cancer patients is significantly reduced.

A multiscale anatomy of cancer cell, e.g., the breast cancer cell, reveals that a single cancer cell contains a number of characteristic elements including specific biomarker receptors (e.g., folic acid ligands) and DNA molecules, as shown schematically in Fig. 12.1. If these cancer-specific molecules can be intercalated by some functional molecules supplied via an implantable patch, then the patch can be envisioned to serve as a complementary technology with current systemic therapy. The patch has the potential to enhance localized treatment efficiency, minimize excess injections/surgeries, and prevent tumor recurrence, provided that the patch is designed such that a quantitative control on the assembly, loading, and release of such functional groups are ensured.

The creation of a fully optimized biocompatible device capable of local delivery has proven to be very challenging, due to a range of chemical and material problems

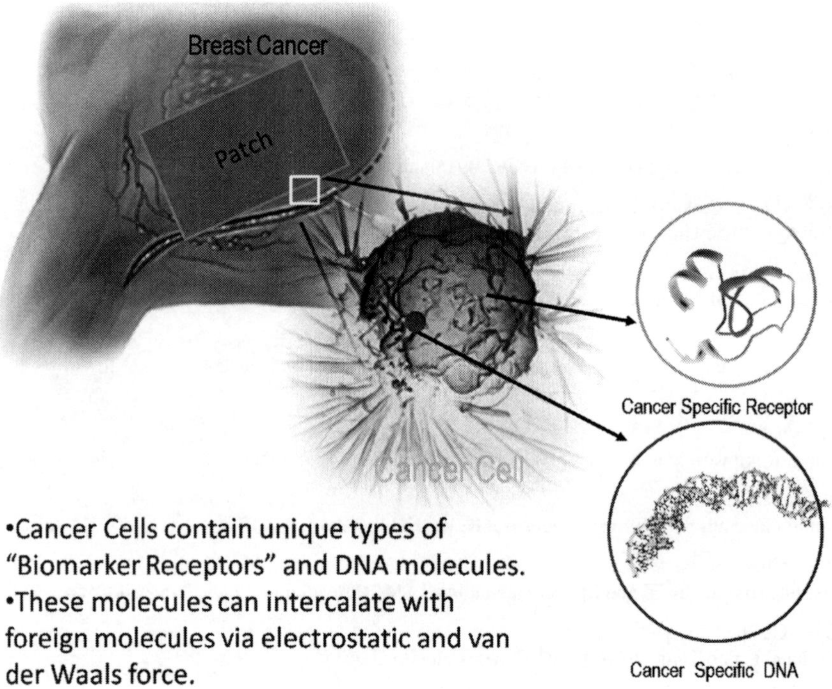

Fig. 12.1 Multiscale anatomy of cancer cell

to be overcome in addition to the pharmacological and biomedical issues. For example, metal-based delivery materials often possess reactive interfaces that can cause cell stress and inflammation, hindering the efficacy of therapy [1, 2]. The selected delivery material must be extremely stable to enable localized release while resisting attack from the immune system, which commonly leads to the breakdown of the device and ultimately to failure of the treatment. Furthermore, systemic nanoparticle-mediated administration of anti-inflammatory compounds is nonlocalized, indiscriminate, dilutes drug efficacy, and lowers drug loading capacities. Therefore, it is important that a technology be developed that can consistently release the cancer drug in a localized and sustained fashion and the fundamental basis of this functionality is understood to drive intelligent device design.

Currently, biodegradable strategies are the preferred option for local drug delivery. This type of device often dissolves between 20 and 30 days after implantation, which is consistent with the timeframe for conclusion of drug release. Sustained treatment using this method requires repeated implantation, which can cause repeated inflammation and side effects that impede cancer treatment [3–5].

Combined imaging/diagnostics and therapy (termed "theranostics") have made important advances over the past several years, and continued work has been sought to design transformative materials to accelerate the fruition of this roadmap [1–42]. Based on their unique mechanical–chemical properties, it can be envisioned that nanocarbons, such as diamond-based nanomaterials, may serve as optimal platforms for the integrative imaging and treatment of cancer cells. Much of the ongoing research in the emerging field of nanomedicine is based on the use of carbon nanotubes and bucky balls, which challenges their underlying translational relevance and innate compatibility to drug delivery. On the contrary, nanodiamonds (NDs) represent a much more viable alternative, because: (a) they have much higher potential for mass production, yet still possess properties common to nanotubes and bucky balls such as the ultrahigh surface-to-volume ratio, (b) they can improve drug administration efficacy by order of magnitude scale (it has been found from some preliminary mouse studies that 20 times less drug is needed for cancer therapy when it is bound to the NDs compared with administration of the drug alone), (c) they are highly biocompatible (noncytotoxic) as opposed to most carbonaceous materials, and (d) they can be functionalized to chemically link with almost any materials. In this chapter, we outline a framework for the development of an ND-enabled drug delivery system, capable of performing both therapeutics and diagnostics functions, via seamless integration of simulation-based engineering and science (SVE&S) and experimental validations. Our vision is to design a system such that it will be capable of releasing drug for prolonged timescale (in the order of months to a year) over a specific residual cancer cell region, thereby preventing the recurrence of tumor growth. This is a key improvement because widespread/nonspecific drug distribution due to material degradation can be avoided, along with the associated medical complications and unpleasant side effects.

12.2 The Drug Delivery Patch

The fundamental requirements for the proposed drug delivery patch are to efficiently detect/diagnose cancer cells and then to deliver therapeutic molecules with a preset dose. In this study, a model drug delivery system is proposed that consists of (a) nanodiamonds (ND), (b) parylene buffer layer, and (c) doxorubicin (DOX) drugs and biomarkers. In its simplest form, an array of self-assembled nanodiamonds is functionalized with DOX and biomarker molecules and is contained inside parylene capsule as shown in Fig. 12.2. Here, parylene polymers serve as the *capsule shell* for the ND complex so that the device integrity is preserved. It can be seen from Fig. 12.2 that the bottom part of the device contains nonporous parylene, and the top part contains porous parylene layers. Parylene is chosen because it is highly biocompatible and also Food Drug Administrations (FDA) approved. The pore size in the parylene layer is in the order of nanoscale and may be fabricated using MEMS-based technology.

It is envisioned that the patch will be implanted in the body after a surgery is commenced to remove residual (and otherwise persistent) cancer cells. The physiological conditions in the vicinity of living healthy cells and cancerous cells are significantly different and can be quantified by different pH levels. The patch is designed such that when it is implanted in the body and comes into contact with cancer cells, the drug molecules will detach from the NDs and flow through the

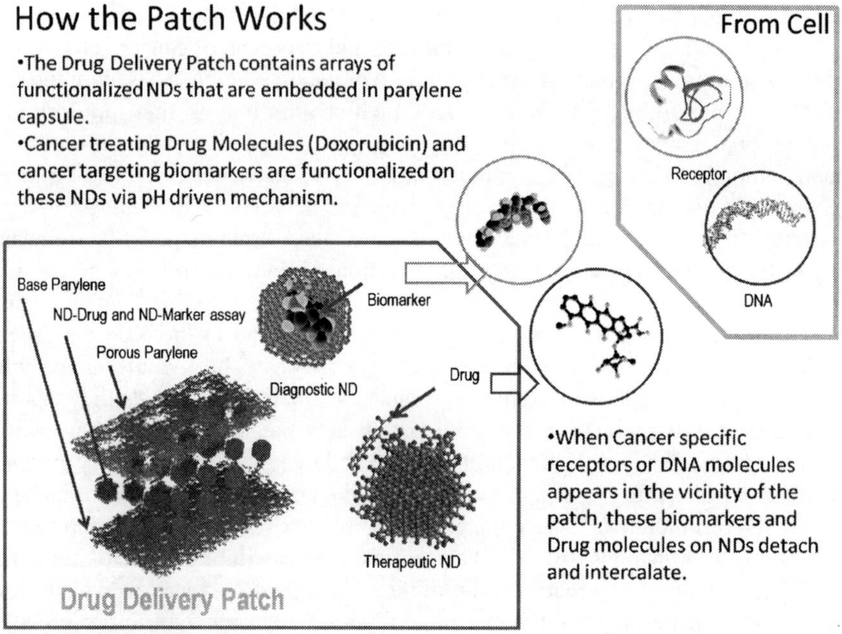

Fig. 12.2 Schematics of the proposed drug delivery patch. The encircled molecules (biomarkers and drug) are the functional elements detached from ND to attach with the cancer specific molecules of cancer cells

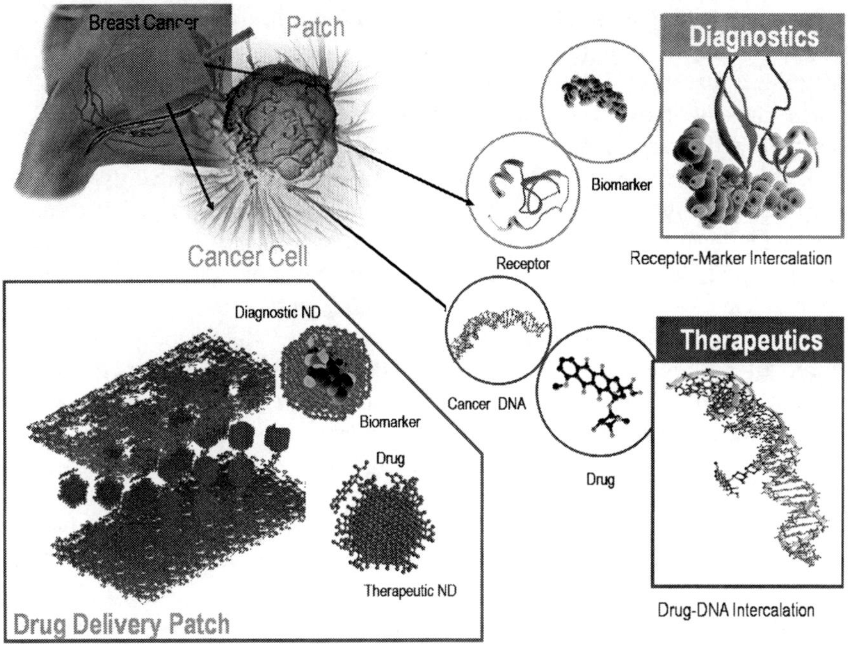

Fig. 12.3 Schematics of the proposed drug delivery patch with therapeutic and diagnostic functionalities

porous parylene toward cancer cells. The characteristic pH level will dictate the drug "detachment", and the degree of porosity in the parylene will control the drug "diffusion rate". Similarly, the ND-biomarker assembly will flow from the device when it comes into contact with cancer-specific receptors. As a result, one device will perform both therapeutic and diagnostic functionalities as schematically shown in Fig. 12.3.

12.2.1 Number Effect in Nanoparticles

12.2.1.1 What is the Number Effect?

A recent finding suggests that a new attribute of nanoparticles exists, which so far has been neglected but could be unique and generally useful for the smaller nanoparticles such as the primary particles of detonation nanodiamond that we are dealing with in this book. The new attribute is the unexpectedly large number of particles in unit volume or unit weight. This was surprising for us because of our prejudice that nanoparticles are much larger than molecules. There are many reasons for us to acquire this prejudice. Nanoparticles are visible under TEM, whereas molecules are not. There are only a few conventional molecules having such a large size as 1 nm, but a nanoparticle having a diameter of 1 nm (e.g., C_{60})

is the smallest (definition of nanoparticle: 1–100 nm in diameter). Molecules are gauged by Ångstrom unit (10^{-10} m), whereas nanoparticles are by nm (10^{-9} m). The truth is that these differences are not really big.

The impact of overwhelmingly large number associated with nanoparticles, especially with single-nano particles, has been noticed by one of the co-authors (E. I. Ōsawa) when he spotted a word '*nanodiamond*' in recent advertisements for sharp pencils (Fig. 12.4). It was interesting to find that a major pencil company claiming to have improved writing performance of their new brand of sharp pencils by *dispersing 400 million particles of nanodiamond into one piece of lead* with 0.5 mm diameter and 60 mm length. *Four-hundred million* particles in such a tiny lead sounded unbelievably large in number. However, the doubt quickly disappeared after simple arithmetic. The number density of the added particles is given by $400\times10^{6}/0.0118 = 3.39\times10^{10}/\text{cm}^{3}$ (the volume of one lead is equal to 0.0118 cm^3). This volume density can be approximately translated into a linear density by taking a cube root of 3230/cm, or 310 nm for average center-to-center distance between the nearest particles. There is certainly a plenty of room to pack 400 million particles in a lead, even if the largest nanodiamond particles with 100 nm diameter were used. In other words, this advertized density sounds great but actually is not really too much for nanoparticles.

Now, if the number of spherical nanoparticles were a decisive factor for the writing performance of lead, how much single-nano buckydiamond (SNBD, the primary particles of detonation nanodiamond) will be needed to pack 400 million particles of them into a lead? It turned out that only 0.5 mg of SNBD per 1 kg of

Fig. 12.4 An advertisement of a new brand of sharp-pencil lead claiming to contain 400 million particles of nanodiamond per piece of lead

Table 12.1 Basic numbers of SNBD particles

Selected numbers for one SNBD particle	Diameter $D(=2r)$	4.8 nm	100 nm[d]
	Volume $v(=4\pi r^3/3)$	$5.79\times10^{-20}\,cm^3$	$5.23\times10^{-16}\,cm^3$
	Weight $w(=\rho v)$[a]	$1.74\times10^{-19}\,g$	$1.64\times10^{-15}\,g$
	PW[b] $(=wN_A)$[c]	104,000	9.88×10^{8}
	Number of C atoms	8,660	8.23×10^{7}
Number of particles in 1 g of SNBD $n(=1/w)$		5.75×10^{18}	6.10×10^{14}

[a] ρ = sp. gr. of diamond = 2.99 g/cm^3 (preliminary results)
[b] "Particulate weight", a concept corresponding to "molecular weight"
[c] N_A = Avogadro number
[d] A model of agglutinated product of detonation nanodiamond

graphite would be needed to achieve the advertized volume density! It is quite impressive to find how enormous number of particles that tiny amounts of SNBD contain. Basic figures necessary to count the number of SNBD particles are summarized in Table 12.1. The number of particles contained in 1 g of SNBD, which is billion times as large as 5 billion is given in the last line. To be more systematic, the number is the order of quintillion (10^{18}).

Actually, such large numbers are beyond our comprehension. We could have some idea on the largeness up to millions (10^6) and billions (10^9); for example, populations of large cities and countries are of these orders of magnitude. Although it is difficult to grasp the largeness for numbers beyond trillion (10^{12}), we should pay attention to the effects of great numbers when dealing with nanoparticles, of which there should be plenty.

Summing up, we learned the following from the above consideration: Nanoparticles are larger than conventional molecules by two to three orders of magnitude. However, as Avogadro Number A_N is so large that small differences in size between molecule and nanoparticles is not as significantly as it looks. The number of particles per unit weight ($=A_N$/PW) still turns out to be an enormous number which is out of our sense on the large number. This is the essence of number effect, unique in nanoparticles. Number effect does not exist in molecules, nor in microparticles, but only in nanoparticles.

In this regard, nanoparticles may be useful for those uses where enormously large number of particles is critical. One such example of such "number effect" is given below.

12.2.1.2 Nanospacers Lubrication

Ever since industrial revolution, oils have been used as practically the sole means for reducing friction between moving parts of all kinds of machines to smoothly transmit power. The major and by far the most critical role of lubricant oil is to provide oil film between touching planes in order to prevent adhesion of the planes by direct or boundary contact under load. However, oil (and grease) film often breaks up when their thickness decreased below a few molecular layers to disastrous

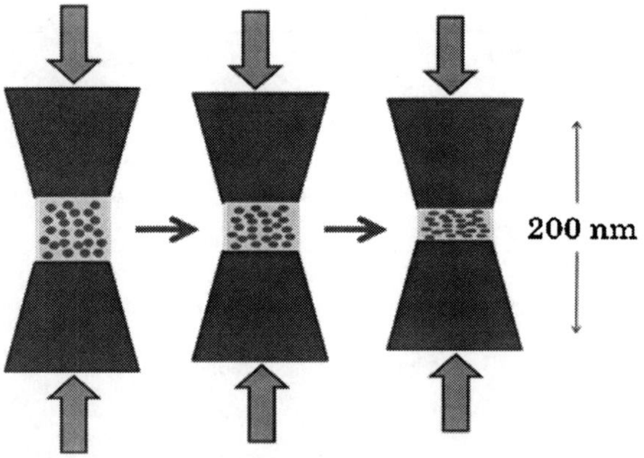

Fig. 12.5 In nanospacers lubrication, colloidal nanoparticles play the role of adhesion preventing spacers under boundary condition

loss of lubrication. In addition, disposal of the used lubrication oil and grease poses a major cause of environmental destruction and increase emission of CO_2 gas. Lubrication oil will be forbidden sooner or later.

An entirely new lubrication method would be to rely on SNBD particles as the spacers to prevent direct contact between shearing planes of moving parts (Fig. 12.5). While it may seem also possible to use SNBD particles as solid lubricant, actually it is not advisable due to very high tendency for these particles to aggregate to give much larger effective diameters unsuitable to penetrate and cover rough microstructure of surface. We evaluate here colloidal solutions of SNBD dispersed in low-viscosity liquid like water.

One of the critical requirements for lubrication by nanospacers is that there must always be enough number of spacer particles in the true contact areas whenever they were beginning to form themselves under the boundary conditions. We examined the prospect of this requirement for SNBD aqueous colloid by using a classical work of Borden and Tabor [43] who estimated area and number of true contact points under various loads between a pair of polished surface of soft-steel plate having an apparent contact are of 2000 mm^2 (Table 12.2). To these data, the number effect of dilute aqueous colloid solution of SNBD (Table 12.1) was fitted, as shown below.

Another critical requirement for the nanospacers lubrication is that the spacer particles stay in a thin space between the approaching pair of true contact area without flowing out of the space with the nonviscous solvent. According to our experimental and theoretical analysis [45], SNBD particles are most likely deformed polyhedra with the facets having high electrostatic potentials and supposed to be tightly hydrated in the colloidal solution. Under the boundary lubrication condition, when the tips of the true contact area on the steel surfaces approach to each other across thin colloidal liquid, the shell of hydrated water on the particle surface in the thin colloidal liquid will be attracted to, not repelled by, the steel surface (Fig. 12.5).

If we assume a surface roughness of 100 nm, then boundary friction condition will start when the thickness of the colloidal solution pinched between the tipped

Table 12.2 Area and number of true contact points and number of SNBD spacers present at a true contact point (TCP) between a pair of shearing soft-steel surface (Fig. 12.5) polished to a surface roughness of 100 nm. Concentration of SNBD is 0.01 wt%[a]

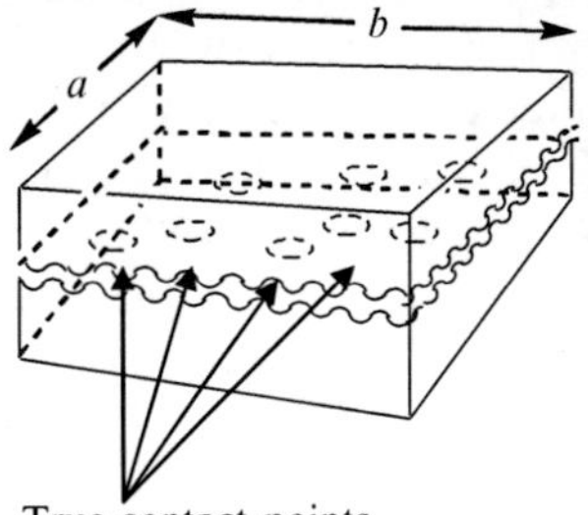

Load W *kgf*	Area of TCP[a]A mm^2	A/S	No. of TCP	No. of SNBD particles in A 10^8
500	5	1/400	35	20.0
100	1	1/2000	22	25.0
20	0.2	1/10000	9	2.1
5	0.05	1/40000	5	0.29
2	0.02	1/100000	3	0.069

[a]Apparent contact area = a × b = 2000 mm^2, thickness of liquid film filling the contact area = average roughness×2 = 200 × 10^{-6} mm, volume of liquid film in apparent contact area = 0.40 mm^3, total number of SNBD particles in 0.01% colloid in this volume = $5.75 \times 10^{18} \times 10^{-2} \times 10^{-5} \times 0.4 = 2.30 \times 10^{11}$

surfaces decreased to about 200 nm. At this point, the number of SNBD particles in the true contact area is equal to a liquid column of 0.01% colloidal solution having a volume of 200 × A nm^3 (Table 12.2). The sum of these values over the true contact points are entered in the last column of Table 12.2 in a unit of 10^8 particles. Even in the very beginning of boundary condition, 7 million SNBD particles participate in the spacer action (the lowest line of Table 12.2). Number of spacers keeps increasing as the load as well as the number and area of true contact points are increased, up to 2 billion under the highest load included in Table 12.2, to take over the load to prevent adhesion. We feel that these numbers are much more than enough, and in practice the concentration of colloid could be reduced to as low as one ten-thousandth % concentration without increasing friction coefficient.

We have already obtained an excellent friction coefficient of 0.02 in our first experiments on nanospacers lubrication, by measuring friction between poly(acrylamide) gel immersed in SNBD colloidal solutions and a reprocating sapphire ball under a small but constant load [44]. However, at that time we have not noticed 'number effect' and vaguely thought that low friction should be realized only in considerably high concentration range of 0.1–5 % based on chemical experience. As clearly shown by the analysis of Bowden–Tabor work in the previous section, the tested concentrations were much too high. We plan to repeat the experiments using much dilute colloidal concentrations.

12.2.1.3 Remarks on the Number Effect

Discovery of nanospacers lubrication reinforced by the number effect will have significantly large impact. In the traditional lubrication oil, the oil itself played dual role of adhesion preventive agent and medium. In contrast, nanospacers lubrication consists of adhesion preventive agent and medium as separate components. The function of medium is to disperse the adhesion preventive agent, and there is no need to form liquid film. Therefore, medium can be of low viscosity like water. For environmental protection, water is the best medium, even though it will induce rust on steel. Rusting consists of a series of electrochemical reactions, wherein Fe is oxidized to $Fe(OH)_2$ and loose $Fe_2O_3{\cdot}xH_2O$. Covering the iron surface with compact oxide layer of Fe_3O_4 is considered to be the cheapest, and most efficient antirusting agent as used in the stainless steel. We would predict that eventually highly diluted aqueous SNBD colloid or lubrication water will replace lubrication oil as lubricant, and all the iron parts of machines will be made of highly stable stainless steel. The lubrication water may contain one more component which replaces the fragile hydration shell on the SNBD surface with more durable and tight oil-like layer to avoid mechanical damage to the true contact areas.

Small mass of nanoparticles brings about large number effect. In other words, each single nanoparticle of SNBD has complex geometric and electronic structure which leads to a variety of functions [45], but they come in large numbers; hence, in many cases, we need a very small mass of nanoparticles. In the past, when the smallest available particles were of the order of microns in diameter, we used to prepare microcomposites containing as much reinforcing components as possible. However, the same results can be realized in much smaller proportions of nanoparticles as reinforcing components in nanocomposites. It is likely that Hall–Petch effect works best for nanocomposites. A logical consequence of number effect is that we may wish to design, prepare, and evaluate nanocomposites in terms of the number of particles rather than the weight. Then, we will be more used to handle large number of like billion-and trillion- like astronomists.

12.3 Materials and Manufacturing

As mentioned, the essential material systems of the patch are nanodiamond, parylene polymers, drug and imaging molecules. Among these, nanodiamond plays the key role in device performance. In the following sections, nanodiamond manufacturing techniques are provided.

12.3.1 High Frequency Plasma-Enhanced Chemical Vapor Deposition (PECVD) for ND Particles and Film Synthesis and Processing

The Nano Particle Technology Laboratory (**NPTL**) is engaged in doing research related to nanoparticle generation, measurement, and some other related applicable research areas such as environment, medical science, energy, new material, element & sensors.

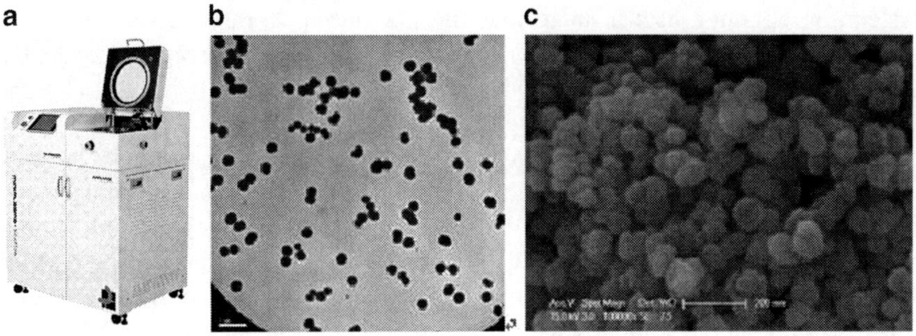

Fig. 12.6 Plasma System (**a**) and Synthesized Nanoparticles (**b** and **c**)

Nanoparticles are widely studied as a building block for broad applications. Several studies have been performed to generate nanoparticles by liquid- or gas-phase synthesis processes. Compare to liquid-phase synthesis, gas-phase synthesis has many advantages such as high purity, continuous process, and size/structure controllability. However, it is still difficult to manufacture nanoparticles smaller than 10 nm due to coagulation.

Recently, NPTL synthesized Si nanoparticles using RF plasma (Fig. 12.6). In which, the size, composition, and growth of nanoparticles can be easily controlled. The Si nanoparticles synthesized, using inductively coupled plasma, have diameter in the range of 3 nm to a few hundred nm and are very uniform in size (relative standard deviation is less than 0.2). These particles are being applied to secondary battery electrode materials.

Recent research articles [62–76] reveal that nanodiamonds (NDs) in particular possess several characteristics including surface functionalization capabilities, biocompatibility, and versatile deposition and processing mechanisms that make them suitable for advanced drug delivery. NDs have formerly been physically immobilized and functionalized in various ways in order to bind with cytochrome *c*, DNA, antibodies, and various protein antigens.

All prior methods of synthesizing diamond are bulk processes. In such processes, new atoms of carbon arrive at the growing diamond crystal structure having random positions, energies, and timing. Growth extends outward from initial nucleation centers having uncontrolled size, shape, orientation, and location. Existing bulk processes can be divided into three principal methods – high pressure, low pressure hydrogenic, and low pressure nonhydrogenic. The high frequency plasma-enhanced chemical vapor deposition (PECVD) is one of the low pressure or CVD hydrogenic metastable diamond growth process.

PLASMA technology (PECVD) is widely used to synthesis and alters the surface properties of few materials without affecting their bulk properties. Plasma deposition involves the addition of a plasma discharge to the foregoing filament process. The plasma discharge increases the nucleation density and growth rate and is believed to enhance ND and ND film formation. There are three basic plasma systems in common use: a microwave plasma system, a radio frequency of RF (inductively or capacitively coupled) plasma system, and a direct current or DC plasma system. The diamond growth rate offered by these systems can be quite modest, on the order of ~ 1 μ/h.

In the case of low pressure hydrogenic method, hydrogen gas mixed with methane is introduced through plasma (discharge), dissociating the methane molecule to form the methyl radical (CH_3) and dissociating the hydrogen molecule into atomic hydrogens (H). Hydrogen is generally regarded as an essential part of the reaction steps in forming ND/ND film during CVD and atomic hydrogen must be present during low pressure diamond growth to: (1) stabilize the diamond surface, (2) reduce the size of the critical nucleus, (3) "dissolve" the carbon in the feedstock gas, (4) produce carbon solubility minimum, (5) generate condensable carbon radicals in the feedstock gas, (6) abstract hydrogen from hydrocarbons attached to the surface, (7) produce vacant surface sites, (8) etch (regasify) graphite, hence suppressing unwanted graphite formation, and (9) terminate carbon dangling bonds. Both diamond and graphite are etched by atomic hydrogen, but for diamond, the deposition rate exceeds the etch rate during CVD, leading to diamond (tetrahedral sp^3 bonding) growth and the suppression of graphite (planar sp^2 bonding) formation. (Note that most potential atomic hydrogen substitutes such as atomic halogens etch graphite at much higher rates than atomic hydrogen). Similarly, dual DC-RF plasma system is used to deposit hydrogenated diamond like carbon (DLC) films form methane plasma. It has the advantages of separately controlling ion density and ion energy by RF power and DC bias, respectively, over conventional simply capacitive-coupled radio frequency PECVD. The sp^3 content, hardness, and Young's modulus of the DLC films increased with increasing RF power at a constant DC bias of −200 V and reached the maximum values at an RF power of 300 W, after which they decreased with further increase of the RF power. The DC bias had a similar but greater effect on the structure and properties of the films, owing to a greater influence of the ion energy on the characteristics of the films than the ion current density. D. M. Gruen group in Argonne National Laboratory has been made ND film w/o hydrogen gas using MWCVD Plasma jet method has high deposition rate so that it is considered alternative method for Microwave CVD (MWCVD) and High frequency CVD (HFCVD). Generally, methane gas is extremely diluted only by 1 vol%. The substrate temperature should be over 700 C for MWCVD & HFCVD film. If we use RF plasma system, we can conduct the same process at lower temperature which be quite advantage for particle deposition for bio device system.

Apart from synthesis, the capability of high-frequency plasma to modify surface physical and chemical properties without affecting bulk properties is advantageous for the design, development, and manufacture of biocompatible polymers. The plasma treated materials have found various applications in automobiles, microelectronics, biomedical and chemical industries. Specific surface properties, such as hydrophobicity, chemical structures, roughness, conductivity, etc., can be modified to meet the specific requirements of these applications. The major effects observed in plasma treatment of polymer surfaces are cleaning of organic contamination, micro-etching, cross-linking, and surface chemistry modification.

Biomaterials that have contact with the human body need an optimal combination of mechanical properties and surface characteristics that result in superior performance in the biological environment. Physicochemical properties of the surface of the material, such as surface free energy, hydrophobicity, and surface morphology, which influence the cell–polymer interaction, determine the choice of the polymer. Since in general all polymers do not possess the surface properties needed for biomedical applications, radio frequency (RF) plasma treatment plays a crucial role in incorporating them. Surface modification in a controlled fashion, deposition of highly cross-linked films irrespective of the surface geometries, formation of multilayer films, eco-friendly nature, and the prospect of scaling-up make the RF plasma treatment extremely suitable for biomedical applications.

12.4 Device Design Roadmap

Over the past couple of years, multiscale modeling and simulations [46–59] have been extensively used to understand properties of solids and structures. We aim to establish a seamless integration of multiple scale computational techniques along with appropriate experimental validation. The design roadmap is schematically shown in Fig. 12.7. In this multiscale analysis, quantum scale (QS) calculation

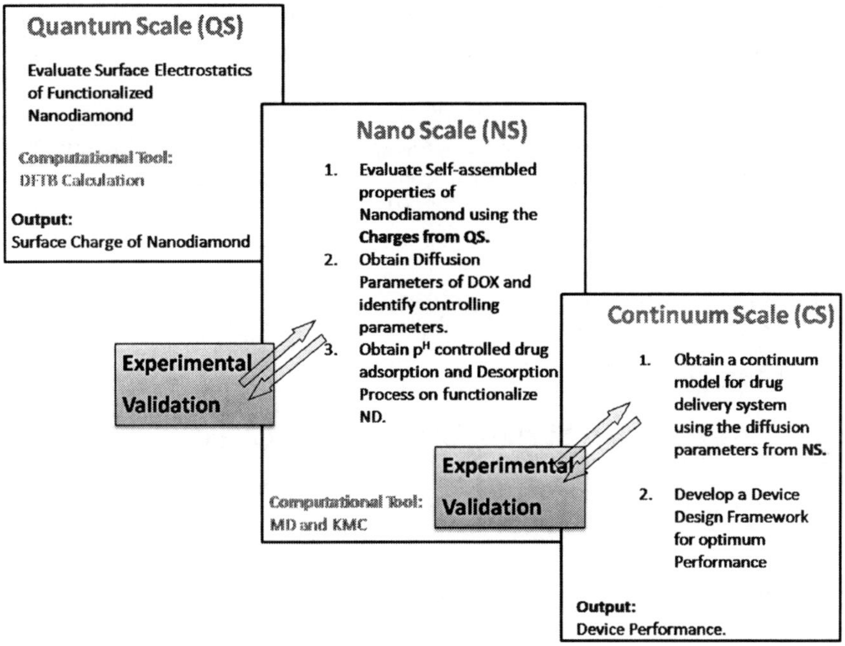
Quantum Scale (QS)
Evaluate Surface Electrostatics of Functionalized Nanodiamond
Computational Tool:
DFTB Calculation
Output:
Surface Charge of Nanodiamond
Nano Scale (NS)
1. Evaluate Self-assembled properties of Nanodiamond using the Charges from QS.
2. Obtain Diffusion Parameters of DOX and identify controlling parameters.
3. Obtain pH controlled drug adsorption and Desorption Process on functionalize ND.
Computational Tool:
MD and KMC
Experimental Validation
Continuum Scale (CS)
1. Obtain a continuum model for drug delivery system using the diffusion parameters from NS.
2. Develop a Device Design Framework for optimum Performance
Output:
Device Performance.
Experimental Validation

Fig. 12.7 Device design roadmap that includes multi-scale simulation and experimental validation

will predict the structure and surface electrostatics of the functionalized NDs. Information from this scale will be utilized in the atomistic calculations at the nanoscale (NS) to evaluate the self-assembly process of functionalized NDs, their electrostatic interactions with drug molecules/biomarkers as well as their diffusion kinetics. Finally, at the continuum scale (CS), the drug delivery process from the device to the targeted area will be modeled and simulated with information provided from sub scale simulations.

Now, in order for the device to function as an innovative drug delivery system with multifunctional capabilities, some fundamental science issues need to be solved. These include quantitative understanding of

1. Structure, surface electrostatics and self-assembly of NDs so that their packing with functional groups can be efficiently facilitated, under a range of conditions.
2. Drug–ND and biomarker–ND interaction process so that they can be controlled with the physiological condition of cancer cells.
3. Drug–ND and biomarker–ND diffusion process so that drug dose rate and cell imaging efficiency can be identified.
4. Overall design optimization with respect to the integrated patch performance.

In addition, the solutions to these scientific issues need to be verified and validated. To do this, a multi-level uncertainty quantification scheme needs to be developed. There are two facets to this uncertainty analysis: (1) comparing the experimental and theoretical results on each scale (quantum, nano, and continuum), and (2) linking these scales such that uncertainty on one scale can realistically propagate into other scales. The knowledge of the uncertainty present in modeling allows for a more robust and reliable model and patch design.

There are many approaches toward uncertainty analysis on a single scale, most involving an iterative process where experiments are repeated and computational models are refined until a realistic calibration between the two can be made (Xiong, 2009; Kennedy and O'Hagan, 2001). However, a multi-level uncertainty quantification scheme that allows for reliable error propagation across many scales does not currently exist. This is another area of this drug delivery system design that needs to be addressed.

12.5 Structure and Surface Electrostatics of Nanodiamonds

12.5.1 Uncertainty Quantification of ND charge

The solutions to these scientific issues also need to be verified and validated to ensure model reliability. To do this, a multi-level uncertainty quantification scheme needs to be developed. There are two facets to this uncertainty analysis: (1) comparing the experimental and theoretical results on each scale (quantum, nano, and continuum), and (2) linking these scales such that uncertainty on one scale can realistically propagate into other scales. The knowledge of the uncertainty present in modeling allows for a more robust and reliable model and patch design.

For example, the surface charges of the NDs discussed in Sect 5.2 can be approached in a variety of ways. A strictly linear projection can be done such that the charge distribution for the larger NDs looks precisely the same as for the smaller NDs. However, this ignores most of the information that is known about the system and assumes a behavior that may not be accurate. As experimental data are limited on this scale, a better understanding of the accuracy of this estimate is important. Errors introduced in this step will propagate throughout the modeled system.

Instead of the this deterministic approach, an estimate of the probability density function (PDF), and therefore the cumulative density function (CDF), of each face of the ND can be generated statistically. The estimated PDF and CDF are developed using a Gaussian-kernel smoothing technique for each of the three truncated octahedral NDs with available DFT calculations. A sample of size n, where n is the number of atoms present on the respective faces of the next larger ND, is generated from the current PDF and the sample PDF is generated. These sample functions are used as a projection of the next larger ND.

This projection scheme allows for a good estimate, but it only utilizes data present on the closest ND to determine the surface charges for unknown NDs. A more rigorous scheme can be developed using a bootstrapped and edgeworth expansion [76]. Bootstrapping refers to resampling a data set to estimate the behavior of the entire system. This ideology carries over well for the use in data projection because of the data sub-steps available (in this case, the surface charge distributions of the different sized NDs). More details of the bootstrap method and Edgeworth expansion can be found in Hall [78], but the application to the ND charge distribution will be outlined here.

Each face of the ND truncated octahedron is considered independently because they have dramatically different charge distributions. For example, face 100 can be isolated into three readily available data sets: the first set from the largest ND with 1639 total atoms, the second set from the next largest ND with 1198 total atoms, and the third set from the next largest ND with 837 total atoms. The two smallest NDs (268 and 548 atoms) will not be considered because their results vary from the larger charge distributions that are of interest for this application.

The surface charges for the 100 faces on the three NDs are extracted, and will be referred to in decreasing order of size, respectively F_1, F_2, and F_3. A random sampling with replacement is done of F_1, F_2, and F_3 to create three more data sets to calculate the projection onto a larger ND, and gives a better projection scheme. From these six data sets, a nested probability distribution can be estimated, and a better understanding of the reliability of the linear projections is available.

The surface charges are only one example of the uncertainty quantification needed in this ND model. There are many approaches toward uncertainty analysis on a single scale, most involving an iterative process where experiments are repeated and computational models are refined until a realistic calibration between the two can be made [79, 80]. However, a multi-level uncertainty quantification scheme that allows for reliable error propagation across many scales does not currently exist. This is another area of this drug delivery system design that needs to be addressed.

12.5.2 Multiscale Analysis of ND charge

The structure of individual nanodiamonds (NDs), described in terms of the degree of crystallinity and characteristic surface chemistry, is crucial to all ND-based technologies. Both crystallinity and surface chemistry affect particle–particle interactions, changing the way individual nanodiamonds self-assembe, and particle–drug interactions responsible for adsorption and desorption of drug molecules. The structure of detonated NDs is still relatively poorly understood, but the past decade has seen a number of seminal computational studies upon which the current understanding is based. [60] Early work utilized thermodynamic (phase stability) theories [63] and first-principles density functional theory (DFT) computer simulations [61] to elucidate the shape of ND, and describe how the shape affects the stability of individual surface facets on small particles (< 2 nm in diameter). While NDs may assume various nominal shapes, including spherical, octahedral, truncated octahedral, cuboctahedral, and cubic morphologies, NDs with truncated octahedral or cuboctahedral morphologies are thermodynamically preferred over other alternative shapes.[63] Similarly, early DFT simulations demonstrated that the preferential graphitization of the (111) surface to form fullerenic sp^2-bonded "bubbles" on the surface, giving rise to a special (all-carbon) core–shell structure known as a bucky-diamond.[65] Bucky-diamonds are characterized by a crystalline diamond sp^3-bonded core, encapsulated by a single- or multi-layer sp^2-bonded fullerenic shell that either partially or fully covers the particle surface. The localized (surface) sp^3 to sp^2 phase transition is spontaneous at room temperature, but can be eliminated by coating the surfaces with suitable passivants [66].

A more recent density functional tight binding (DFTB) study [67] examining the crystallinity of somewhat larger NDs (2–3 nm in diameter) confirmed the thermodynamic preference for the truncated octahedral shape, and the localized graphitization of the octahedral (111) facets. The (100) and (110) surface facets were shown to resist graphitization, but exhibited reconstructions containing sp^{2+x} hybridized carbon atoms. In addition to this, these new results predicted an anisotropic (facet dependent) pattern of charge distributions on ND surfaces. The resultant surface electrostatic potential revealed that the (100) surface facets possess a strong positive electrostatic potential, while the (111) facets possess a negative electrostatic potential depending upon the degree of surface graphitization. This suggests that Coulombic interparticle interactions may be responsible for nanodiamond agglomeration, and that a preferred orientation for particle–particle interactions would produce self-assembled ND agglutinates. This hypothesis was validated via a systematic DFTB study in 2008 [68], which elegantly demonstrated preferred configuration of ND interparticle interactions based on strong, long-ranged Coulombic attractions.

While these studies give some insight into ND electronic structure for ND between 1–3.3 nm in diameter, experimentally available detonation NDs range in size from 4–10 nm [81]. Unfortunately, it is not yet possible to determine the equilibrium structure of charged NDs over this entire size range because of computational constraints. Moreover, experimentally results indicate that detonated NDs are heavily functionalized with various functional groups including carboxyl (–COOH) and carbonyls (CO). As conventional DFT and DFTB calculations will be too computationally expensive to treat entire ensembles of structures at experimentally relevant sizes, we have implemented a multiscale approach to nanodiamond simulation, building upon the ab initio foundation and extending this knowledge to larger systems. In the first instance, we have employed a projection method to identify charge distributions for larger nanodiamonds, beyond the sizes explored in previous works.

In this method, effective surface electrostatic potentials for smaller NDs that are obtained from DFTB calculations [67] are first analyzed. Figure 12.8 shows normalized surface potentials for truncated octahedral and cuboctahedral geometry [67]. The surface potentials per atom are calculated using the classical Coulomb's law that utilizes the Mullikan charge distributions from DFTB calculation:

$$\Psi_i(r) = \frac{q_i q_j}{4\pi\varepsilon\varepsilon_0 r_{ij}}, j \neq i \tag{12.1}$$

where ψ is the surface potential, q's are the atomic charges, ε is the dielectric permittivity and r is the distances between atoms. Here, i is a free index and j is the dummy index and the relation follows Einstein's summation convention.

In the case of the truncated octahedral subset, as shown in Fig. 12.8a, it can be observed that there is a strong correlation between the particle shape and structure of the surfaces, and the sign of electrostatic potential. The (100) surfaces, and the (100)/(111) edges, exhibit a strong positive potential, whereas (like the octahedral particles), some of the graphitized (111) surfaces exhibit a strongly negative potential.

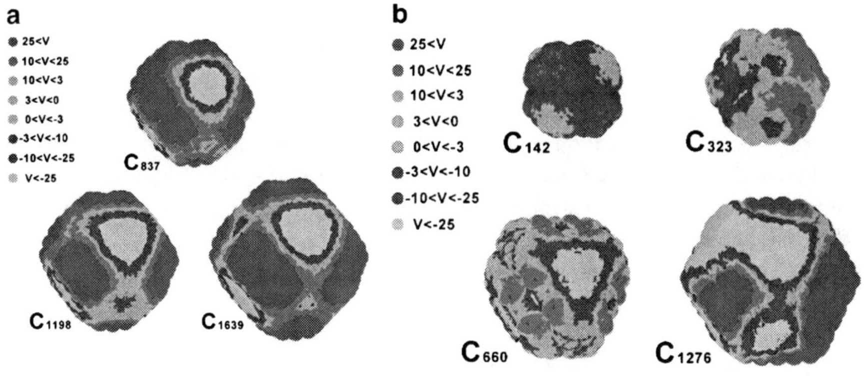

Fig. 12.8 Normalized surface electrostatic potential for the relaxed structures of the (**a**) truncated octahedral and (**b**) cuboctahedral subsets. Here C_{xxx} refers to xxx carbon atoms in structure, and higher the number the larger is nanodiamond size. The figure is obtained from [67]

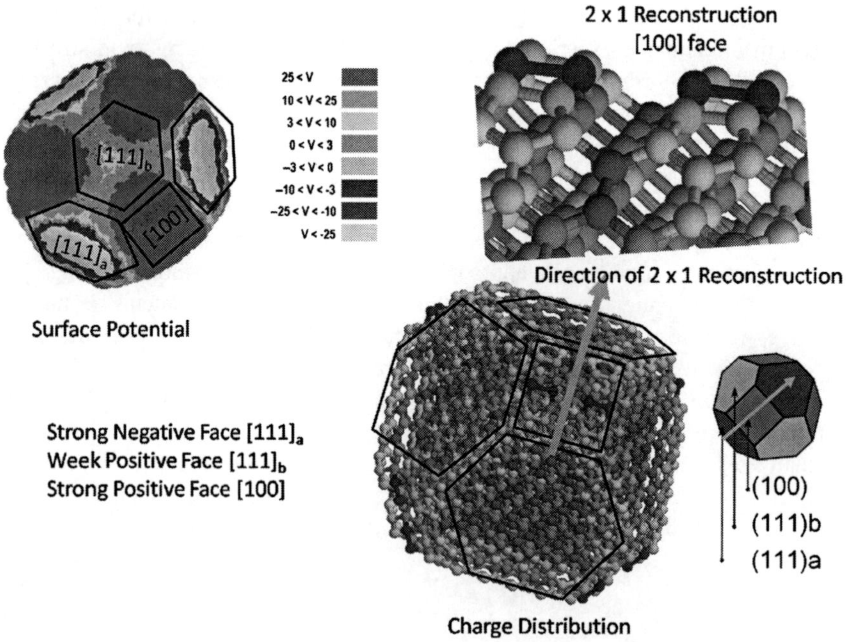

Fig. 12.9 Charge distribution and corresponding surface potential on truncated octahedral nanodiamond. Note the geometric asymmetry in the overall structure induced by 2×1 dimer on the (100) face that in turn causes two distinct distributions of charges and potentials on (111) faces

It is interesting that only half of the (111) surfaces exhibit the strong negative V, while the remaining (111) surface exhibit much more variation in potential (depending on the position relative to the facet center). This is due to some "artificial" asymmetry in the overall particle shapes. This asymmetry has been introduced on purpose to ensure that the (100) surfaces have the opportunity to form two ideal 2×1 dimer rows, but results in rectangular (100) facets, and some (111) facets being slightly larger than others. This is shown in Fig. 12.9.

Since the asymmetry in the (111) surface potentials is not due to cross-linkers, or variations in the core–shell structure [66], but due to geometric asymmetry and imperfect graphitization alone, it can be anticipated that a symmetric geometry would yield a similar potential field on all (111) surfaces if graphitization is homogeneous. With this assumption, the charge and potential distribution with respect to ND size has been analyzed and the relation between charge and surface potential has been explored.

12.5.2.1 Charge and Potential on (100) Surface

Figure 12.10(a) shows the charge distribution on (100) surface of a truncated octahedral nanodiamond. In this figure, the surface atoms of the (100) faces are only colorized to identify the charges more clearly. The atoms colored in gray represent

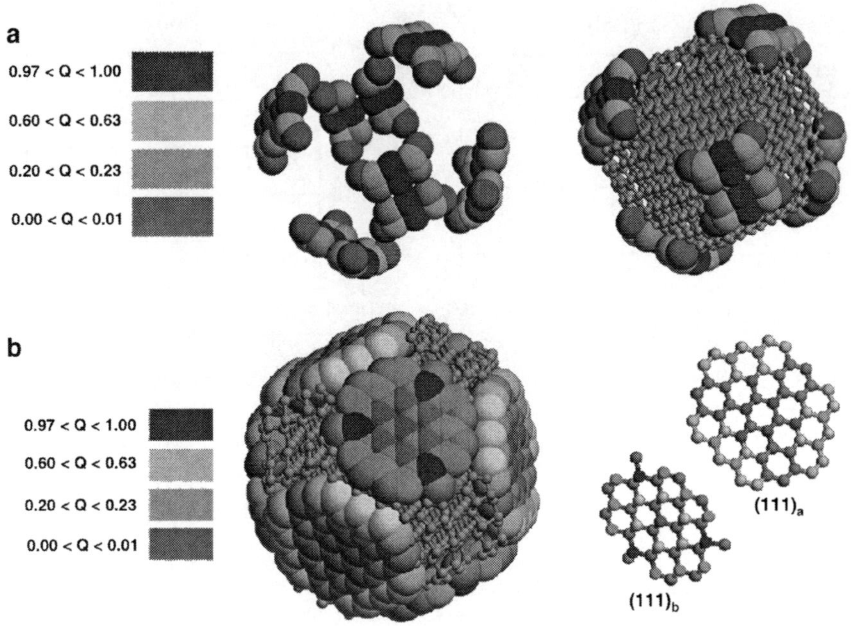

Fig. 12.10 Charge distribution on (**a**) (100) and (**b**) (111) faces of a truncated octahedral nanodiamond

all remaining atoms that do not participate in the (100) terminal atomic layers. It can be noted that there are eight (100) facets on this nanodiamond shape. A closer look at any particular (100) surface reveals that there are four distinct types of positive point charges present. Of these charges, the four corner atoms (colored red) have negligible charges, partly because they are sitting beneath the (100) surface plane. A dramatic reduction in charge is observed for other atoms that are not lying on the surface [67]. Conversely, the rows of atoms perpendicular to the 2×1 missing row directions (colored blue) exhibit a considerably higher positive potential. These high values are attributed to the reconstruction itself. This charge projection scheme, we ignored these two extreme charge values, assuming instead that the strong positive charge on the (100) facets of larger (>5 nm) NDs will not be dominated by these atomic-scale reconstructions and negligible charges due to out-of-plane atom sites. With this assumption, we evaluated the average charge $Q_{average}$ using the relation:

$$Q_{average} = \frac{1}{N}\sum_{i}^{N} q_i \tag{12.2}$$

Now, we employed the same procedure to evaluate the average charge for three different nanodiamonds, and the results are plotted in Fig. 12.11a. It can be seen from Fig. 12.11a that charge on (100) faces approaches a constant value (+0.4378 ec) as ND size increases.

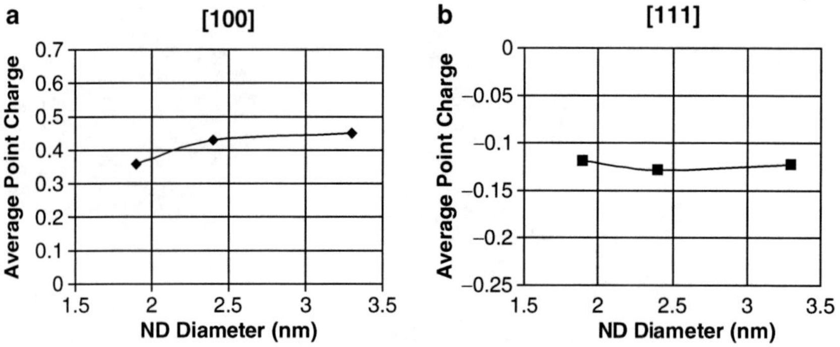

Fig. 12.11 Projected average charge distribution on (**a**) (100) faces and (**b**) (111) faces

12.5.2.2 Charge and Potential on (111) Surface

Using a similar approach to the (100) surface facets, the charge distribution on a (111) surface of a truncated octahedral nanodiamond is shown in Fig. 12.10b. In this figure, only the surface atoms on the (111) faces are colorized. As before, the atoms colored in gray represent all remaining atoms that do not participate in the (111) terminal atomic planes. It can be noted that there are six (111) faces on this nanodiamond, and it can be identified that the surface geometries of all (111) is not identical. In contrast, the half of the graphitized (111) surfaces exhibits a strongly negative potential (denoted as a type $(111)_a$ facet), and half exhibits much more variation in potential (denoted as a type $(111)_b$ facet). This is due to asymmetry in the overall particle shape introduced to preserve the ideal 2×1 reconstruction on the [67] facets, but resulting in imperfect graphitization of the $(111)_b$ facets (as mentioned in Sect. 5.2.1) [67]. In our charge projection analysis, we ignore the contributions from $(111)_b$ faces with the assumption that the charge distributions of $(111)_b$ faces would be similar to $(111)_a$ if the geometric asymmetry in the structure is absent, and graphitization is geometrically homogenous.

Employing an identical procedure to evaluate the average charge for all three nanodiamonds, and the results are plotted in Fig. 12.11b. It can be seen from Fig. 12.11b that the charges on (111) faces approaches a constant value (−0.1239 ec) as ND size increases. Using these two approximated constant charges on (100) and (111) surfaces, the corresponding surface potential is computed and compared with original DFTB calculated data, as shown in Fig. 12.12. In the future, to simulate large nanodiamonds, these constant charges will be utilized to extend this work well beyond the modest length scale accessible to DFT and DFTB simulations.

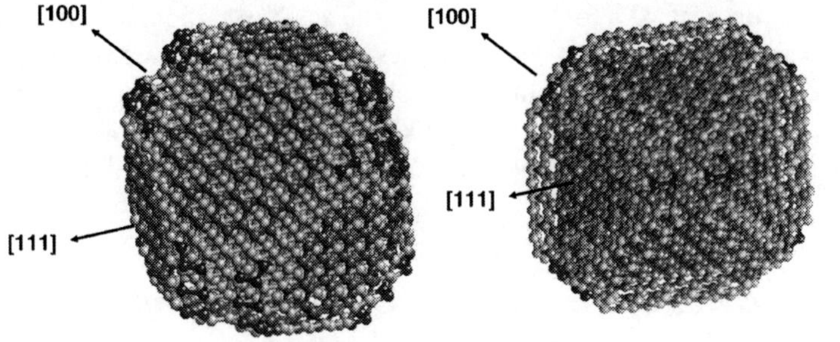

Fig. 12.12 Comparison of charge distribution and surface potential between the facet dependent constant charge approximated and DFTB calculated ND surface

12.6 ND–Drug Interactions

12.6.1 Functionalization of Nanodiamonds

The next phase of calculation involves binding of the DOX to the functionalized NDs. The understanding the drug absorption and desorption profile to and from NDs is the most critical portion of the device design. There is no clear understanding on how many or what types of functional group are present on any ND surface. Some experimental observations [77–85] suggest that the types of functional groups are greatly dependent on the available dangling bonds on ND, whereas the degree of functionalization is dependent on functional group size, shapes, and orientations. In general, it is extremely difficult to quantitatively predict either the types or the number of functional groups present on any surface. It is, however, possible to obtain the upperbound estimate of the functional group by counting the number of potential sites with respect to nanodiamond size. Moreover, it is also possible to identify potential functional elements based on their available valences. For example, it is likely that carboxyl [–COOH] functional group that has valence equals one would attach on (111)/(111) edges because such edges have one dangling bond. Similarly, the carbonyl groups [–CO] with two valence electrons are more likely to attach on (100) faces because each site has two dangling bonds. It can be noted that either of the functional group are negatively charge implying they both should attract positively charged ions. In the future, some phase stability studies will be conducted in the future to quantify more on the possible functional groups.

12.6.2 Conjugation With Single-Stranded DNA Through π-Stacking

The base of single-stranded DNA (ssDNA) is known to conjugate with sp^2 hybridized carbon structures through π-stacking [84, 85]. A similar mechanism can be adopted to attach the base of ssDNA to NDs. Once the ssDNA-carbon nanostructure complex is obtained, these can be used as probes to identify target ssDNA with particular sequence [86]. Also, biotinylated ssDNA can be conjugated with carbon nanostructures to provide a platform to link with a variety of biomolecules, including antigen and antibody, through the strong, versatile bioton–avidin interaction [87].

12.6.3 Electrostatic Interactions Between Doxorubicin and Functionalized Nanodiamond

Typical functionalized NDs and drug molecules contain opposite charges at their surfaces, it has been a natural interpretation that interactions between ND and drug molecules should be straightforward – NDs should attract to drugs as they come into contact. However, recent experiments, however, suggest that NDs usually do not interact with drug molecules in the presence of neutral solutions. The addition of NaCl in the solution improves the interaction dramatically. Moreover, some experiments [6], without any exhaustive conclusion, suggest that drug loading/unloading process may be connected to the pH content of the solution. It has been observed from Dean Ho's experiment [6] that ND and DOX in aqua solution do not adhere to each other. However, a dramatic change in drug adsorption on ND surfaces has been observed with the addition of NaCl (or NaOH) in the solution. The process can be explained from electrokinetic theories involving anionic–cationic interactions. It can be argued that in aqueous solution, without the addition of any other perturbation ions, the repulsive interaction of DOX–DOX and ND–ND is greater than the cohesive interaction of ND–DOX and thus an insignificant amount of DOX can be adsorbed onto the NDs. With the addition of NaCl, the increase of Cl^- ions may shift the balanced interactions toward the formation of DOX–ND complexes because cationic DOX is also balanced with anionic Cl^- ions. The actual mechanism can be modeled via molecular dynamics simulations and can be contrasted with an electrokinetic theory. A well-validated theory will essentially ease the development of a seamless continuum model for drug delivery.

12.6.4 Fundamental Theory of Electrostatic Interactions Between Charged Particles

The interactions of charged particles or ions evolve over time, while maintaining a self-balanced total charge in the system of interested must be self-balanced. How the charges are distributed locally in the system and what factors control the local environment are unclear. The Gouy–Chapman double layer model addresses this issue [89].

Considering an infinitely spanned planner interface with a surface potential $\psi_0(V)$ and a surface charge density $\sigma_0(Cm^{-2})$, the Gouy and Chapman model attempts to describe the distribution of charge and potential in the solution as a function of distance from the surface.

A fundamental result from electrostatics gives the relationship between charge density $\rho(Cm^{-3})$ and the potential $\psi_0(V)$ at any point in the domain of interest. This is the classical *Poisson Equation*:

$$\nabla^2\Psi = \frac{\partial^2\Psi}{\partial x^2} + \frac{\partial^2\Psi}{\partial y^2} + \frac{\partial^2\Psi}{\partial z^2} = -\frac{\rho}{\varepsilon} \tag{12.3}$$

where ε is the permittivity of the medium.

In addition, in an electrolyte solution, the charge density depends on the local concentrations of anions and cations, which in turn depends on the location potential through the *Boltzmann Equation:*

$$n_i = n_{i0}\exp\left(-\frac{z_i e\Psi}{kT}\right) \tag{12.4}$$

where n_i is the number concentration of ion i at a point where the potential is ψ, n_{io} is the corresponding concentration in the bulk solution ($\psi = 0$), z_i is the valence of ion i, e is the electron charge, k is the Boltzmann constant. The equation clearly suggests that when the potential has a sign opposite to the sign of charge, then the location concentrations will be higher than the bulk value and vice versa.

Now combining (12.3) and (12.4), the classical *Poisson–Boltzmann* Equation for electrostatic interactions can be obtained as:

$$\nabla.\varepsilon(r)\nabla\Psi(r) = \rho_{bulk} + \sum_i z_i e n_{i0}\exp\left(-\frac{z_i e\Psi}{kT}\right) \tag{12.5}$$

For a one-dimensional system with homogenous dielectric medium, the PB equation can be written as

$$\frac{d^2\Psi}{dx^2} = \frac{2zen_0}{\varepsilon}\sinh\left(\frac{z_i e\Psi}{kT}\right) \tag{12.6}$$

For a system with small potential:

The *sinh* term can be expanded and linearized as:

$$\sinh(\Psi) = \Psi + \frac{\Psi^3}{3!} + \frac{\Psi^5}{5!} + \ldots.. \approx \Psi \tag{12.7}$$

The linearized approximation makes it easier to solve the PB equation.

12.6.5 pH-Dependent Adsorption–Desorption

At the ionic level, variation in pH simply means variation in anionic–cationic concentration. As a result, it should be possible to develop a correlation between

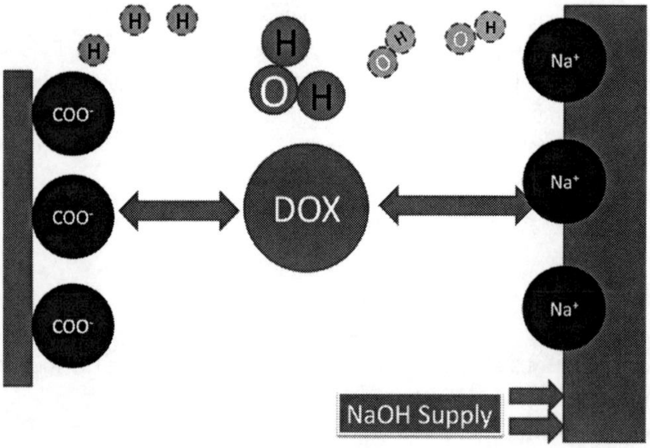

Fig. 12.13 (**a**) Basic components of ND–DOX–Parylene drug delivery system. (**b**) Major diffusion process involved in the drug delivery system. The basic device consists of a non-porous parylene substrate, ND–drug assembly and porous parylene media. Process 1: Drugs are adsorbed on the ND surface via pH-mediated electrostatic interactions between ND–drug and solutions. Once adsorbed on the surface, the centrifuged-out ND-drug assembly is put on the non–porous parylene bath to ensure device integrity. Process 2: Diffusion of ND-drug system is controlled by varying the porous parylene thickness. Controlling is possible between the time frame of 10 min and 2 days. Process 3: pH-controlled drug release

pH level and adsorption–desorption criterion. One such approach is described in following schematic diagram (Fig. 12.13). The notes on Fig. 12.13 will help understand the approach.

12.7 ND-Drug Diffusion Through Porous Parylene

12.7.1 Fundamental Theory

In principle, diffusion of ND–drug system can be modeled using Fick's Laws of Diffusion which states that:

$$\frac{\partial C}{\partial t} = D\left(\frac{\partial^2 C}{\partial x^2} + \frac{\partial^2 C}{\partial y^2} + \frac{\partial^2 C}{\partial z^2}\right) \tag{12.8}$$

where, D is the diffusion coefficient. At the continuum level, ND–drug system diffusion through parylene layer should obey this Fick's Law. As the parylene pore radius is at the order of nanometer, it is expected that VDW interaction force will impart contribution on the diffusion process. This could be verified by performing MD/MC simulation on drug diffusion through parylene layers.

Now, once we have the diffusion coefficient determined, we can employ them to develop a continuum model using immersed finite element method.

12.7.2 Diffusion Through Porous Parylene and Continuum Model

The focus of this section is on the description and numerical modeling of drug release from the porous parylene matrices. Polymeric matrices can be thought as coherent systems that contain mechanical properties in between those of solids and liquids. The delivery system at the continuum is simply a three-dimensional network comprised of a liquid medium with embedded high molecular weight molecules, similar to that of sponge. Porosity of such matrix systems becomes crucial, as the drug diffusion may occur not only through network meshes but also through the pores. Drug release kinetics may be affected not only by the physical and chemical characteristics of the drug/ND/polymer matrix, but also by many factors including polymer swelling and erosion, drug dissolution and diffusion characteristics from the ND, drug/ND distribution inside the polymer matrix, drug/ND/polymer ratio, and overall system geometry. Upon contact with the release fluids, physiological media, the polymer swells, and release can take place. This implies the transition process of the polymer from the glassy dry state to the rubbery swollen one, where molecular rearrangements of polymetric chains reach a new equilibrium condition. This glassy–rubbery transition increases polymer chain mobility, enabling the drug to dissolve and diffuse through the gel layer. The time required for this transition depends on the relaxation time of the given polymer/solvent system, which in turn is a function of both solvent concentration and temperature. If the relaxation time is much lower than the characteristic time of diffusion, the solvent absorption may be described by means of Fick's law with a concentration-dependent diffusion coefficient. On the contrary, Fickian solvent absorption with constant diffusivity takes place. If both the relaxation time and the characteristic time of diffusion are about the same, solvent adsorption does not follow Fick's law of diffusion. Hence, drug release becomes non-Fickian where solvent absorption and drug release depends on the polymer/solvent viscoelastic properties.

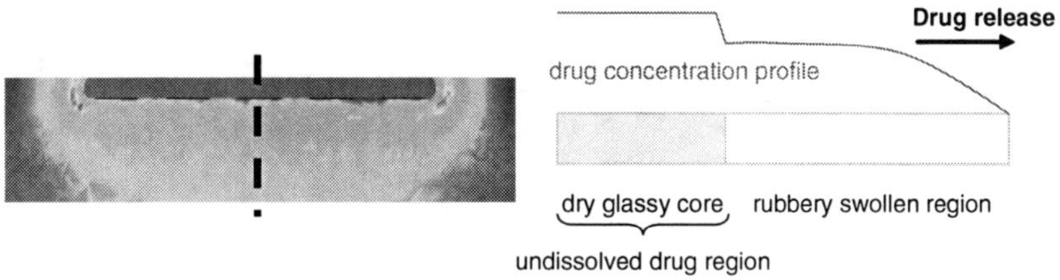

Fig. 12.14 The external fluid update of the drug molecules

At the continuum level, the drug release mechanism from the ND/polymer matrices can be divided into three main fronts: the eroding front separating the release environment from the matrix, the swelling front separating the dry glassy core from the swollen matrix portion, and the diffusion front, which is found between the outer portion of the swollen matrix and the inner part where the drug is not yet completely dissolved. It should be noted that at the matrix/release environment interface, uniform distribution of drug may lead to a burst effect in the release profile followed by a slow release and must be avoided.

Most of the modeling effort at the continuum will be invoked within the immersed finite element method (IFEM). IFEM becomes an ideal platform to treat these kinds of problems due to its capabilities of solving fluid–structure interaction problems. Furthermore, within this framework, new models can be developed to describe drug release from the ND/parylene matrices.

Many simple empirical and semi–empirical approaches have been developed to describe drug release from matrix systems; among them is the one by Higuchi that applies to a planar system:

$$M_t = A\sqrt{D(2C_0 - C_s)C_s t} \qquad C_0 > C_s \tag{12.9}$$

where M_t is the amount of drug released until time t, A is the release area, D is the drug diffusion coefficient, C_0 is the initial drug concentration in the matrix, while Cs is the drug solubility. This model, despite its simplicity, may be extended to more complex geometries and porous systems.

Consider an incompressible three-dimensional deformable structure Ω^s (as shown in Fig. 12.15), i.e., drug delivery system or a cancer tumor, completely immersed in an incompressible fluid domain Ω^f. Both, the fluid and the solid occupy the domain Ω, but do not intersect

$$\Omega^f \cup \Omega^s = \Omega$$

$$\Omega^f \cap \Omega^s = 0 \tag{12.10}$$

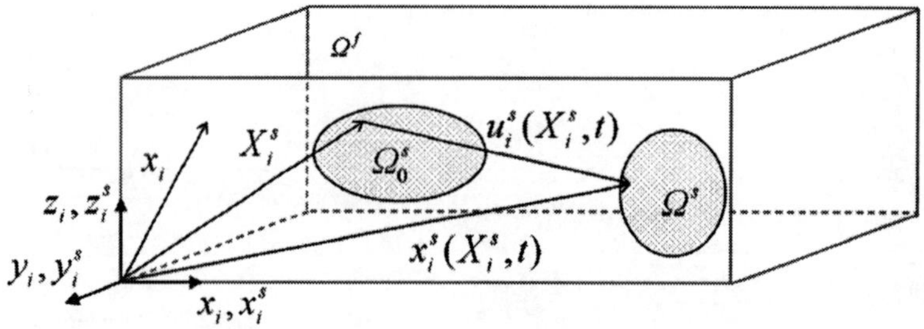

Fig. 12.15 The Eulerian coordinates in the computation fluid domain $\Omega^f$are described with the time invariant position vectorx_i. The solid positions in the initial configuration Ω_0^s and the current configuration $\Omega^s$are represented by X_i^s and $x_i^s(X_i^s,t)$, respectively

With these assumption, Eulerian fluid mesh is adopted which spans the entire domain Ω and a Lagrangian solid mesh is constructed on top of the Eulerian fluid mesh. The coexistence of fluid and solid in Ω^srequires some deliberation when developing the momentum and continuity equations. As we know, the inertial force of a particle is balanced with the derivative of the Cauchy stress σ and the external force f^{ext}exerted on the continuum

$$\rho \frac{\mathrm{d}v_i}{\mathrm{d}t} = \sigma_{ij,j} + f_i^{\mathrm{ext}} \tag{12.11}$$

Noting that the solid density $\rho^s$is different from the fluid density ρ^f, i.e., $p = p^s$ in Ω^s and $p = p^f$ in Ω^f , the inertial forces can be divided into two components within the solid domain Ω^s in the following manner:

$$\rho \frac{\mathrm{d}v_i}{\mathrm{d}t} = \begin{cases} \rho^f \dfrac{\mathrm{d}v_i}{\mathrm{d}t}, \mathbf{x} \in \Omega / \Omega^s \\ \rho^f \dfrac{\mathrm{d}v_i}{\mathrm{d}t} + \left(\rho^s - \rho^f\right)\dfrac{\mathrm{d}v_i}{\mathrm{d}t}, \mathbf{x} \in \Omega^s \end{cases} \tag{12.12}$$

Following the same concept, both the external force $f_i^{\mathrm{ext}}$and the Cauchy stress is decomposed as

$$f_i^{\mathrm{ext}} = \begin{cases} 0, \mathbf{x} \in \Omega / \Omega^s \\ \left(\rho^s - \rho^f\right)g_i, \mathbf{x} \in \Omega^s \end{cases}$$

$$\sigma_{ij,j} = \begin{cases} \sigma^f_{ij,j}\ \mathbf{x} \in \Omega / \Omega^s \\ \sigma^f_{ij,j} + \sigma^s_{ij,j} - \sigma^f_{ij,j}, \mathbf{x} \in \Omega^s \end{cases} \tag{12.13}$$

It is important to note that the fluid stress in the solid domain in general is much smaller than the corresponding solid stress. Furthermore, since the computational fluid domain is the entire domain Ω, hydrostatic pressure is ignored. The fluid–structure interaction (FSI) force within the Ω^s is defined as

$$f_i^{\mathrm{FSI,s}} = -(\rho^s - \rho^f)\frac{\mathrm{d}v_i}{\mathrm{d}t} + \sigma^s_{ij,j} - \sigma^f_{ij,j} + (\rho^s - \rho^f)g_i, \mathbf{x} \in \Omega^s \tag{12.14}$$

The fluid–structure interaction force is naturally calculated with the Lagrangian description, where a Dirac delta function δ is used to distribute the interaction force from the solid domain onto the computational fluid domain.

$$f_i^{\mathrm{FSI}}(\mathbf{x},t) = \int_{\Omega^s} f_i^{\mathrm{FSI,s}}(\mathbf{X}^s,t)\delta(\mathbf{x} - \mathbf{x}^s(\mathbf{X}^s,t))\mathrm{d}\Omega \tag{12.15}$$

Hence, the governing equation for the fluid domain can be derived by combining the fluid terms and the interaction force as:

$$\rho^f \frac{dv_i}{dt} = \sigma^f_{ij,j} + f_i^{\mathrm{FSI}}, \mathbf{x} \in \Omega \tag{12.16}$$

As we consider the entire domain Ω to be incompressible, we only need to apply the incompressibility constraint ($v_{i,i} = 0$) once in the entire domain Ω. To define the Lagrangian description for the solid and Eulerian description for the fluid, different velocity field variables, v_i^s and v_i, are introduced to represent the motions of the solid in the domain Ω^s and the fluid within the entire domain Ω. The coupling of both velocity fields is accomplished with the Dirac delta function

$$v_i^s(\mathbf{X}^s, t) = \int_\Omega v_i(\mathbf{x}, t)\delta(\mathbf{x} - \mathbf{x}^s(\mathbf{X}^s, t))\mathrm{d}\Omega \tag{12.18}$$

The coupling between the fluid and solid domains is enforced via the Dirac delta functions. The nonlinear system of equations is then solved using the standard Petrov-Galerkin method and the Newton–Raphson solution technique. Moreover, to improve the computational efficiency, GMRES iterative algorithm is employed to compute the residuals based on the matrix-free techniques.

12.8 Diagnostic Nanodiamonds

12.8.1 Imaging Applications as BioMarkers

Organic fluorophores have been widely used to tag antibodies in typical immunological assays including an immuno-fluorescent staining. However, photobleaching of the chromophore limits the life time and long-term stability, resulting in the disability of researchers to investigate specimens repeatedly and over long periods. Nanoscale particles, such as NDs, can be employed to overcome this hurdle. NDs show persistent, nonphotobleaching Raman scattering enabling the detection of the signal from a small amount of molecules. Two distinctive Raman modes, so-called D and G modes, correspond to sp^3 and sp^2 hybridized carbon structures. The NDs will be engineered from the synthesis step to optimize these Raman signals for better sensitivity. Also, a variety of dopants can be incorporated to induce florescence in the visible range. These dopants will be optimized to provide high quantum yields.

12.8.2 Imaging Functionality of Nanodiamond

Figure 12.16 shows the confocal fluorescence image of 5-nm ND and A549 lung cancer cell. The fluorescence images shown are after 30-min incubation with the

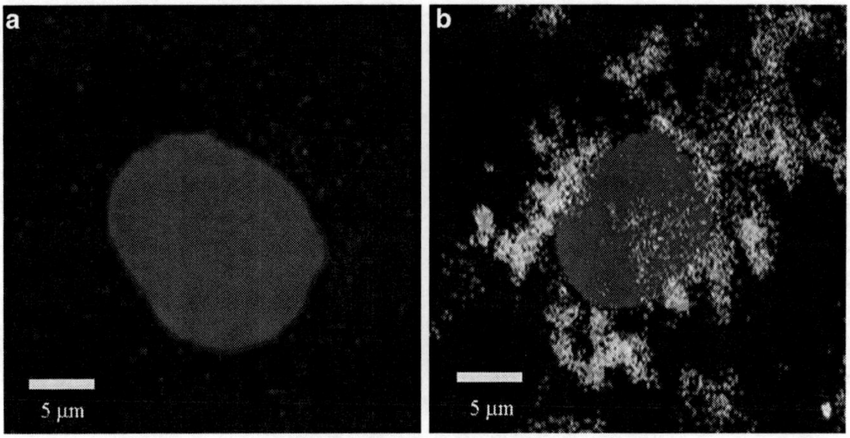

Fig. 12.16 Fluorescence images of (**a**) ND only and (**a**) ND–GH on A549 cancer cell. (**a**) The nonlabeled ND can be stripped after 30-min incubation with A549 cancer cell. (**b**) However, the growth hormone-labeled ND can recognized A549 cancer cell specifically

nonlabeled nanodiamond (a) and growth hormone-labeled nanodiamond (b). The right image indicates the labeled nanodiamond can recognized cancer cell specifically. Meanwhile, the non-labeled nanodiamond can be uptake freely after 12 h incubation (data not shown). Therefore, ND may be a good candidate as drug or gene delivery shuttle and indicator in biomedical usage.

The right figures are interactions of growth hormone-labeled NDs (ND-GH) with A549 cell. The A549 cancer cell contains growth hormone receptors on the surface of its cell membrane. Therefore, the NDGHs are located on the cell surface and cannot be observed inside the cell. This images indicated that the modified ND can be used as cancer cell targeting molecules and the optical properties of ND can help us to identify those cancer cells. Namely, functionalized ND can be used in bio-image of cancer research [88].

As shown in Fig. 12.17, the topic of this study in ND-related cancer targeting and treatment is based on the impurity, nitroso, of ND. The nitroso is the residue of the TNT and hexogen during diamonds detonation. This impurity is embedded in the structure of nanodiamonds. However, this molecule can be photo activated and release NO by LASER irradiation.

These accumulated NO, as depicted in Fig. 12.18, will be heated and increase extremely high pressure and facilitate the diamond transform into graphite. This process will enlarge the size of ND tenfolds, as a nanoblast. Therefore, this conformation transition can be used as nanoknife in nanomedicine. This is model to interpret the mechanism of functionalized ND complexes recognize specific target of cell membrane and can be triggered blast as nanoknives for nano medicine.

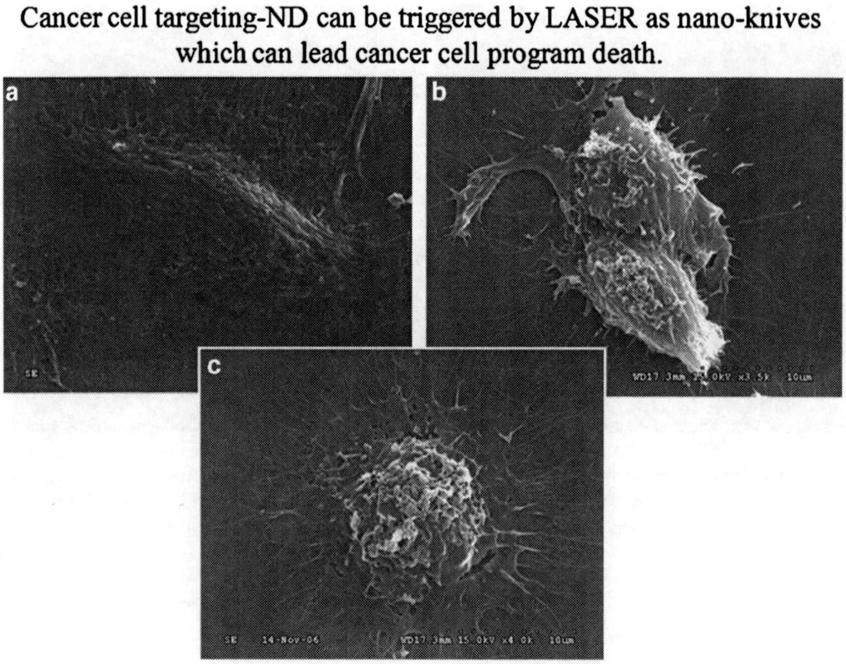

Fig. 12.17 Cancer cell targeting

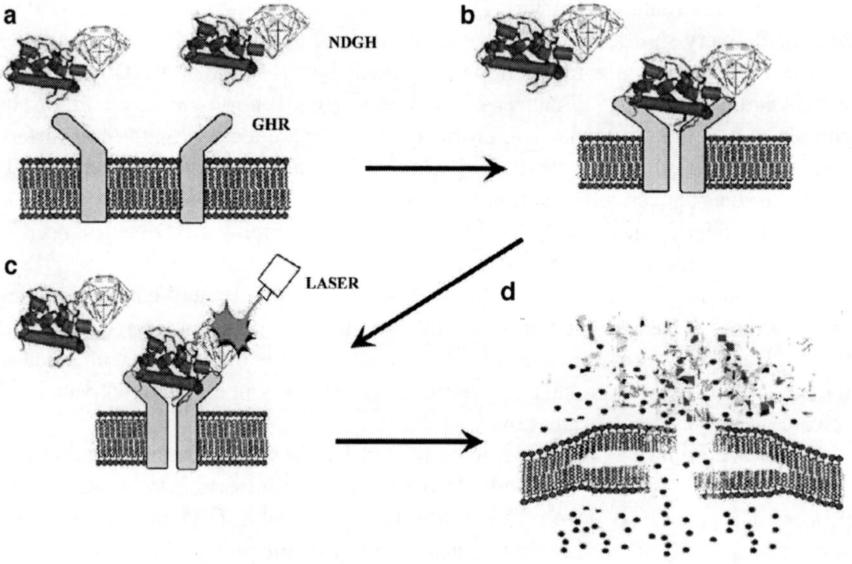

Fig. 12.18 Proposed mechanism of nano-surgery

12.9 Summary and Conclusions

In this book chapter, a broad range of materials have been explored as candidates for the imagery/diagnosis and therapeutic release toward cancer. The development of a platform approach toward rationally designed nanocarbon-enabled imagery and drug delivery that is broadly applicable would be an important advance towards material-driven enhancements in therapy. This research unites an emerging global effort towards the application of novel ND platforms, a model nanocarbon system, as imaging agents and drug carriers due to their comprehensively and quantitatively demonstrated biocompatibility and lack of cytotoxicity, as well as their ability to mediate the sustained release of nearly any type of therapeutic. It is imperative that in addition to experimental observation, additional work is required to fundamentally understand the material properties that govern these characteristics so that mechanistic approaches towards optimizing and tuning drug release parameters, and defect-based ND fluorescence can be manipulated to generate scalable, on-demand performance from these technologies [90–93]. It is this marriage of modeling/simulation and experimental validation to design (modeling) and develop (experimental) a multifunctional microfilm device that is simultaneously capable of releasing moeity-functionalized fluorescent NDs for targeted cancer cell imagery and drug–ND hybrids for sustained drug delivery to these cancer cells.

Over the several sections of this book chapter, we have briefly introduced a novel concept to numerically design a multifunctional drug delivery patch system for cancer therapeutic and diagnostic applications. New parameters for coulombic interactions in nandiamonds have been developed from DFTB calculated results. An uncertainty quantification study has also been performed to demonstrate the multiscale capability of the charge prediction method. These parameters can be employed in large-scale MD simulations to capture the fundamental physics of nanodiamond structure and self-assembly. An upper bound estimate for the loading capacity of functional groups on nanodiamonds is outlined. This estimate will pave the way to understand drug loading capacity of nanodiamond. Research is underway [90–101] to address several fundamental science issues on drug delivery system. For instance, using the newly developed coulombic interaction parameters, the self assembled morphology [95, 99] of ND-clusters will be determined. Such morphology will be helpful in designing efficient patch system. Using ND-drug charge interactions in various physiological environment (characterized by pH), the mechanics of pH-mediated drug loading and unloading from ND will be investigated [94, 97]. Using classical Fick's law of diffusion and van der Waals potentials, drug diffusion through parylene nanopores will be investigated [100]. Finally, a continuum model based on Immersed Finite Element Techniques will be developed to study the overall patch performance.

Acknowledgment Financial Support from the US National Science Foundation and the World Class University program (R33-10079) under the Ministry of Education, Science and Technology, Republic of Korea, are greatly appreciated.

References

1. Niemeyer CM (2001) Nanoparticles, proteins, and nucleic acids: biotechnology meets materials science. Angew Chem Int ed 40:4128–4158
2. Michalet X, Pinaud FF, Bentolila LA, Tsay M, Doose S, Li JJ, Sundaresan G, Wu AM, Gambhir SS, Weiss S (2005) Quantum dots for live cells and in vivo imaging. Diagnostics and beyond. Science 307:538–544
3. Jeong B, Bae YH, Lee DS, Kim SW (1997) Biodegradable block copolymers as injectable drug-delivery systems. Nature 388:860–862
4. Jeong B, Bae YH, Kim SW (2000) Drug release from biodegradable injective thermosensitive hydrogel of PEG-PLGA-PEG triblock copolymers. J Control Rel 63:155–163
5. Gombotz WR, Pettit DK (1995) Biodegradable polymers for protein and peptide drug delivery. Bioconj Chem 6:332–351
6. Huang H, Pierstorff E, Osawa E, Ho D (2007) Active nanodiamond hydrogels for chemotherapeutic delivery. Nano Lett 7:3305–3314
7. Huang H, Pierstorff E, Osawa E, Ho D (2008) Protein-mediated assembly of nanodiamond hydrogels into a biocompatible and biofunctional multilayer nanofilm. ACS Nano 2:203–212
8. Osawa E (2007) Recent progress and perspectives in single-digit nanodiamond. Diamond Relat Mater 16(12):2018–2022
9. Dolmatov VY (2006) Applications of detonation nanodiamond. Ultrananocryst; Diamond, 477–527
10. Yeap WS, Tan YY, Loh KP (2008) Using detonation nanodiamond for the specific capture of glycoproteins. Anal Chem 80:4659–4665
11. Sakurai H, Ebihara N, Osawa E, Takahashi M, Fujinami M, Oguma K (2006) Adsorption characteristics of a nanodiamond for oxoacid anions and their application to the selective preconcentration of tungstate in water samples. Anal Sci 22:357–362
12. Krueger A, Stegk J, Liang Y, Lu L, Jarre G (2008) Biotinylated nanodiamond: simple and efficient functionalization of detonation diamond. Langmuir 24(8):4200–4204
13. Krueger A, Liang Y, Jarre G, Stegk J (2006) Surface functionalization of detonation diamond suitable for biological applications. J Mater Chem 16(24):2322–2328
14. Barnard AS (2006) Theory and modeling of nanocarbon phase stability. Diamond and Related Materials 15(2–3):285–291
15. Barnard AS (2008) Self-assembly in nanodiamond agglutinates. J Mater Chem 18(34): 4038–4041
16. Mielke SL, Belytschko T, Schatz GC (2007) Nanoscale fracture mechanics. Annu Rev Phys Chem 58:185–209
17. Paci JT, Belytschko T, Schatz GC (2006) Mechanical properties of ultrananocrystalline diamond prepared in a nitrogen-rich plasma: a theoretical study. Phys Rev B 74:184112-1–184112-9
18. Osawa E (2008) Monodisperse single nanodiamond particulates. Pure Appl Chem 80(7): 1365–1379
19. Barnard AS, Vlasov II, Ralchenko VG (2009) Predicting the distribution and stability of photoactive defect centers in nanodiamond biomarkers. J Mater Chem 19(3):360–365
20. Barnard AS, Sternberg M (2008) Vacancy induced structural changes in diamond nanoparticles. Journal of Computational and Theoretical Nanoscience 5(11):2089–2095
21. Chang Y-R, Lee H-Y, Chen K, Chang C-C, Tsai D-S, Fu C-C, Lim T-S, Tzeng Y-K, Fang C-Y, Han C-C, Chang H-C, Fann W (2008) Mass production and dynamic imaging of fluorescent nanodiamonds. Nature Nanotechnology 3(5):284–288
22. Fu C-C, Lee H-Y, Chen K, Lim T-S, Wu H-Y, Lin P-K, Wei P-K, Tsao P-H, Chang H-C, Fann W (2007) Characterization and application of single fluorescent nanodiamonds as cellular biomarkers. Proc Natl Acad Sci USA 104(3):727–732
23. Sumant AV, Grierson DS, Gerbi JE, Birrell J, Lanke UD, Auciello O, Carlisle JA, Carpick RW (2005) Toward the ultimate tribological interface: surface chemical optimization and nanoscale single asperity properties of ultrananocrystalline diamond. Adv Mat 17:1039–1045

24. Naguib NN, Elam JW, Birrell J, Wang J, Grierson DS, Kabius B, Hiller JM, Sumant AV, Carpick RW, Auciello O, Carlisle JA (2006) The use of tungsten interlayers to enhance the initial nucleation and conformality of ultrananocrystalline diamond (UNCD) thin films. Chem Phys Lett 430:345–50
25. Sumant AV, Gilbert PUPA, Grierson DS, Konicek AR, Abrecht M, Butler JE, Feygelson T, Rotter SS, Carpick RW (2007) Surface composition, bonding, and morphology in the nucleation and growth of ultra-thin, high quality nanocrystalline diamond films. Diam Rel Mat 16:718–24
26. Lee W, Jang S, Kim MJ, Myoung J-M (2008) Interfacial interactions and dispersion relations in carbon-aluminium nanocomposite systems. Nanotechnology 19:285701-1–285701-13
27. Steager EB, Kim C-B, Kim MJ (2008) Temperature effects on swarming flagellated bacteria in microfluidic environments. J Heat Transfer 130:080908–1
28. Kim YS, Liao KS, Jan CJ, Bergbreiter DE, Grunlan JC (2006) "Conductive thin films on functionalized polyethylene particles. Chem Mat 18:2997–3004
29. Jan CJ, Walton MD, McConnell EP, Jang WS, Kim YS, Grunlan JC (2006) Carbon black thin films with tunable resistance and optical transparency. Carbon 44:1974–1981
30. Ho YP, Chen HH, Leong KW, Wang TH (2006) Evaluating the intracellular stability and unpacking of DNA nanocomplexes by quantum dots-FRET. J Control Rel 116:83–89
31. Zhang J, Zimmer JW, Howe RT, Maboudian R (2008) Characterization of boron-doped micro- and nanocrystalline diamond films deposited by wafer-scale hot filament chemical vapor deposition for MEMS applications. Diam and Rel Mat 17:23–28
32. Discher BM, Won YY, Ege DS, Lee JC, Bates FS, Discher DE, Hammer DA (1999) Polymersomes: tough vesicles made from diblock copolymers. Science 284:1143–1146
33. Yang Y, Zeng C, Lee LJ (2004) Three-dimensional assembly of polymer microstructures at low temperatures. Adv Mat 16:560–564
34. Qi L, Gao X (2008) Quantum dot – amphipol nanocomplex for intracellular delivery and real-time imaging of siRNA. *ACS Nano* , DOI: 10.1021/nn800280r
35. Discher DE, Ahmed F (2006) Polymersomes. Ann Rev Bio Eng 8:323–341
36. Geng Y, Discher DE (2005) Hydrolytic shortening of polycaprolactone-block-(polyethylene oxide) worm micelles. J Am Chem Soc 127:12780–12781
37. Ahmed F, Discher DE (2004) Controlled release from polymersome vesicles blended with PEO-PLA or related hydrolysable copolymer. J Control Rel 96:37–53
38. Discher DE, Eisenberg A (2002) Polymer Vesicles Science 297:967–973
39. Lam R, Chen M, Pierstorff E, Huang H, Osawa E, Ho D (2008) Nanodiamond-embedded microfilm devices for localized chemotherapeutic elution. ACS Nano 2:2095–2102
40. Cui Y, Wei QQ, Park HK, Lieber CM (2001) Nanowire nanosensors for highly sensitive and selective detection of biological and chemical species. Science 293:1289–1292
41. Baughman RH, Zakhidov AA, de Heer WA (2002) Carbon nanotubes–the route toward applications. Science 297:787–792
42. Bianco A, Prato M (2003) Can carbon nanotubes be considered usefull tools for biological applications? Adv Mat 15:1765–1768
43. Bowden T, Tabor D (1961) Friction and lubrication in solids, Japanese edition. Maruzen, Tokyo, p 27
44. S Mori, A Kanno, H Nanao, I Minami, E Ōsawa, Tribological performance of nanodiamond for water lubrication, Proceedings of the 3 rd International Symposium on Detonation Nanodiamonds: Technology, Properties and Applications, July 1–4, 2008, St. Petersburg, Russia, p. 21–28, Ioffe Physico-Technical Institute
45. Ōsawa E, Ho D, Huang H, Korobov MV, Rozhkova NN (2009) "Consequences of strong and diverse electrostatic potential field on the surface of detonation nanodiamond particles,"Diam Rel Mater, 18, doi.org/10.1016/j.diamond.2009.01.025.
46. Qian D, Liu WK, Zheng Q (2008) Concurrent quantum/continuum coupling analysis of nanostructures. Comput Methods Appl Mech Eng 197(41–42):3291–3323
47. Kopacz AM, Liu WK, Liu ShuQ (2008) Simulation and prediction of endothelial cell adhesion modulated by molecular engineering. Comput Meth Appl Mech Eng 197(25–28): 2340–2352

48. Liu Y, Kieseok Oh, Bai JG, Chang C-L, Yeo W, Chung J-H, Lee K-H, Liu WK (2008) Manipulation of nanoparticles and biomolecules by electric field and surface tension. Comput Meth Appl Mech Eng 197(25–28):2156–2172
49. Liu WK, Jun S, Qian D (2008) Computational nanomechanics of materials. Journal of Computational and Theoretical Nanoscience 5(970–996):2008
50. Liu WK, Kim Do Wan, Tang S (2007) Mathematical foundations of the immersed finite element method. Computational Mechanics 39(3):211–222
51. WK Liu, Liu Y, Farrell D, et al (2006) Immersed finite element method and applications to biological systems. Comp Meth Appl Mech Eng, 195(1722-1749),
52. Liu WK, Sukky J, Qian D (2006) Computational nanomechanics of materials, Handbook of theoretical and computational nanotechnology, M Reith and W Schommers (Eds). 4, 132–191, American Scientific Publishers
53. Liu WK, Park HS, Qian D, Karpov EG, Kadowaki H, Wagner GJ (2006) Bridging scale methods for nanomechanics and materials. Comput Meth Appl Mech Eng 195(13–16):1407–1421
54. Liu WK, Park HS, Qian D, Karpov EG, Kadowaki H, Wagner GJ (2006) Bridging scale methods for nanomechanics and materials. Computer method in applied mechanics and engineering 195:1404–1421
55. Park HS, Karpov EG, Liu WK (2005) Non-reflecting boundary conditions for atomistic, continuum and coupled atomistic/continuum simulations. Int J Numer Meth Eng 64:237–259
56. WK Liu, HS Park, D Qian, EG Karpov, H Kadowaki, GJ Wagner. Bridging Scale Methods for Nanomechanics and Materials", accepted for publication in Comput Meth Appl Mech Eng 2005, special issue in honor of the 60th birthday of Prof. T.J.R. Hughes.
57. Park HS, Karpov EG, Klein PA, Liu WK (2005) The bridging scale for two-dimensional atomistic/continuum coupling. Phil Mag 85(1):79–113
58. Park HS, Karpov EG, Liu WK, Klein PA (2005) The bridging scale for two-dimensional atomistic/continuum coupling. Phil Mag 85(1):79–113
59. Park HS, Karpov EG, Klein PA, Liu WK (2005) Three-dimensional bridging scale analysis of dynamic fracture. J Comput Phys 207:588–609
60. Applied Mechanics and Enineering. 198 (15-16), pp 1327–1337, March 2009
61. Barnard AS, Russo SP, Snook IK (2005) J Comput Theor Nanosci 2:180
62. AS Barnard, SP Russo, IK Snook (2003) J Chem Phys 118, 5094; AS Barnard, SP Russo, IK Snook (2003) Phys Rev B 68, 073406
63. AS Barnard, SP Russo, IK Snook (2003) Philos Mag Lett 83, 39; AS Barnard, SP Russo, IK Snook (2004) Diamond Relat Mater 12, 1867
64. Barnard AS, Zapol PJ (2004) Chem Phys 121:4276
65. AS Barnard, SP Russo, IK Snook (2003) Phys Rev B 68, 073406
66. AS Barnard, SP Russo, IK Snook (2003) Int J Mod Phys B 17 (21) 3865
67. Barnard AS, Sternberg M (2007) J Mater Chem 17:4811
68. Barnard AS (2008) J Mater Chem 18:4038
69. Harold P Bovenkerk, Thomas R Anthony, James F Fleischer, William F Banholzer, CVD diamond by alternating chemical reactions, US patent 5,302,231, 12 april 1994
70. Dieter M Gruen, Thomas G McCauley, Dan Zhou, Alan R Krauss, Tailoring nanocrystalline diamond film properties US patent, 6,592,839, 15 July 2003
71. Li H, Tao Xu, Chen J, Zhou H, Liu H (2003) Preparation and characterization of hydrogenated diamond-like carbon films in a dual DC-RF plasma system. J Phys D: Appl Phys 36:3183–3190
72. Synthesis and Application of Nano-crystalline Diamond Thin Film, KISTI reports, 2005
73. Baik E-S, Baik Y-J, Lee SW, Jeon D (2000) Fabrication of diamond nano-whiskers. Thin Solid Films 377–378:295–298
74. Remes Z, Kromka A, Vanecek M, Grinevich A, Hartmannova H, Kmoch S (2007) The RF plasma surface chemical modification of nanodiamond films grown on glass and silicon at low temperature. Diamond & Related Materials 16:671–674
75. Hirakuri KK, Minorikawa T, Friedbacher G, Grasserbauer M (1997) Thin film characterization of diamond-like carbon films prepared by r.f. plasma chemical vapor deposition. Thin Solid Films 302:5–11

76. Krueger A (2008) New carbon materials. Chem Eur J 14:1382–1390
77. www.kurzweilai.net/meme/frame.html?main=/articles/art0632.html: How to make nano diamond.
78. Hall P (1992) The bootstrap and edgeworth expansion. Springer Series in Statistics. Springer-Verlog, New York
79. Kennedy MC, A O'Hagan (2001) Bayesian calibration of computer models. Journal of the Royal Statistical Society Series B 63, 425–464.
80. Xiong Y, Chen W, Tsui K-L, Apley D (2009) A better understanding of model updating strategies in validating engineering models. Comput Meth Appl Mech Eng 198(15–16):1327–1337
81. Krüger A, Kataoka F, Ozawa M, Fujino T, Suzuki Y, Aleksenskii AE, Ya. Vul A, Osawa E (2005) Unusually tight aggregation in detonation nanodiamond: Identification and disintegration. Carbon 43:1722–1730
82. Panich AM, Shames AI, Vieth HM, Osawa E, Takahashi M, Ya Vul A (2006) Nuclear magnetic resonance study of ultrananocrystalline diamonds. Eur Phys J B 52:397–402
83. M Gruen, Shenderova O (2005) Synthesis, properties and applications of ultrananocrystalline nanodiamond. Springer, USA. 217–230
84. Zheng M, Jagota A, Semke ED, Diner BA, Mclean RS, Lustig SR, Richardson RE, Tassi NG (2003) Nature mater 2:338
85. Zheng M, Jagota A, Strano MS, Santos AP, Barone P, Chou SG, Diner BA, Dresselhaus MS, Mclean RS, Onoa GB, Samsonidze GG, Semke ED, Usrey M, Walls DJ (2003) Science 302:1545
86. Hwang ES, Cao C, Hong S, Jung HJ, Cha CY, Choi JB, Kim YJ, Baik S (2006) Nanotechnol 17:3442
87. Chengfan Cao, Jung Heon Kim, Ye-Jin Kwon, Young-Jin Kim, Eung-Soo Hwang and Seunghyun Baik (2009) An immunoassay using biotinylated single walled carbon nanotubes as Raman biomarkers, Accepted to Analyst.
88. Cheng C-Y, Perevedentseva E, Tu J-S, Chung P-H, Cheng C-L, Liu K-K, Chao J-I, Chen P-H, Chang C-C (2007) Direct and in vitro observation of growth hormone receptor molecules in A549 human lung epithelial cells by nanodiamond labeling. Appl Phys Lett 90:163903 (SCI)
89. Hunter RJ (1981) Zeta potential in colloidal science. Academic, London
90. Wing Kam Liu, Ashfaq Adnan, Adrian Kopacz, Roadmap for nanodiamond-based drug delivery design for cancer therapeutics and diagnostics, 10th US National Congress On Computational Mechanics, Columbus, OH, 2009
91. Wing Kam Liu, Ashfaq Adnan, Adrian Kopacz (2009) Design of nanodiamond-enabled drug delivery system by simulation based science & engineering, ASME International Mechanical Engineering Congress & Exposition, Lake Buena Vista, Florida – November 13–19.
92. Wing Kam Liu, Multiscale design of nanodiamond-based drug delivery system for engineered medicine, coupled problems 2009, Computational Methods For Coupled Problems In Science And Engineering, 8–10 June 2009, Ischia Island, Italy.
93. Wing Kam Liu, Ashfaq Adnan, Adrian Kopacz, Nanoscale Science In Therapeutic And Diagnostic Applications 2009 ASME International Mechanical Engineering Congress & Exposition, Lake Buena Vista, Florida – November 13–19, 2009.
94. Adnan A, Kam Liu Wing (2009) Mechanics of pH mediated adsorption/desorption of doxorubicin drug from functionalized nanodiamond. 10th US National Congress On Computational Mechanics, Columbus, OH
95. Kopacz A, Adnan A, Kam Liu W (2009) Functionalized & self-assembled nanodiamonds for diagnostic and therapeutic applications. 10th US National Congress On Computational Mechanics, Columbus, OH
96. Wing Kam Liu, Ashfaq Adnan (2009) Nanoscale science in therapeutic and diagnostic applications. 10th US National Congress On Computational Mechanics, Columbus, OH
97. Ashfaq Adnan, Wing Kam Liu (2009) Mechanics of pH controlled loading and release of chemotherapeutics from functionalized nanodiamond, 2009 ASME International

Mechanical Engineering Congress & Exposition, Lake Buena Vista, Florida – November 13–19, 2009
98. Michelle Hallikainen, Ashfaq Adnan, Wing Kam Liu, Predicting nanodiamond structure and surface charge distribution using molecular dynamics and Bayesian statistics. 10th US National Congress On Computational Mechanics, Columbus, OH, 2009
99. Adrian Kopacz, Ashfaq Adnan, Wing Kam Liu (2009) Equilibrium functionalization and self-assembly of nanodiamond as a platform for engineered medicine", 2009 ASME International Mechanical Engineering Congress & Exposition, Lake Buena Vista, Florida – November 13–19, 2009
100. Paul Arendt, Wei Chen, Wing Kam Liu, Ashfaq Adnan, Multiscale design of a simplified nanodiamond based drug delivery system. 10th US National Congress On Computational Mechanics, Columbus, OH, 2009
101. Wing Kam Liu, Ashfaq Adnan, Adrian Kopacz, Young-Jin Kim, Moon Ki Kim, Multiscale design of nanodiamond-based drug delivery system for cancer therapeutics and diagnostics. 2nd International Symposium On Computational Mechanics (ISCM II) and 12th International Conference On Enhancement And Promotion Of Computational Methods In Engineering And Science (EPMESC XII), November 30 – December 3, 2009, Hong Kong – Macau

Index